Synthesis Lectures on Operations Research and Applications

This series focuses on the use of advanced analytics in both industry and scientific research to advance the quality of decisions and processes. Written by international experts, modern applications and methodologies are utilized to help researchers and students alike to improve their use of analytics. Classical and cutting-edge topics are presented and explored with a focus on utilization and application across a range in practical situations.

Robert Handfield · Daniel J. Finkenstadt

Supply Chain Immunity

Overcoming our Nation's Sourcing Sickness in a Post-COVID World

Springer

Robert Handfield
North Carolina State University
Raleigh, NC, USA

Daniel J. Finkenstadt
Naval Postgraduate School
Monterey, CA, USA

ISSN 2770-6303 ISSN 2770-6311 (electronic)
Synthesis Lectures on Operations Research and Applications
ISBN 978-3-031-17183-3 ISBN 978-3-031-19344-6 (eBook)
https://doi.org/10.1007/978-3-031-19344-6

This Springer imprint is published by the registered company Springer Nature Switzerland AG
The registered company address is: Gewerbestrasse 11, 6330 Cham, Switzerland

To my best friend and my lifelong partner, Sandi. I am so lucky to be able to live this life with you by my side. And to the tireless work of our public sector workers, first responders, and healthcare workers, as well as defense, federal and state agencies. Your silent work did not go unnoticed.

Robert Handfield

I dedicate this first book to my childhood sweetheart, Amy. You make all things possible. Also, to my late grandma Peak who instilled in me a desire to learn. Finally, to the men and women who worked so diligently to respond to the unprecedented demand of 2020. Many will never fully understand what you have done. Thank you.

Daniel J. Finkenstadt

Prologue: How We Got Involved

One of the first signs of COVID-19 is a distinct and overwhelming loss of smell. The patient loses all ability to distinguish between familiar smells, fails to pick up on strong or pungent odors, and in many cases also completely loses their sense of taste. In 2020 COVID-19 hit the world full force. Sadly, the world had either lost its ability to smell (e.g., detect) that a global pandemic was coming and could not taste what was coming and prepare for it. We didn't pick up on the signals broadcast in public health intelligence sources, the reports in medical social media, or rumors of disruptions in the manufacturing and supply chains abroad. We left open critical airline corridors into major cities such as Chicago, LA, New York, and Atlanta that allowed the disease to spread like wildfire once infected passengers disembarked into these population centers. This was followed by geopolitical ideological and economic opportunism; at the same time countries were shutting borders and halting shipments of critical medical supplies. Indeed, it seemed as if no one was in charge, and chaos reigned. Command and control government actions were completely inadequate, and the general populace was left in the dark, wondering what was happening at every level. Irrational demand and supply behaviors exacerbated an already untenable situation of social distancing, lockdowns, and crushing medical system overloads, as patients streamed into ICUs and death rates climbed.

This set of authors had a front-row seat during the national pandemic response, including government, health care, supply chain, and manufacturing at all levels from local municipalities to federal and global efforts. Unfortunately, it wasn't pretty. We were privileged to work alongside a multi-functional team of experts and professionals trying to manage an overwhelming threat on the fly. Many times it was simply a coalition of the willing taking on roles of massive importance for the health of the world. We are eternally grateful to the men and women who supported this response, and we feel the need to write a book which seeks to pass on critical lessons learned (and we believe there are) and hopefully to be better prepared for this type of disaster to ever happen again.

COVID-19 is expected to cost the world in excess of $12 trillion through 2024 based on international monetary fund estimates.[1] As of this writing the pandemic has already cost the USA alone nearly $3.6 trillion.[2] More importantly, it has cost over 5.7 million lives to date.[3] We are not claiming that this book could have prevented any monetary or health impacts from COVID-19. But we remain hopeful that the lessons learned and insights conveyed within these pages can mitigate (or perhaps even attenuate) this level of impact in the future for the next global catastrophe. And there will be one. Many of the issues that have occurred with COVID-19, especially within supply chains, are not the result of "black swan" events as much as they are from previous "ostrich"-based strategies.[4] That is to say, they are not "one in a million years" events any more, but can be determined on a much more probabilistic basis and with better preparedness by making supply chain risk management an on-going part of business strategy and planning. A caveat is we don't know what disruptions will occur; but we know there will be more disruptions because of the world we live in.

It is important that we qualify our ability to speak on these topics before we get too far along in our observations, studies, opinions, and advice. The lead author, Dr. Rob Handfield, is a global expert in supply chains and has worked for several decades studying healthcare supply chains. He has over 30 years of experience studying many of the areas production and logistics that are discussed in this book, which is one of the main reasons we asked him to join the various federal task forces advising the initial supply chain and acquisition response in 2020. Dr. Dan Finkenstadt is a life-long defense procurement expert with over 18 years of experience sourcing supplies and services all over the world from Texas to Afghanistan and a professor of enterprise sourcing and category management within the DoD. Pete Guinto who supported the two co-authors is also a defense procurement expert with degrees and practice in healthcare law and respiratory therapy. Pete was assigned as the market intelligence lead for the Air Force and on DoD and federal-level supply chain task forces from February to July of 2020. He engaged Dr. Finkenstadt and Dr. Handfield as advisory experts to aid in the response. During this period the authors participated in hundreds of meetings and interviews with federal agencies, state agencies, local officials, non-profit organizations, manufacturing organizations, defense suppliers, healthcare suppliers, hospitals, healthcare workers, logistics personnel and firms, data experts, journalists, legislative staffs, and others. The authors kept rigorous notes of these events and interviews and have published extensively on these experiences

[1] https://www.reuters.com/business/imf-sees-cost-covid-pandemic-rising-beyond-125-trillion-estimate-2022-01-20/#:~:text=WASHINGTON%2C%20Jan%2020%20(Reuters),global%20lender%20said%20on%20Thursday.

[2] https://www.usaspending.gov/.

[3] https://covid19.who.int/.

[4] Steenkamp, J.B.E.M. (2020) Time to Lead: Lessons for Today's Leaders from Bold Decisions that Changed History. Fast Company Press.

in peer-reviewed journals and professional outlets. This book summarizes those experiences and lessons (some published to date and some not) in an easy to read, consolidated book for everyday citizens, practitioners, students, and managers. We hope it can shed some light on these unprecedented events and ways to better manage them in the future.

This book is divided into three major parts: (1) The Disease, (2) Symptoms and Treatments, and finally, (3) Immunity. We found it both interesting and appropriate to align the sections of this book with the stages of the pandemic—as you will see, the metaphors are completely aligned with the events that happened to this nation and others around the world. In Part I—*The Disease*, we discuss the specific attributes of the COVID-19 virus that made it so difficult to contain and manage and introduce another metaphor—the "commons"—which captures the public health system designed to control a pandemic. Unfortunately, the government did not manage the commons well, resulting in many irrational behaviors when it came to healthcare supplies and services (e.g., hoarding and obfuscation), which further inhibited the ability to respond to the deadly spread of the virus. In Part II—*Symptoms and Treatments*, we expand on the woeful national response that we observed, including gaps in data availability, communications, and overall governance of the event. We discuss the original design of the Strategic National Stockpile, based on the original intent of its creators, and discuss the massive supply visibility problems we observed, using personal protective equipment (PPE) as a representative case. Also in Part II we discuss specific recommendations derived from our observations, focusing on improved preparation and response to massive supply chain disruptions in the future using five major areas of supply chain immunity concern we developed during our experiences with the COVID-19 response. Finally, in Part III—*Immunity*, we highlight example cases of supply chain disruptions since 2020 to demonstrate that the principles in this book do not just apply to the pandemic. Finally, we provide a new model for decision-making in uncertain, data saturated environments for use in concert with principles discussed in Part II to bolster supply chain immunity.

But before we get into the solution, let's describe the problem and review the timeline of events surrounding COVID-19 in 2020.

Contents

Part I

The Disease

COVID-19, A Rapidly Evolving Disaster in 2020

1

The following timeline is based on a collection of events and observations by a team of researchers including the three authors of this book. These events were captured and cataloged by members of the research team in support of a peer-reviewed article entitles, "A Commons for a Supply Chain in the Post-COVID-19 Era: The Case for a Reformed Strategic National Stockpile" published in the Milbank Quarterly in 2020.[1] Portions of the timeline of events were also captured from the American Journal of Managed Care.[2]

January

In January of 2020 COVID-19 hit the world stage. On 9 January the World Health Organization (WHO) announced a mysterious coronavirus-related pneumonia in Wuhan, China. At this point the WHO still had doubts about the origin of the disease and there we less than 60 cases. Travel precautions were started to be discussed by experts. By 20 January 2020 there were additional cases popping up in Thailand and Japan. The Centers for Disease Control (CDC) in the US started to screen for the disease at major airports that had flights from Wuhan, including JFK International, San Francisco and Los Angeles. The next day the first case of COVID-19 was confirmed by the CDC in Washington state using polymerase chain reaction (PCR) testing. The CDC then deployed a team to help investigate using contact tracing. The same day, in China, Dr. Zhong Nanshan confirms person to person transmission. At his point in China, four are dead and there are more than 200 cases in China alone. The WHO is still unsure about a public health emergency

---

[1] Handfield R, Finkenstadt DJ, Schneller ES, Godfrey AB, Guinto P. A Commons for a Supply Chain in the Post-COVID-19 Era: The Case for a Reformed Strategic National Stockpile. Milbank Q. 2020;98(4):1058–1090. https://doi.org/10.1111/1468-0009.12485.

[2] https://www.ajmc.com/view/a-timeline-of-covid19-developments-in-2020.

R. Handfield and D. J. Finkenstadt, *Supply Chain Immunity*, Synthesis Lectures on Operations Research and Applications, https://doi.org/10.1007/978-3-031-19344-6_1

declaration. Two days later, on January 23rd, Wuhan is put under Quarantine. In just a short 48 hour period 13 more deaths have occurred and the an additional 300 people have the virus. China closes off all of Wuhan and its 11 million citizens. Restriction protocols are placed on Huanggang, China 30 miles east of Wuhan, bringing the lockdown figure to 18 million Chinese citizens. By the end of January the WHO declares a global health emergency. There are more than 200 dead and 9800 cases. There are cases in the US, Germany, Japan, Vietnam and Taiwan.

February

Finally, on February 2nd, global air travel restrictions are put in place. By that afternoon those en route to the United States are given a choice to leave China or face a 2-week home-based quarantine if they had been in Hubei province. Mainland visitors are forced to undergo health screenings upon their return, and foreign nationals can even be denied admittance. Australia, Germany, Italy and New Zealand begin to impose similar air-travel restrictions. A day later the US declares a public health emergency. The Trump administration declares a public health emergency due to the coronavirus outbreak. By February 10th, the COVID-19 deaths in China have exceeded the severe acute respiratory syndrome (SARS) outbreak with 908 deaths in a month versus 774 deaths during all of SARS. By the end of February (25th) the CDC's director of Immunization and Respiratory Diseases states that COVID-19 meets two of the three necessary factors for pandemic status; deadly illness and person-to-person spread. It has not quite reached the third requirement for worldwide spread.

March

On March 6th a cruise ship carrying 3500 people off the coast of California tests 46 passengers and finds 21 of them positive, including 19 crew. The ship is held at sea outside of San Francisco while testing is conducted. This event leads to mass lawsuits against Carnival Corp. for gross negligence in passenger safety. Finally, on March 11th, the WHO declares COVID-19 a full pandemic. Tedros Adhanom Ghebreyesus, WHO Director states at a briefing in Geneva, that the WHO is "deeply concerned by the alarming levels of spread and severity" of the outbreak. He also expressed concern about "***the alarming levels of inaction***." Two days later President Trump declares COVID-19 a national emergency, unlocking billions of dollars in federal funding to fight the disease's spread. The same day (13 March) the administration issues a travel ban on non-Americans who visited 26 European countries within 14 days of coming to the United States. People traveling from the United Kingdom and the Republic of Ireland are exempt.

A day later, on March 14th, our task force starts to form. Two members of our team are initially contacted by leaders within the DoD to begin strategizing ways to advise the United States Air Force in methods and sources to support the supply chain relief efforts for DoD, FEMA and DHHS requirements. The team would be tasked to accomplish several aims in the direct response to COVID. First, act as the supply procurement backstop for other federal agencies, if they were stretched too thin or were unable to buy what was needed, the team would execute contracts to purchase goods for any national purpose. Second, prepare the Air Force Research Lab (AFRL) team for potential capacity expansion efforts that were being missed in the mainstream coverage of the event, execution would be in AFRL, the executive agent for Defense Production Act (DPA) title III funding. March 17th brings a flurry of events. The University of Minnesota launches clinical trials to investigate whether hydroxychloroquine can prevent an individual exposed to COVID-19 from becoming ill or reduce the severity of the infection. The trial is limited to those at high risk of exposure and aims to enroll 1500 individuals. The Centers for Medicare & Medicaid Services (CMS) expands telehealth rules to allow for telehealth visit coverage as a regular visit in order to protect older patients from exposure. The same day the Trump administration asks Congress to send emergency relief checks to stimulate the economy. This happens just as the US is reporting its 100th COVID-related death. Two days later California becomes the first state to issue a stay-at-home order, mandating residents to stay at home unless part of an essential job or shopping for essential items such as groceries. This order also directs priority to sickest COVID patients.

The reality during this time is that there were many more cases in the US by this point. Gottlieb in this book covers the nature of this spread, indicating that the US's inability to control the borders and perform tests for those coming into the United States resulted in a rapid spread of the virus throughout the United States. He also notes that there were a massive shortage of tests and not enough labs required to track the genetic fragments of the virus in individuals. This occurred because of the lack of a deliberate strategy to deploy the different kinds of tests, how they should be used, and when they can be made available after a new pathogen emerges. Instead, every testing company stepped into the market at different times, with little meaningful coordination. A real pandemic playbook would have mapped out these decisions in advance, and seeded the development of the required testing layers. In addition, the nation's installed base of testing equipment was a hodgepodge of testing instruments. The US lacked the establishment of a strategic network of labs with a uniform base of instruments that can be used to process the largest possible pool of test kits.

On March 23, 2020, the first major supply chain event takes place. The Wall Street Journal reports that the SNS had run out of critical medical supplies for the states. Overwhelmed hospitals are keeping out everyone who does not need to be there, including delaying the start of new clinical trials. The Center for Biosimilars® reported that drugs with fresh FDA approvals are not likely to launch, as their chances of making it into circulation are dim with hospitals struggling just to find enough PPE. Reports come out two days later showing that longer, tighter social distancing in Wuhan China may be flattening the curve. The same day (25 March) our initial COVID-19 acquisition task force team formally stands up within the Air Force Lifecycle Management Center and the Office of the Air Force's Deputy Assistant Secretary for Contracting. This task force include various lines of effort teams. Pete Guinto and Dr. Finkenstadt are initial members, and they request support from Dr. Rob Handfield, a professional colleague of Dr. Finkenstadt. The team's goal is to prepare for acquisitions that FEMA/DHHS were unable to handle or for capacity expansion efforts that the Air Force Research Laboratory needed assistance with. Within a day the early team has delivered a list of potential critical items derived from discussions with clinical personnel's including the Ohio Board of Medicine Members, nurses, physicians, and first responders, supply chain professionals, and acquisition personnel. Critical items list included vent consumables, intubation supplies, bag valve masks, ventilator/intubation meds, expanded PPE list, high flow oxygen devices (particularly high flow nasal cannulas), and other items that were at that time either unnoticed or underpublicized. At the same time the team identifies major data shortfalls.

The graph below, taken from Resilinc, shows how the impact on supply chains spread, originating in China, but quickly spread into Italy, the Philippines, Spain, France, Malaysia, Israel, Singapore, and California. One can trace the shut downs of countries and resulting supplier disruptions as they occurred. The entire world went into lockdown. The second chart shows how this occurred across multiple industries—no one was immune.

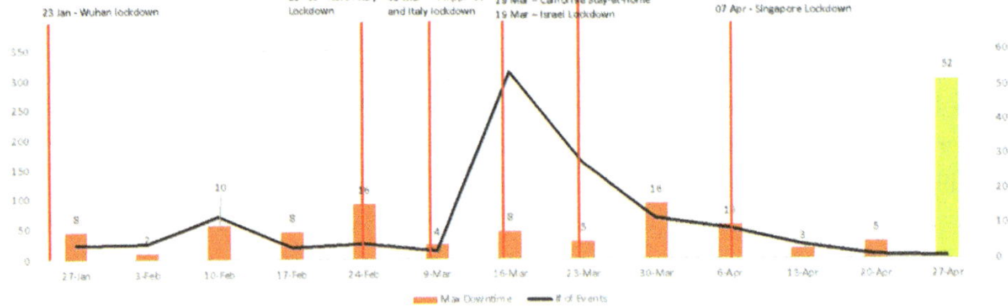

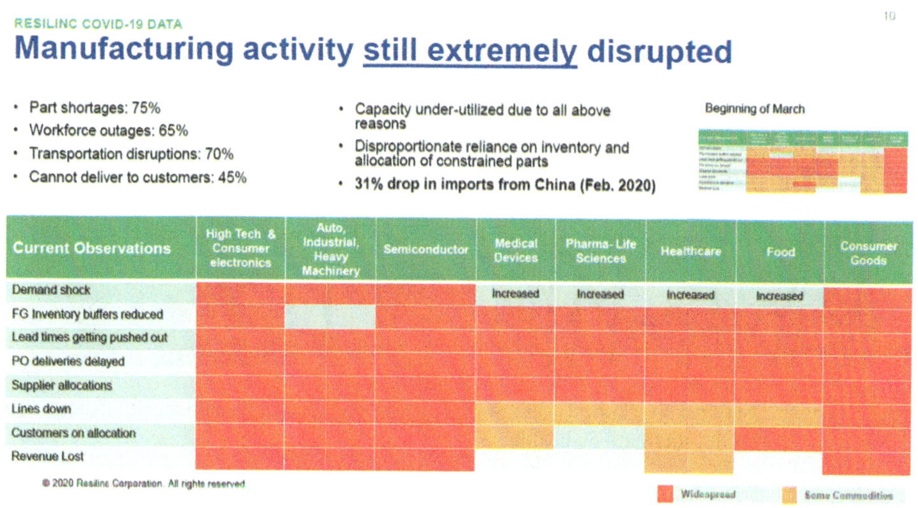

Discussions with various Air Force team members, Army, Navy, and Defense Health Agency personnel revealed that there was no system to track inventory or usage of critical items. Our primary concern was raised—we needed data to determine what items were being used most frequently, what items were experiencing the greatest supply shortfalls, and what points of care have the greatest need. We needed this information because this effort was not a *supply and demand* exercise. Due to irrational demand, it was, instead, a situation of *supply and need*. There seemed to be no mechanism for monitoring need and usage at points of care. As a result, the team was prepared that shortfalls would be identified with very little lead time, identifying shortfalls in time to identify supply chain constructures or anticipating shortfalls in time to enable meaningful capacity expansion efforts was virtually impossible. We begin to add informal members to advise our team from nonprofit agencies and private industry. The same day (March 26) the Senate passes the Coronavirus Aid, Relief, and Economic Security (CARES) Act, providing $2 trillion in aid to hospitals, small businesses, and state and local governments, while including an elimination of the Medicare sequester from May 1 through December 31, 2020. A day later, on March 17th, it was signed into law by President Trump. The bipartisan legislation provides direct payments to Americans and expansions in unemployment insurance.

Our team had its first discussion with DPA Title III players on 27 March to discuss capacity expansion for critical items. It became abundantly clear that DHHS was not prepared or resourced to spend billions in capacity expansion efforts. Our team began to prepare for the potential of capacity expansion efforts rather than procurement of goods and services. On 30 March the FDA issues an emergency use authorization (EUA) for "hydroxychloroquine sulfate and chloroquine phosphate products" to be donated to the SNS and donated to hospitals to treat patients with COVID-19. The EUA would be rescinded June 15, except for patients in clinical trials, in the wake of reports of heart rhythm problems among some patients. By the end of March it is apparent that COVID

Fig. 1 Strategic National Stockpile Image Source https://blogs.cdc.gov/publichealthmatters/2016/04/sns-key-to-protecting-the-nations-health/

can spread through the eye. A report in JAMA Ophthalmology creates a stir with the finding that patients can catch the virus that causes COVID-19 through the eye, despite low prevalence of the virus in tears. The coverage of the study involving 38 patients from Hubei Province, China, drew some of AJMC.com's highest readership of 2020, as the findings contradicted assumptions by leading professional societies. The same day our team identifies critical data gaps and disparity in supplies in the Ohio health system. Calls with Summa Health, the Cleveland Clinic, the OH board of medicine, Green County Emergency Management, Kettering Health and the state legislature reveal that significant disparities in supply availability were present within the state. Large health systems and small health systems in the same region were presented with drastically different clinical supply settings. No data providing insight into inventories or usage was available to the state or to federal agencies and as a result all emergency supply requests were routed through county emergency management agencies that each had different policies in regards to when a request to the state and then federal emergency management agencies would be contacted. One county even indicated a policy to not report to the state until inventories of items were completely depleted. Reporting was generally limited to ventilators, PPE, and available beds.

April

On April 1st meetings begin between our team and members of Helena Group to discuss their COVID Network Project. This tool used a variety of sources and machine learning to connect vital medical supplies to areas of need. The system tracks cases, 1-week projected cases, deaths, 1-week projected deaths, population, poverty, comorbid conditions, social distancing, medical facilities, and facility risk to the county level across the entire continental US. Our team begins to coordinate meetings with FEMA and DHHS data manager/analysts to consider the Helena Group COVID Network Project as a means of matching supply and need versus simply supply and demand. We had observed hoarding behaviors by medical facilities and state and local governments and predicted that the hoarding would impact efficient resource allocation without using this or a similar tool to adjudicate true need versus stated demand.

The next day we began communications with a McKinsey consultant team. Their team had utilized internal resources and also government funded efforts to attempt to map supply and demand of critical items. Immediately, it was clear that no federal agency was collecting the necessary data to understand what needs existed. The possibility that FEMA was constructing a "data tower" was brought to the team's attention, however the thought processes behind the data tower appeared to be deeply flawed and management by a very inexperienced FEMA team was deeply troubling.

On April 4th, the team engaged with Amazon to explore the potential for teaming to allow for rapid delivery of goods to the point of need with little notice based on observed necessities. Amazon personnel, located in Detroit and as a result deeply focused on the mission, were excited to assist but indicated that their distribution networks were not sought by federal agencies. In regards to data, the use of Amazon data to determine demand and attempt to correlate need was determined to be completely ineffective as their queries and requests for supplies were so disproportionate to rational needs and varied so widely that though it could be made available, would be essentially worthless for planning rational allocation models.

Based on observed issues with market analysis to date, we recommend and receive approval (on 5 April 2020) for a devoted **market intelligence team** to coordinate across all previous task force lines of effort. Our primary charter was to develop supply, demand and need site pictures for COVID-19 current and future response. We focused on current sourcing capacities and constraints, distribution (to include import/export) challenges and materials risks. We also focused on demand in terms of prioritization regimes, assessing need versus demand, and explored means to identify hoarding behaviors across the medical demand base. It is at this time that our team expands. We brougt in additional advisors from academia. The team was approved and Pete Guinto took the lead role. This team was given expanded responsibility to work directly with the joint DoD acquisition task force supply chain team.

On 7 April our market intelligence team attended the first DoD Joint Acquisition Task Force Supply Chain meeting. The exceptional quality of DoD personnel involved was immediately evident. However, **the absence of meaningful data to determine gaps between supply and demand** was also immediately evident, further solidifying team view that inventory and usage data was an absolute necessity during a response of this magnitude. Decision-making without data is like driving without a speedometer, an odometer, or a gas tank indicator. You have no idea how fast you are going, when you will run out of gas, or how far you have traveled. Similarly, this team had no idea what the current levels of inventory were across healthcare networks, how much was on order, how much of it was available, and what the expiration dates were of materials that were in inventory, and how many days of supply were left before hospitals started running out. They didn't know which hospitals had the greatest number of cases, where to send supplies if they became available based on the most urgent need, or where the "hot spots" were where people were dying. The team explored various sources: the CDC National Hospital Safety Network NHSN data, White House Model, and the FEMA Data Tower model. None of them had sufficient data to answer these questions, but there were indications that the CDC data was likely the best source yet still highly subject to hospital manipulation since "days of inventory" were reported instead of actual inventory and usage. The White House Model and FEMA models were so dramatically different (magnitudes different for some items) that the clear consensus was that both models must be fundamentally flawed. Additionally, models were at this point severely limited in regards to the depth of supply tracking, only basic PPE, Ventilators, and beds were being modeled. The same day the JATF leadership requested that Pete Guinto and Dr. Finkenstadt integrate into the JATF team to assist with transition of FEMA "War Room" efforts from FEMA to the DoD. This point marked an enormous shift in effort for the authors to assure that secure sources were utilized for procurement of goods and that procurement planning became a priority.

A day later Mr. Guinto and Dr. Finkenstadt agreed to support the FEMA efforts as co-advisors between the DoD and USAF task forces. They were subsequently assigned to support the DoD joint acquisition task force supply chain efforts as "individual sourcers". The role included providing data, managing the vetting process and escalating high quality sources of medical supplies to manufacturer site visit organizers (primarily in China). We are told that the previous volunteers worked to find sources, qualify, or disqualify them, speak to them about capacity and pricing, organize site visits and then, if they seemed like quality sources, pass them to FEMA procurement with the price, quantity and logistics "laid out for final approval". Our team members started to receive emails from members of the FEMA volunteer group that was working to source PPE and other vital supplies. Most sources provided by these volunteers did not pass the initial supply chain risk assessment vetting and our team recommended that we abandon their use. Our team started to provide recommended sources using other market research and intelligence methods outside of the volunteer network to reduce supply chain risk. Two days later we become aware that there was a congressional inquiry into the FEMA "war room" based on an NBCNews.com article by Jonathan Allen.

On April 11th, upon a request from DHA Medical Logistics, members of the our team provided a set of early lessons learned from COVID-19 supply chain struggles. This early report provided six general recommendations (inclusive of short and long-term strategies) that DHA could consider as lessons learned for future supply chain crises: (1) develop a comprehensive supply chain risk management program, (2) reposition and redesign supply chain as a strategic competency for pandemic/disruption response, (3) design new roles for a responsive and efficient supply chain, (4) require prime vendors to have pandemic and natural-event strategic plans, (5) draw on industry/academic best practices, and (6) leverage relationships with civilian health care systems to share supplies. The same day we engage with the Joint Artificial Intelligence Center (JAIC). The Market intelligence team engaged with the JAIC in regards to projects aimed at ensuring stability of critical infrastructure—most critically public utility, food and clothing supply. Assessments from their analysts was that it is likely already too late to meaningfully influence the data enterprise to assure timely response to the near term impacts of COVID. Instead, their focus was to ensure national stability if the crisis raged through the summer and further intensified in the fall.

By April 16th the Trump administration released broad guidelines for how people could return to work, to church, and to restaurants and other venues. The plan outlined the concept of "gating criteria," which called for states or metropolitan areas to achieve benchmarks in reducing COVID-19 cases or deaths before taking the next step toward reopening. On April 23rd, after assisting the Iowa Public Utilities sector with an urgent request for supplies that if unmet could have led to regional blackouts in Iowa due to COVID infections of control room personnel, the Market Intelligence Team engaged with the Department of Energy (DOE) and was asked to present and facilitate a discussion with the Department of Energy and Electric Subsector Coordinating Council (ESCC). CEO engagement from the three major trade industries followed and it was clear that all members were experiencing severe shortfalls with regards to any PPE, disinfectants, or testing necessary to prevent outbreaks among essential personnel. Our market intelligence team prepared and submitted requests to fund and authorize the Air Force to create contract instruments to assure delivery of all supplies and hospital aseptic maintenance level cleaning of facilities rapidly to ensure the continued security of the power grid. This request was never approved.

To support the analysis of market intelligence research, Dr. Handfield reached out to a handful of graduate students, including Nikhil Singh, Beena Thomas, and a number of others. These students combed library resources and online databases to identify potential sources of PPE and other critical medical products. The students developed a comprehensive database of potential providers of PPE across the world.

By the end of April as the pandemic lingers, economic impacts are mounting and beginning to work against public health. More and more cases of deferred care are occurring (patients avoiding doctor's offices and hospitals for procedures that could wait). According to Gallup polling this includes 1 in 7 Americans who report they would not seek care for a fever or dry cough—the classic symptoms of COVID-19. Their main justification is cost concerns. Those most likely to avoid medical treatment for symptoms are younger than age 30 and make less than $40,000 a year. By the end of April, 26.5 million Americans have filed for unemployment since mid-March. There starts to be early discussions about the potential benefits of Remdesivir. The National Institutes of Health (NIH) trial data, which are not peer reviewed, show that remdesivir, made by Gilead Sciences, is better than placebo in treating COVID-19. Patients with advanced COVID-19 and lung involvement who received the antiviral had faster recovery times. At the same time our team develops and provides a system for adjudicating rapid 72-hour emergency medical supply requirements through a rapid market intelligence process/network for use by the JATF. These were requests in which supply emergencies validated by the National Response Coordination Center involved supply shortages that would likely result in loss of life in 72 hours or less. Multiple requests were fielded across the country for items including bag valve masks, various PPE items, medication necessary for intubation/ventilation, and other various needs.

May

On 1 May, shortly after the trial data are published, FDA grants an EUA to Remdesivir after preliminary data from an NIH trial found the treatment accelerated recovery in individuals with advanced COVID-19 and lung involvement. By 9 May the FDA broadens authorization of a saliva-based test to detect COVID-19 infection; the EUA is granted to Rutgers Clinical Genomics Laboratory. The test makes it possible for those who cannot get to a collection center to get tested, including those who are home because they are ill, quarantined, or at high risk of infection due to their age or comorbidities. On 12 May Anthony Fauci, MD, director of the National Institute of Allergy and Infectious Diseases, testifies before the US Senate that the US death toll of 80,000 is likely an underestimate. He warns against the relaxation of social distancing and says he is "cautiously optimistic" that a vaccine will be effective and achieved within 1 or 2 years.

A day later (13 May) our team provides the initial governance concept and National Contingency Supply Chain Cell (NCSCC) structure for an "SNS 2.0" strategic plan to the JATF supply chain lead in response to a request for information to support a review of Peter Navarro's Op-Ed regarding the President's visit to Allentown, PA. The team voices concern over the focus on increasing the SNS inventory size and heavy reliance on firms such as Cardinal Health, Owens and Minor, McKeeson and AmerisourceBergen to maintain larger inventories of PPE as a key feature of an SNS 2.0. The next day, our team interviews a member of the Johns Hopkins team supporting the current SNS. They indicate alarming management practices regarding item expiration management. These include large lot buying for items that expire simultaneously and are not scheduled for replenishment in sufficient time to prevent a shortfall, manual data entry and personal cell phone camera use for tracking items in the stockpile, loss of traceability into the shelf-life values lost between item production and when it actually shows up at the SNS site, and tendency to wargame contingencies based on historic versu forward-looking risks. By 19 May 2020 our team briefs a SNS 2.0 strategy on behalf of the U.S. Air Force COVID-19 Acquisition Task Force to the JATF. No immediate decisions are made by the JATF, but feedback is highly positive.

On May 21, 2020 the Trump administration and AstraZeneca announce a collaboration to speed development of a COVID-19 vaccine called AZD1222. Health and Human Services says it expects the first doses to be available as early as October 2020. Six days later the use passes 100,000 COVID deaths. The CDC says surpassing 100,000 deaths is a "sobering development and a heart-breaking reminder of the horrible toll of this unprecedented pandemic." It asks that Americans continue following local and state guidance on prevention strategies, such as social distancing, good hand hygiene, and wearing a face mask while in public.

June

On June 4th both the New England Journal of Medicine and The Lancet retract studies on the use of hydroxychloroquine in COVID-19, after the study authors said they could not vouch for the data used. A private database of medical records compiled by a little-known firm called Surgisphere was used in both studies. By June 10th US COVID-19 cases reach 2 million as cases rise in 20 states under relaxed social distancing rules. HHS announces on 16 June that vaccines under Operation Warp Speed would be provided for free to elderly patients and other vulnerable populations who cannot afford it. By 18 June the WHO has ended the study into hydroxychloroquine as a treatment for COVID-19. The data from the Solidarity Trial show the drug did not reduce mortality. Two days later the NIH announces it is halting a clinical trial examining the safety and effectiveness of hydroxychloroquine as a treatment for COVID-19. The study indicates that the treatment does no harm, but also provides no benefit. A study in Science Translation Medicine suggests on 22 June 2020, that as many as 80% of Americans who sought care for flu-like illnesses in March were actually infected with the virus that causes COVID-19. According to the research, it may have amounted to 8.7 million infections. On 26 June, the White House Coronavirus Task Force holds a briefing for the first time in two months. The focus of the discussion is the rising number of cases and growing positive test rate in some states. As cases rise, Texas and Florida both decide to halt the re-openings as each state records growing numbers of cases. By the end of June Gilead Sciences sets a price for Remdesivir, which can shorten hospitalization stays for patients with COVID-19, at $520 a vial. With a treatment course of 6 vials, the typical treatment course will be $3,120 per patient for people covered with private insurance. Critics of the price point are quick to point out that taxpayers funded the COVID-19 Remdesivir trial through the National Institute of Allergy and Infectious Diseases. And finally, on 30 June, Dr. Fauci testifies before the Senate Health, Education, Labor, and Pensions Committee, that while the current daily number of new cases in the United States is hovering around 40,000, that could reach as high as 100,000 new cases per day given the outbreak's current trajectory.

As the month of June 2020 closed out, our team transitioned management of the market intelligence team to new members. Pete Guinto and Dr. Finkenstadt transitioned to new roles within the DoD. All three of us have continued to research, advise, assist, and teach local, state, federal and non-profit organizations surrounding the pandemic and various other supply chain issues resulting from this global event. This chapter provides a brief overview of critical events. In part it shows the types of efforts in which we were involved, but mainly it shows the rapid fashion of an evolving global contingency, the disparate nature of global coordination and the dramatic gap in necessary purchasing and supply chain management for future catastrophes. We hope this book can be a start to filling that gap for our readers and their stakeholders.

2 Tragedy of the New Commons

For any disease to flourish and spread, conditions have to be right; or more appropriately, things have to go wrong in the right sequence. During the early periods of the pandemic, market conditions in place perfectly allowed for inefficient supply and demand supply chain behaviors; helping COVID thrive. Every distribution system seemed to encounter a stone wall consisting of self-interested human behavior, in what Garret Hardin dubbed in 1968 as "The Tragedy of the Commons".[1] This phenomena was first noted in 1832 by Oxford economist William Forster Lloyd while observing common pastures in England. He observed that the livestock on the common pastures was smaller and in poorer condition while the land itself was more bare than the land around it. In this situation the individual gain of a herdsman adding more livestock to the commons outweighed their individual stake of communized costs, at least in the short run. Sadly, however, all mismanaged commons end similarly; as each individual calculates their gain relative to their share of costs, they take part in opportunistic and self-serving behaviors. Collectively all individuals in the commons acting out of self-interest deplete the commons of resources, thereby rendering it useless for all.

[1] Hardin, Garrett. "The Tragedy of the Commons." Science 162 (1968): 1243–1248.

R. Handfield and D. J. Finkenstadt, *Supply Chain Immunity*, Synthesis Lectures on Operations Research and Applications, https://doi.org/10.1007/978-3-031-19344-6_2

Source https://www.econlib.org/library/Enc/TragedyoftheCommons.html

We began to see self-interested, myopic and short-sighted behaviors proliferate the supply chain landscape almost immediately after COVID-19 arrived in the U.S. These behaviors began to impact supply chains of toilet paper, yeast, milk, eggs, meat, masks, personal protective equipment (PPE), ventilators, and polymerase chain reaction (PCR) and antibody tests. Eventually it extended into every area of demand during the pandemic. Additional contingencies that arose from 2020 to present have introduced more hoarding and irrational buying behaviors. Multiple fuel network disruptions from weather and cyber intrusions have led to a run on the pumps with citizens going so far as to use plastic trash bags and coolers to collect what they perceived as a safety stock of fuel. Such tragedies have to be expected and planned for to keep from depleting resources inefficiently and unnecessarily in times of crisis.

Source https://poole.ncsu.edu/thought-leadership/article/save-me-a-roll-understanding-consumer-purchasing-behaviors-in-the-midst-of-covid-19/

The US health sector failed to properly manage need versus demand signals and collective management of items that became a new commons during the pandemic. Need signals would have provided insight into the requirements to equitably maintain a healthy population. Demand signals are not appropriate for managing resources during a mass contingency that reaches global "commons" proportions because demand can be influenced by self-serving goals that do not consider impacts on others and long-term effects. We observed and experienced shortages in the available supply of beds, pharmaceuticals, critical devices, and human resources. The disruptions were not confined to a single purchased material, a specific region, or a specific cross-border export restriction. They were everywhere, driven by a mass contagion affecting the entire global supply chain. Not only were these disruptions widely dispersed, they were also long-lasting.

During a global event such as COVID-19, critical items need to be abundant, credible, useable and accessible in a way that can meet the needs of the greatest proportion of the population. This is especially true when it comes to our first responders and healthcare providers – that are normally prioritized within the population as they are on the front lines of helping others to survive the virus. Though COVID-19 resulted in enormous loss of life, it was not nearly as bad as it could have been. The worst of the health care supply chain woes were fortunately short lived, months instead of years, and lethality was high, but not as high as our worst fears. Even so, the healthcare supply delivery infrastructure immediately created haves and have-nots. Large hospitals with larger purchasing volumes and financial resources hoarded supplies. At the same time, within the Government, the absence of quality data to assess need made it impossible to equitably distribute finite supplies to maximize social utility. The scale of that population and the question of how to allocate distribution of those items is of paramount importance. Stewardship of commons, both new and old, is required. Countries need a "commons-based strategy". This cannot simply be a stockpile repository, it must include a network of repositories, fluid inventories, and analytic monitoring governed by the experts. We need coordinated "commons" management plan(s) that will direct both conventional and new suppliers to meet demands, equitably address needs for basic life support and sustainment goods and services and to eliminate hoarding and other opportunistic behaviors.

We have offered up the concept of "certainty satiation marketing" as a means of dealing with the increased desirability and demand that builds when goods are perceived as scarce.[2] Researchers have found a plethora of contributing factors to irrational buying behavior such as social media and social cognitive biases, pandemic stress, fear of contagion, and particular personality traits (conscientiousness) (Labad et al., 2021). The current set of supply chain disruptions is being set off by a combination of COVID cases in manufacturing hubs around the world, labor shortages, lack of capital infrastructure investment, and surges in demand. This has led to a massive set of supply chain shortages, including a scarcity of microchips and massive container logjams in ports brought on by a combination of manufacturing loss during the pandemic shutdowns and increased

[2] Finkenstadt, DJ, Handfield, R and Eapen, TT. (2022). *Certainty Satiation Marketing for Disrupted Supply Chains.* California Management Review—Insights. https://cmr.berkeley.edu/2022/01/certainty-satiation-marketing-for-disrupted-supply-chains/.

demand for consumer goods by a world forced to sit at home. Free market price levels alone have not been able to satiate demand.

The global logistic system is not prepared to process the flow of goods based on irrational buying-selling behaviors that we observe during mass contingencies like pandemics and other supply chain influences. The supply chain has to be viewed in a form of "systems thinking", that views the end to end supply chain as a series of nodes which must all flow without friction. On-line ordering and delivery has made it all to easy to continue to buy as government's continue to deliver economic aid packages that provide cash to consumers who are looking to spend as a relief from the pressures of lockdowns and perceived scarcities. We propose the concept of certainty satiation marketing that involves demand moderation and creates a satiation effect that is an important approach for managing the supply chain disruption realities we face today. We define Certainty Satiation Marketing as: Marketing via the transparency of supply for demand moderation to create a satiation effect. It is focused primarily on managing the information and advice available to consumers so that they can reach satiation at the appropriate time, given environmental market conditions.[3]

Consumer and business demand management have been handled in various ways. Natural price increases should manage demand based on economic theory, but we have seen that price increases can reach a point of hyperinflation. The World Bank is projecting a period of stagflation at levels not seen since the 1970s.[4] Rationing, which has also been used to impact demand, has downsides as well. Buyers and sellers develop strategies for consumption and production based on asymmetric information. Over time the market is supposed to reach a rational equilibrium. Since 2020 we have seen all sorts of erratic buying behaviors for things like PPE, toilet paper, fuel and, most recently, baby formula. In these cases, demand should be managed with a concerted marketing strategy that reduces asymmetric information and increases market certainty until things level out. There is a gap in consumer certainty that needs to be filled. Standard approaches used to inhibit irrational demand (i.e. product rationing and tiered pricing) can increase the perception of scarcity, which in turn leaves customers unsatiated. Figure 2.1 visualizes how irrational demand is influenced by a lack of satiation that aggravates the supply chain and creates a vicious demand satiation uncertainty loop.[5]

Satiation is defined in economics as the point at which additional consumption beyond such a point would yield diminishing marginal utility relative to the cost to consume it. In the marketing literature, satiation is defined as "the process whereby consumers enjoy a stimulus less as they consume more of it" (Coombs and Avrunin 1977; Redden

---

[3] Finkenstadt, D.J., Handfield, R., and Eapen, T.T. (2022) Certainty Satiation Marketing for Disrupted Supply Chains. California Management Review—Insights. https://cmr.berkeley.edu/2022/01/certainty-satiation-marketing-for-disrupted-supply-chains/.

[4] https://www.worldbank.org/en/news/press-release/2022/06/07/stagflation-risk-rises-amid-sharp-slowdown-in-growth-energy-markets

[5] Finkenstadt, et al. (2022) Ibid.

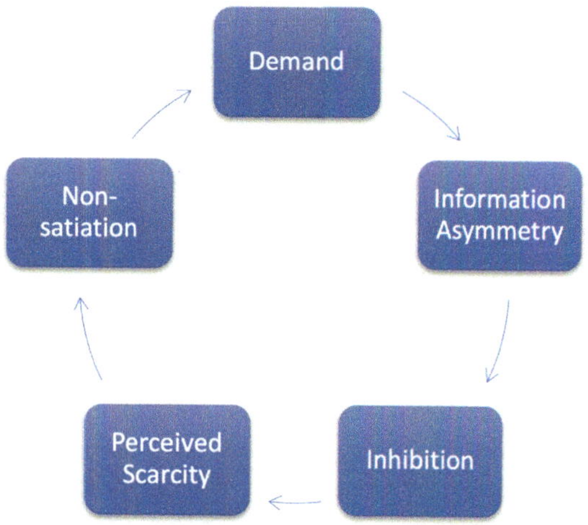

Fig. 2.1 Demand satiation uncertainty loop. Adapted from Source: https://cmr.berkeley.edu/2022/01/certainty-satiation-marketing-for-disrupted-supply-chains/

2008)" (Sevilla and Reddin, 2014, p. 206) something which most firms would like to avoid! In general, most marketers would prefer that consumers avoid reaching a point of satiation, and prefer continuous consumption to provide a stream of revenue and on-going profit maximization. Marketing research also suggests that people reach this satiation point as they consume additional quantities of an item or service and, once the satiation point is reached, they move on to consume other types of products or services (Coombs and Avrunin, 1977; Herrnstein and Prelic, 1991). Research suggests that creating supply scarcities or perceptions of limited availability can prolong the point of consumer satiation and improve customer enjoyment over time (Sevilla and Reddin, 2014).

But how do we apply these concepts to the turbulent levels of supply chain disruption faced in our global economy today? Does it make sense to continue to try to reduce satiation rates, and continue to encourage consumers to consume as much as they want? Conversely, should firms continue to promote the scarcity of its products, sending a message of actual limited availability, and allow prices and thus revenues to continue to grow?

We suggest that these are not the optimal long-run paths for marketers, supply chain managers, firms, and governments to follow. Rather, both private and public sector organizations have a social responsibility to employ transparent and factual messaging, to create a new level of satiation for consumer certainty that may help to reduce strains on a global system of goods and services by increasing awareness for resource availability. This is consistent with the emerging view of a 'commons' during times of mass contingency and global healthcare disruption. Our research suggests that the COVID-19 pandemic has created many new forms of supply chain commons for many public health goods and services, which should be based on transparency and equitability in supply chains to achieve a desired point of stabilization (Handfield et al., 2020; Finkenstadt et al., 2021). We suggest that the supply chain commons metaphor should be married with social marketing concepts via certainty satiation. Certainty satiation marketing and supply chain commons planning can help consumers reach a point of satiation in resource certainty as a type of market contingency demand management during a period of extreme supply chain disruption.

Demand management is a technique used in category management by supply chain and procurement professionals to prevent the buying of unneeded goods and services. The US government employs demand management as part of its category management program and considers it one of its five key recommendations for category management. The US government defines demand management as the practice of "analyzing what and how an agency is buying, with a specific focus on reducing the total cost of ownership" (p. 9) by standardizing requirements, coordinating agencywide buying events, sharing demand strategies, and seeking volume-based (tiered) pricing (Weichart, 2019). The primary purpose of demand management in public procurement is to buy things smarter in an effort to increase value for dollar spent, reduce waste and increase public trust in the use of fiscal appropriations. The government has used these methods to improve sourcing in everything from milling machines for aircraft maintenance to working dogs for explosives and drug detection. These conservation mandates, which have been in use since World War II, can be equally applied to the private sector, and can have an important implication for containing the supply chain disruptions with similar intent to government category management. The government communicates demand management messages through policy and regulation, but such mandates cannot be as easily applied to free market consumers. All parties should consider alternative communication channels to inform consumers about the realities of available goods and services to reach a form of informed satiation; reducing the impacts of hoarding behaviors and/or logistical logjams. Even today we are experiencing massive disruptions in almost every industrial and commodities sector. Beef, poultry, and produce prices are rising, due to production and demand cycles; oil prices surpassed $100/bbl, and resin prices have also surged. Steel, aluminum, and copper are impacting the housing market, and the cost of electronics, appliances, and consumer packaged goods are all going up by 5% or more as a function of raw material shortages and disruptions. In recent conversation a Chief Purchasing Officer of a major CPG company shared how she was in the process of "educating" her Chief Marketing Officer on how raw material prices could not be made up through productivity, and that the only way to continue operating a profitable business was to pass on these price increases to consumers. In a similar vein, we need to be thinking of how to educate consumers of the reality of what is happening in global supply chains, that is leading to increased leadtimes, increased pricing, and reduced product availability on retail and e-commerce channels. The irony of this point cannot be missed: we need to inform a broad consumer base that their commercial buying behaviors, resulting in a demand surge are largely responsible disrupting global supply chains!

Public Service Announcements (PSAs) have been around in America since World War II as a form of social marketing. Social marketing is defined as "process that applies marketing principles and techniques to create, communicate and deliver value in order to influence target audience behaviors that benefit society (public health, safety, the environment, and communities) as well as the target audience" (Kotler, Lee and Rothschild, 2006 personal communication cited in Cheng, Kotler and Lee, 2009). Researchers have cited the use of social marketing for health promotion, injury prevention, environmental protection and community mobilization (Kotler and Lee, 2008). These marketing communications are intended to influence consumer behaviors to get them to (1) accept a new behavior such as buying electric vehicles, (2) reject a potential undesirable behavior such as starting to use drugs, (3) modify current behaviors such as reducing water usage or (4) abandon an undesirable behavior such as driving without a seatbelt. We have also seen the government use rationing as a means to control public demand for consumer goods. A downside is that this method has been found to increase the development of black markets. Black markets increase the risk of counterfeiting and we have seen that behavior take hold in dangerous and unregulated ways for items like N95 masks, vaccines and vaccine cards during COVID19. So, we assume that we would prefer not to ration entire swaths of consumer goods and services. And this leads us back to social marketing. How can we use social marketing as a means to reduce market demand and convince consumers that their needs are and will be satiated in a reasonable time, without impulsively hoarding or scrambling for goods that seem to be scarce during supply chain disruptions?

We suggest that firms consider our concept of certainty satiation marketing. We are NOT suggesting that we always tell customers to STOP consuming, as is done to prevent environmental harm (i.e., reducing coal or petroleum fuel consumption). Instead, we are asking them to consider the impact on global commons (i.e. for the common good of humanity and the community). Why is certainty satiation marketing different than traditional social marketing? The rationale driving certainty satiation marketing run counter to traditional marketing principles, and are not focused exclusively on modifying consumer behavior for a traditional social good. The social benefit in this case is related to the need to ease constraints facing global supply chain operations. Supply chains are pivotal, as we have seen, to the flow of everything from food markets to defense markets, and shortages can create havoc in healthcare, air travel, transportation, and even a trip to the supermarket. Massive disruptions in the processing capabilities of supply chain systems are often triggered by irrational buying and selling behaviors; as these behaviors spread, the equivalent is a bullwhip effect on steroids, leading to chaos in global supply chain system performance. The imbalances currently facing supply chains is having major impacts on public health and welfare, but even disrupted parts delivery for harvesting equipment for farmers, putting the 2021 fall harvest at risk when equipment cannot be maintained in a functioning state. But satiation marketing is also about clearly articulating the availability and shortages of goods to consumers. Many consumers have the public good in mind, and the knowledge of scarcity will guide most consumers to make rational buying decisions within the spirit of that public good. This is not about stating how much or how little they should buy, but rather involves imploring consumers to only buy what they need, and not to hoard at the expense of others in their own communities. This line of thinking may turn the marketer's job a bit on its head. Instead of focusing on delaying satiation by manipulating perceived scarcities for the purpose of continuous demand and revenues, the strategy involves increasing supply chain transparency and equitability for the purposes of reaching rational satiation levels commensurate with market realities. In as sense, creating alignment between demand and supply constraints is a means of informing consumer expectations, which will in most cases lead to measures of conservation.

Conservation is not a bad thing. As mentioned, conservation measures during World War II led many private sector companies like GE and Ford to create a new engineering tool called Value Analysis and Value Engineering. The goal of VA/VE was straightforward: to simplify the design and manufacture of products, to achieve the equivalent or better functional outcome, at a lower or equal price. One of the most important tools in VA/VE was simplification and standardization, which involved using whatever material was in plentiful supply, and avoiding unique or difficult to order materials, including those that were being rationed by the US government like steel, cotton, etc. VA/VE proved to be an important option in the stable of tools in satiation marketing. Conservation was also applied during COVID in healthcare, when many hospitals began by examining their clinical practices, and advising physicians, nurses and staffs on new recommendations to reduce the overall demand for masks, gowns, and gloves. An example was to re-use the gown one was wearing when visiting multiple patient rooms, and not to change into a new gown every time one entered a new room. Another example was to sterilize N95 masks using hydrochloric acid using approved FDA technologies. These policies made a huge dent in the volume of goods used by hospitals during a period when supply for PPE was stretched thin, and allowed many hospitals to scrape by without running out of PPE.

There is also risk that by not partaking in certainty satiation marketing, entities may choose to enact demand management strategies that seem helpful but are actually harmful. We have mentioned that rationing can lead to things like black markets. But it can also lead to opaque markets. A non-profit PPE provider we interviewed informed us that they were unable to advertise domestically produced N95 masks on Google. Google had stated that the US government was asking them to not advertise N95s anymore due to supply shortages. The irony was that there was actually an abundance of domestic products. The shortage was only for traditional name brand products like 3M. Google was essentially rationing advertising information. Google, it seems, was not aware of a host of domestic respirator firms stood up using government funding in 2020. They were also not aware that these firms were at risk for bankruptcy because they could not overcome the competition from cheap Asian products and couldn't get their advertising posted to take advantage of the market when those competitors were low on stock. Using stronger certainty satiation marketing strategies, Google and the government could, instead, increase transparency into the types of products that were available for advertising (i.e. domestic respirators). Interviews we conducted with 3M revealed that their normal production levels are approximately 20 million masks. In March 2020, the company received orders for more than 2 billion N95 masks—a number that was of course far more than they could even produce in a year! By rationing the information on available stock, Google and the government had inadvertently contributed to the feeling of resource uncertainty. As Fig. 2.1 shows, this leads to more irrational ordering behaviors which further impacts available stocks of critical supplies.

We have considered various ways in which demand satiation can occur. Satiation focused strategies leverage information to build trust and should be considered in the future. Simply providing further visibility into supply chains reduces uncertainty surrounding future availability of goods and services and can help to rationally balance demand. Of course, this assumes that suppliers and manufacturers have this visibility themselves. Social proof and quantity anchoring can be used to give a clearer picture of others consumption rates, thereby influencing consumer's reference points for rational consumption in uncertain climates. Similar to social marketing, consequence extrapolation can be tailored to provide information about consumer demand and potential choices on the health and well-being of others. For example, COVID-19 vaccine booster shots are currently an ethical dilemma when half the world has not received their first shot. Yet human behavior in this uncertain time has caused many to seek their booster shots illicitly, impacting the global commons (Gutman, 2021). How might they behave if they new the actual impact of jumping the line?

Incentive alignment works in tandem with supply chain visibility and has been shown to improve outcomes for all parties (Narayanan and Raman, 2004). Better information leads to stronger incentive-alignment. Suppliers could use methods that align price savings features of a product to its actual usage versus simply a promotional period of time. At the same time suppliers and retailers can look to supply-based promotions as a way of moving excess stock of substitution products. For example, in the case of N95 masks, producers could offer better pricing for domestic product during a time when traditional, cheaper suppliers are short of supply, and advertising platforms like Google can help highlight them during such a period and thus increase the overall benefit to the greater commons supply chain and consumer.

In a similar manner time-shifted promotions can be used to provide discounts into a future period when normal consumption would be assumed to have taken place. Price promotion need not be the only mechanism shifted to rebalance demand. Buy-now deliver later (BNDL) and subscription warranties provides a mechanism for consumers to pre-order items and simply wait on them to arrive at a called upon date in terms of BNDL or a specified delivery schedule in terms of subscription warranties. This creates a means of delaying delivery until a normal, more rational, consumption period expires and reduces hoarding incentives. However, both of these methods require a great degree of trust on the part of the consumer that firms and governments can support them through strong enforcement governance and supply chain transparency.

Building trust may not occur in the chaos of a pandemic or other mass calamity without getting the consumers, firms, and governments involved in collaborative activities that build on information transparency and incentive-alignment. Google Trends shows that from Oct 2019 to Nov 2021 searches for supply chain "shortage", "crisis", and "management" are up 900%, 500% and 250% respectively. The public is aware there is a problem, and they are more curious than ever in recent history. It is an exceptionally opportune time to use trust-building engagements with the public to discuss efficient, resilient, and ethical supply chain management and consumer demand behavior during times of global crises. This can be used to encourage actions as simple as group purchasing with capped quantities at lower prices, to engaging in crowdsourcing solutions to supply and demand issues plaguing these new market commons in a post-pandemic world (Table 2.1).

Table 2.1 Summary of example demand management (DM) Strategies (Eapen and Finkenstadt, 2021)

DM driver	Strategy	Summary and limitations
Inhibition	Demand based pricing	Customer pay a higher price when aggregate demand goes up. This is the standard economic response to increased demand and is associated by the customer with inflation
	Product rationing	Each customer or household is allowed to purchase only a fixed quantity. Customers are sometimes required to show identification
	Tiered pricing	The consumer pays regular price for a threshold number of units. Purchase beyond this quantity is charged a higher price. Theoretically, there is no limit on purchase quantity
Information	Supply chain visibility	Consumers are made aware of specific information about the supply chain to assuage concerns regarding availability
	Social proof and quantity anchoring	Consumers are shown information relating to suggested quantity that can potentially satiates demand. For example, a message next to the product might read, "An average person in the US consumes an equivalent two units of the product per week"

(continued)

Table 2.1 (continued)

DM driver	Strategy	Summary and limitations
	Consequence extrapolation	People are provided information about needs of other users or provided with dynamic information on how their potential choice (if extrapolated) can impact the global commons or other individual consumers
Incentives	Consumption-linked pricing/promotion	Promotions or coupons only visible to customers once the product is opened or consumed. For example, a coupon that is printed on one of the tissues roll cores for a discount[6]
	Supply-based promotions	Promotion of products that are excess in supply
	Time-shifted promotions	Promotions that are redeemable only at future date. Time shifted promotions are designed to even the customer demand. Let's say each unit of product is given a single coupon for 10% discount. However, the coupon is only redeemable after 1 month from the purchase of the original product[7]
	Buy-now deliver later (BNDL)	Products can be ordered today for a future delivery at current price. Consumers are guaranteed of obtaining product at today's price and save on storage space[8]
	Subscription warranties	Consumers can buy vouchers for future delivery of products which if unfulfilled will result in penalty payout by the company[9]

(continued)

[6] This approach partly solves the dilemma that irrational demand is both a good problem and a bad problem to have. The supplier needs to even demand.

[7] For example, for every toilet roll box purchased, the customer is given a discount coupon with a future redemption date that is 2–3 months away from current purchase date. The redemption date may be randomly staggered to further even future demand.

[8] This strategy responds to consumers' concern about future increase in prices.

[9] This strategy responds to consumers' concern about future availability of product.

Table 2.1 (continued)

DM driver	Strategy	Summary and limitations
Involvement	Trust-building	Consumer groups are encouraged to participate in discussions about supply chain to build trust within communities
	Group purchases	Group discounts where groups are allowed to buy products for a lower price with limitations on overall items purchased by a group
	Conservation and education campaigns	Customers are involved in supply chain conservation and management efforts. For example, crowdsourcing contests could be used to encourage customer participation in the global supply chain. Education campaigns on supply chain conditions, demand management, and risks to the global commons provided via webinar, open-source literature and social media and audio platforms

Part II

Symptoms and Treatments

3 Slow Response

3.1 Vaccine Hesitancy

Disasters can change the course of history. The goal of the organizations that provide humanitarian assistance and disaster relief (HADR) is to reduce pain and casualties. Effective logistics in the case of a pandemic necessitates getting the necessary emergency supplies and services to the affected population in time. Disasters have been classified as shown in Fig. 3.1.

Though pandemics are of slow onset, they can disperse quickly as the virus spreads or mutates. It is known that 284,000 died due to H1N1 in 2009, and Ebola consumed 28,616 in a West Africa outbreak in 2014. COVID-19 now has a vaccine. Yet at this writing, the deaths due to COVID-19 are more than a million in the United States (US) and 6.5 million in the world. The deaths in US before the emergence of a vaccine are understandable but not after the vaccine became available, given that there are many underdeveloped and developing nations in the world that have either marginal or no access to the vaccines. It is unfortunate that a rich country like US with fairly smaller population has more deaths than Brazil or India. As the variants such as the Omicron BA. 5 continue to evolve and become more easily transmissible, hospitalization is less likely to occur, but the continued increase in new outbreaks for those who have been vaccinated and have the disease will increase the likelihood of other medical problems associated with long COVID.

During a pandemic such as COVID-19, the distribution and delivery of critical supplies and services to the affected population is a major undertaking. Providing relief depends on the speed of the response and matching the supply with the need assessed. Response supply chains for vaccines and Personal Protective Equipment (PPE) must be established in anticipation of pandemics and need to be robust. "The challenges in a response supply chain such as demand surges, uncertainty of supplies, [and] critical time windows in face of vulnerable infrastructure (physical as well as social), raise several serious questions.

R. Handfield and D. J. Finkenstadt, *Supply Chain Immunity*, Synthesis Lectures on Operations Research and Applications, https://doi.org/10.1007/978-3-031-19344-6_3

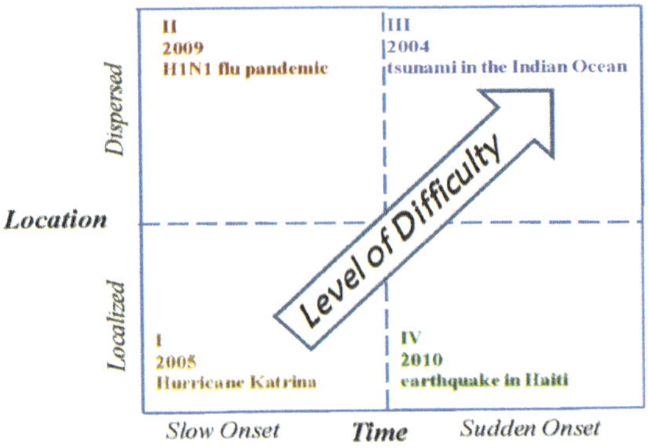

Fig. 3.1 Classification of disasters. *Source* Apte (2009)

For example, how can adequate pre-positioned disaster relief inventory be established and maintained? How reliable are the potential, virtual, as well as real supply lines?" (Apte, 2014). Other questions, especially for pandemics that come to mind are, should the supplies be sourced locally, or should they be imported? Is pre-positioning the best strategy for pandemics?

These questions beg answers. Especially when COVID-19 descended upon US, the lack of information about the impending pandemic was detrimental. Knowing this ahead of time might have resulted in planning for various issues. The vaccine that was created, which was one of the greatest feats in medical history, could have been anticipated. The disbursement of information and the steps taken to create the vaccine could have reduced, if not eliminated, mis/disinformation. Unfortunately, none of this happened, and with the proliferation of social media, the lies spread. This resulted in hesitancy and in some cases downright denial for the vaccine. When people started refusing the vaccine, all systems were strained. The occupancy in hospitals increased to the point of lack of ICU beds. The healthcare community was stressed out and downright shattered. This happened not just in the US but elsewhere in Europe and Asia as well. Many of us remember the first pictures from Italy and India. Such stress on the healthcare system could have been avoided if the necessary personnel and PPEs could have been strategically prepositioned in terms of supply chain.

Now that the vaccines are available, there is a new problem. There is huge hesitancy and downright refusal to get vaccinated. The numbers tell the story. Figure 3.2 shows the percentage of fully vaccinated people in various countries across the world, at this writing. One in five Americans still state that they would refuse the vaccine (Hoffman, 2021).

There are at least two reasons for lack of vaccination: one, there are no vaccines available and two, vaccine hesitancy or denial. In poor countries, lack of vaccines leads to lack of vaccination. However, when vaccines have been available for more than a year in most of the richer countries for everyone 12 years and older, the underlying reason most likely is vaccine hesitancy or denial. It is commendable to see that poorer countries like

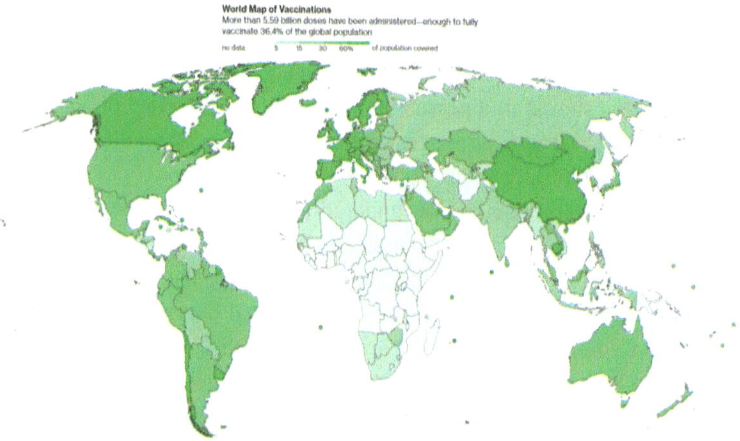

Fig. 3.2 Percentage of fully vaccinated people in various countries across the world, at this writing

Malaysia and Cambodia have a majority of the population fully vaccinated. This brings us again to the problem of vaccine hesitancy or denial during a season when boosters are needed to stave off additional waves of infection.

Unfortunately, the virus does not recognize national boundaries. It seeks to grow and find new ways of attacking the human body through twists and turns, transforming itself into variants. This then becomes a major threat to developing or underdeveloped nations especially on the African and Asian continents. Mutation of the virus strains the humanitarian organizations in their efforts to mitigate the pandemic. Their resources are depleted faster, and hence the scarcity of medical supplies—including vaccines, PPE, and personnel to deliver the supplies, creating another crisis. Providing vaccines to poorer nations is feasible since there are many types of vaccines available now. What is harder is changing people's minds about receiving the vaccine.

COVID-19 has become the most illuminating event in modern humanitarian disaster history, highlighting the importance of supply chains. We have heard that expert opinions can impact the public's vaccine acceptance. However, most of this assertion is focused on medical experts. Given that supply chain commentary surrounding the vaccine has become ubiquitous in the press, we wanted to explore the perceptions of the vaccine's supply chain effectiveness from experts in the field. We do not test the impact of those perceptions on actual vaccine acceptance rates. Our interest is in the stated confidence and intent of these experts, as that is what is most likely to be communicated to the general population and have the most influence beyond the individual experts themselves. We do offer findings to suggest that traditional Supply Chain Operations Reference (SCOR) model perceptions may not be the best measure of confidence for the global vaccine program and individual expert willingness to accept the vaccine. This finding calls for an additional means of measuring supply chain expert perceptions that can be indicative of influence on the larger public. We posit a potential scale developed to measure perceptions of supply chain immunity. Given the impact of vaccine hesitancy on the pandemic's ability to spread/mutate and otherwise chip away at the greater global good, experts in every area of the vaccine process should find means to take stock of their perceptions, measures, and public influence. This research is the beginning of a series of studies into supply chain expertise, perceptions, and influence.

Disaster response is either helped or hindered by information flows. This is especially true in pandemic disasters where information can impact confidence in and willingness to participate in vaccination programs. Such impacts can turn the tide for the entire world. Research has shown that misinformation and public health expert influences can impact vaccine inclination among respondents. Previous research has shown that trust in public health experts and non-experts impacts COVID-19 vaccine inclination (Loomba et al., 2021; Callaghan et al., 2020). However, research has not explored experts in related fields such as supply chain management. Supply chain (SC) experts are a relevant expert group to explore given the overwhelming volume of and credence given to SC discourse and testimony surrounding the COVID-19 pandemic. In 2021, one of the authors testified in front of the Homeland Security and Government Affairs committee surrounding supply chain issues directly attributed to COVID-19. Two of the authors have been advising various government and non-government organizations on US national supply chain response lessons learned and implications for future legislation as part of their experience with the COVID-19 response. An informal poll of professionals conducted in September 2021 on a social media platform shows that, while most persons rely on the expertise of public health officials to provide accurate details about the COVID-19 vaccine program, 25% would rely on supply chain experts. This indicates that there may be significant influence on the wider population by the positions of supply chain professionals when it comes to the COVID-19 vaccine program.

However, as the world seeks to reach herd immunity to curb this humanitarian crisis, we must consider all areas of influence to maximize vaccine acceptance rates, especially in countries with remarkably low vaccination rates to date. We measure the supply chain management confidence for the COVID-19 vaccine program using aspects of the supply chain operations reference (SCOR model) within and between two sample groups of supply chain professionals.

3.2 Testing: The Missing Link to Stop the Spread of COVID

As noted earlier in the book, testing represents one of the most important components of a preparedness strategy for COVID. During the throws of the COVID pandemic, a team of researchers at NC State University conducted interviews with 5 manufacturers of COVID-19 testing kits (Abbott, Becton Dickinson, Biomedomics, Diacarta, and Siemens), two suppliers of test kit raw materials (IDT, Thermo Fisher), 5 laboratories

(Discover, Helix, LabCorp, Quest Diagnostics, and Toledo Labs), 3PLs and distributors (UPS, Cardinal Health), as well as several government organizations (White House Supply Chain Task Force, National Governor's Association, and the National Association of State Procurement Officers. We also interviewed a number of Chief Procurement Officers from several states that were particularly hit hard by the COVID pandemic (Utah, Montana, North Carolina, California, and Louisiana, amongst others). We also met with the Department of Health and Human Services for the state of North Carolina, and a number of NGOs (360RX, USP, Project N95, ECRI, ACLA), and a large national long term care facility (Genesis). The interviews yielded a number of important insights that were critical in helping us to understand the nature of the challenges faced in the end-to-end supply chain for COVID-19 testing.

The good news was the speed at which employees at these organizations diverted their focus from current technologies to focus exclusively on delivering testing for COVID-19 response. New tests were quickly developed, and production was ramped up across the industry to meet the escalating demand for COVID-19 test kits and testing. The demand for testing remained high, and continued to escalate throughout most of 2020 and into 2021.

> "A common story – some employees lived at the plant and didn't leave the plant and kept production going..."

With the start of the pandemic, all providers sought to leverage existing technologies and ramped up their workforce to meet the need causing a series of bottlenecks that sprang up for critical components within the supply chain, including reagents, adjuvants, plastic components, transport materials, and other elements. In some cases, organizations traveled to Asia to develop new suppliers of these components, to ramp up critical capacity for consumables used in test kit development. Shortages of equipment for testing also popped up, as existing equipment was used at capacity levels far beyond what they were designed for, and in some cases, the equipment began to break down.

> "When the pandemic hit we changed our focus to address the need for additional testing and developed an assay and find and stabilize a supply chain for collection materials that gave us the ability to take a large footprint and bring a good mass of tests to the market and to continue to grow that.... and our genetics business hasn't gone away – and we have to balance our lab throughput to maintain our core business and address the needs".
>
> "We don't have a crystal ball to predict the next bottleneck (in the supply chain) – say for the next week. We need to plan continuously. We don't know what will hit us next."
>
> "When I get products – we are also on allocation model with our customers – and everything is a fraction – and we get limited production – and we are second or third on allocation and we have to allocate to our customers."

Our data suggest that the number of weekly antigen tests that became available in January 2021 was approximately 48 M per month, and that annual capacity was in the range of 1.3 B tests. Although demand today is much less than this volume, we noted that the total volume of testing in 2020 was much less than it should be, in order to halt the spread of the virus. If the recommended volume of testing were to be conducted, (the equivalent of the testing requirement established in the Harvard study of 20 M tests per day), there would be a shortfall of about 100 M tests per week. Using the Brown University analysis, the numbers are more promising, and the overall production levels could potentially be expanded enough to meet a more robust and proactive test strategy of 4.4 M tests per day. However, when one considers the likelihood of additional component and machine capacity constraints outlined previously, we are more likely to see a capacity of perhaps 20 M per week, which is considerably lower than what is needed for each of these testing approaches.

- **PCR Test Kits:** By and large, existing capacity was able to keep up with demand, especially for PCR, although lead times were extended to 8–10 days during peak periods due to bottlenecks in laboratories. Manufacturers or test kit equipment are still backed up, especially for consumables. Laboratories we spoke with now have ramped up their capacity to conduct tests, and larger testing labs are able to move capacity around from lower demand areas to hot spots.
- **Antigen Test Kits:** There was a large shortage of antigen tests early in the pandemic. The Abbott BinaX test originally deployed with only 200 M tests that were procured by the government first for distribution. Other antigen tests, including the Quidel and BD Veritor tests which required equipment for testing became more widely available during the summer of 2020. Our interviews suggest that the Quidel and BD tests are more challenging to administer and therefore not as easy to deploy. Newer antigen "rapid" tests are coming into the market, thus making antigen tests largely widely available. "At home" tests are also being introduced, which are easier to use and these have less difficulty in collecting samples.
- The false negative percentage for antigen tests remained at approximately 10%, and for that reason, there is some hesitancy in how to use antigen tests vs. PCR (molecular tests). However, the probability of positively identifying an infection with an antigen test can be improved by repeating the test A variety of approaches to use the antigen and PCR testing in combination strategies have emerged across states and hospitals. A majority of these use antigen tests to initially test for positivity. If positive, the individual is tested with a PCR and is quarantined.

> "The landscape changes and what we learned: there are so many variables and kinds of tests being accumulated and the inexpensive tests for antibodies and antigens have been dismissed as not important - doesn't have the required sensitivity."

- **Antibody Test Kits:** Capacity for antibody test kits remains well above the demand for these kits. Initially, 200 M tests were distributed by the US Government and allocated to states. In fact, several manufacturers we spoke with note that the expiration date on some kits were potentially going to render their stock unusable, as so many had been produced and demand has been very low. There remains a great deal of confusion as to the role of antibody tests during the COVID-19 pandemic, but we believe their role will become much more important as the vaccine is rolled out.
- Supply bottlenecks have existed for manufacturers, and many of these have been shifting dynamically during the course of the current pandemic. There has also emerged a shortage of trained staff to collect and process tests, which has led to increased salaries and overtime rates in testing labs. Although smaller companies currently have available capacity, they are more dependent on manufacturers for equipment, who are backlogged. The smaller companies are also more likely to be more impacted by component supply shortages.
- Manufacturers of PCR tests experienced a number of supply bottlenecks, including allocation backlogs of several months for plastic custom plates (Hamilton equipment) as well as reagents and adjuvants. Hologic's custom pipette tips also had backlogs, with many of these suppliers located in Asia.

Prior to the pandemic, most healthcare organizations did not consider strategic sourcing or management of PPE to be a core capability essential to their business continuity. But the pandemic exposed the flaw in that logic. The spread of the disease across much of the globe came as a shock to business and government leaders, and there were widespread problems in the U.S. federal response, including outdated tracking systems and inadequate planning. In effect, FEMA's message to businesses and the states was: "*We will try to be there for you, but you really need to take care of yourselves*". One state chief procurement officer told us that at some point he realized a bitter truth: that "*federal looks out for federal, and leaves the states to their own devices.*" Many organizations had no choice but to scrounge for their own supplies. Some got lucky: The state of Utah discovered that an importer of goods for the 2002 Salt Lake City Olympics had maintained contacts with Chinese suppliers of PPE, and this company became a go-between, ensuring that products were sourced from reliable suppliers. Other organizations were not so fortunate. Lacking experience in sourcing from overseas, they resorted to scouring the internet, chasing rumors, or responding to email offers. But opportunists were way ahead of them: Price gougers, counterfeiters, and scammers had appeared almost overnight to take advantage of the desperate need. State governments were inundated by so-called "distributors" who demanded payment up front for masks from "friends" in China. We had seen similar scams for construction equipment spring up after major hurricanes, but this was on a much larger scale.

There were other problems at the state level too, such as organizational snafus at procurement offices. States operated independently, in some cases hoarding PPE or being reluctant to share information with one another on their sources. As a result, no one had a complete picture of where critical PPE supplies were located, needed, or should be coming from. With time, some of the worst shortages have abated. As infections declined in China, companies there resumed making and shipping PPE, and a number of domestic companies, including 3M and a number of smaller firms, ramped up production. A few nonprofits, such as Project N95, have sprung up to help organizations source supplies. Many state procurement offices finally feel they have enough PPE on hand for their needs.

Yet even now, two years into the pandemic, shortages remain either an immediate problem or a looming worry, particularly in healthcare, a heavy user of PPE. Some hospitals are still foraging for equipment and are forced to decontaminate disposable masks and gloves for reuse. In November 2020, Baystate Health system in Massachusetts reported being able to purchase "just enough" N95 masks "*to get by, but not enough to store many away*," according to *The New York Times*. Dr. Adam Brown, president of Emergency Medicine at Envision Healthcare, one of the largest medical groups in the U.S., told us that the organization does not have enough N95 and KN95 masks to follow the normal single-use protocol for airborne illnesses, so it started disinfecting and reusing masks as needed. Shortages are often worst at organizations that lack funds and access, such as historically black colleges, disadvantaged communities, and Indigenous reservations, according to Project N95 Executive Director Anne Miller. Shortages of gloves and gowns will likely be exacerbated by distribution of the vaccine, as federal kits will not be able to include medical gloves due to shortages in the market.

The capacity bottlenecks that exist because manufacturers are reluctant to invest significant capital in major scaleup, and the demand for testing will likely go back to previous levels. Scaling up of equipment/technologies have significant capital expense and long-term cost implications for manufacturers, as indicated by this quote:

> "Pre-Covid, the market was perfectly sized for PCR testing and corresponding reimbursement rates, but post-covid if you can't get paid to run molecular tests it is hard convincing. Roches $1M equipment is difficult for hospitals to invest in although ok for big academic centers. Pre-Covid 200 units were being made now scale up is for 1000 units."

The limited availability of instruments created a unique conundrum. Labs purchased multiple testing platforms to address the test capacity shortage, however, that has led to allocation of consumables. There was also a shortage of the ready use antigen test kits (Abbott), and demand was far outstripping supply. Due to the upstream constraints in component manufacturing for test kit equipment, companies like Roche, Hologic went on allocation strategies for making available test kits and/or reagents to various testing centers and healthcare agencies. This has also disturbed the ability of these centers to expand their testing capacities. For allocation strategies, priority has been given to the "big" labs such as Quest Diagnostics, Lab Corp, and others, that have long-term supplier relationships with these manufacturers/suppliers and can purchase large order quantities. This allows them to be placed at the front of the queue, even though allocation of these orders is still going on across the globe. Quality control of products is also suffering in some cases.

The organizations we met with agreed that there is a lack of understanding on how to use the antibody, antigen, and PCR tests in a manner to improve surveillance and prevent spread of the virus. PCR tests continue to be considered as the "gold standard" due to its higher reliability, and most universities and schools rely on PCR tests exclusively. Only a small group of organizations have begun using the combined tests in a rational manner. For example, a large nursing home IDN we met with understands that the goal was to use antigen tests for all employees who come in, and if they test positive, (what they call a "suspicious test"), the employee was told to immediately take a PCR and to go home. This nursing home also favors using the Abbott tests ($5) as they are faster and less expensive than the BD ($30) and Quidel ($24)) antigen tests. However, the company invested in a large volume of BD and Quidel equipment and supplies (and were told by the federal government how this equipment would be allocated to which of their 360 nursing centers). The chief pharmacist fears that these tests could expire first before the Abbott tests, (which they also received a limited quantity from the federal government). The company is running about 62,000 tests a week, the majority of which are BD Veritor tests.

3.3 The Need for a Testing Strategy

To summarize, it was our opinion that not only was the U.S. not doing enough testing, but that if hospitals and schools were using the right combination of antigen, PCR, and antibody tests, then demand would outstrip supply. The bottlenecks for manufactured equipment and consumables continued to get worse, not better in the next few months. There also needs to be some thought given to the role of antibody tests vs. PCR tests in rolling out the vaccine.

We also found that testing and isolation based on COVID-19 symptoms alone would not be sufficient to control the spread of the SARS-CoV-2 virus.[1] Effective suppression depends critically on testing asymptomatic individuals as part of a comprehensive strategy. Throughout the first months of the pandemic in the U.S., however, testing was limited to the diagnosis of symptomatic individuals. Confirmation that symptomatic individuals are infected by the SARS-CoV-2 virus is done most accurately by using nucleic acid-based tests such as those that require analysis that uses a laboratory technique called polymerase chain reaction, or PCR. These tests, however, are quite expensive (~$100), require special laboratory resources, and have processing times, from sample collection to the reporting of the results, of at least 24 hours to two days or more, depending on the availability of lab processing capacity.

[1] https://scm.ncsu.edu/scm-articles/article/designing-a-test-kit-supply-chain-in-response-to-covid-19.

Alternatives to diagnostic testing of symptomatic individuals are screening testing, (intended to identify infected people who may be asymptomatic), and surveillance testing, (which involves the systematic collection, analysis, and interpretation of testing outcomes for ongoing public health reporting).

As surveillance testing is typically performed on de-identified specimens, this makes it difficult to use the results for individual decision-making. The SCRC team's research led to the conclusion that screening testing of asymptomatic individuals in order to detect those who are likely infectious has been underutilized throughout the course of the COVID-19 pandemic. An important supply chain-related benefit of screening testing is that it allows a more direct estimate of the number of tests needed in any given region, as it is based on planned screening protocols to be deployed in specific locations. This is in contrast to PCR-based diagnostic testing of only symptomatic individuals, which is driven by the highly uncertain and hard-to-predict incidence of symptomatic cases and the additional uncertainty regarding whether, and where, symptomatic individuals will seek out a test. This is a prime example of medical intelligence informing sourcing strategies.

In contrast, screening via antigen tests—also called point-of-care tests or rapid-response tests—costs as little as $5 per test for antigen kits purchased in bulk and provides results in 30 min or less. The kits also include codes that can be used to share the test results digitally with public health agencies. Antigen tests do, however, suffer from low analytical sensitivity, meaning that they require greater amounts of viral material to register a positive detection of COVID-19. This means that these tests may be less capable of accurately registering a "true positive" than a more sensitive testing method, such as lab-based PCR. The lower sensitivity of antigen test kits resulted in hesitancy in using them at the outset of the COVID-19 pandemic. Lower sensitivity, however, can be overcome by increasing the frequency of testing. For example, a test that is 85% sensitive has a 15% chance (100–85%) of failing to detect someone who actually is infectious, or a "false negative," but a negative result in each of two, back-to-back applications of this test indicates only a 2.25% chance of a false negative (given by 15% × 15%), effectively raising the sensitivity to approximately 98%. Moreover, the antibodies used on antigen test cards to detect the presence of SARS-CoV-2 are directed at the proteins that are not variant-specific, allowing them to be used effectively even as the SARS-CoV-2 virus continues to mutate.

The teams we worked with determined that widespread deployment of antigen tests (i.e., point-of-care, rapid tests) for screening at key congregate settings (e.g., schools, large business facilities) would be highly effective for both supply planning and protecting public health. Antigen test-based screening simultaneously achieves positive public health outcomes by identifying infected individuals, perhaps even before they show symptoms, while also allowing a clear definition of demand for tests. By contrast, not only is the demand for symptom-driven collection of samples to be processed via PCR highly uncertain and difficult to predict, but the delay in producing the test results leaves the public potentially exposed to virus spread while infected individuals await their results and continue to go about their business. Of course, in the broader testing environment, extensive screening via antigen tests would coexist along with PCR testing for symptomatic individuals and/or those who are able to wait in isolation for lab results.

Unfortunately, we have found that these types of strategies are not well understood nor utilized in the US. This remains an open problem, and testing is critical to speed up the response to a major pandemic event like COVID. There is also a need for improved data quality and timeliness of testing results—this data is imperative for allocation, visibility, and planning of material distribution.

3.4 The Strategic National Stockpile (SNS)

Disaster and emergency events in the US have typically been regional and limited in duration; examples include hurricanes, floods, fires, earthquakes, industrial accidents, or terrorist attacks. In all such cases, such events have generally evoked an immediate response on the part of emergency responders, including FEMA, the Strategic National Stockpile, DHS, the American Red Cross, private sector organizations, and state agencies. Humanitarian logistics services were generally required one or two weeks, after which time local parties became involved. The response to large scale disasters typically involve distribution of critical medical and disaster response supplies to surrounding regions, often following an established procedure for procuring and distributing supplies (e.g., food, shelter, water, and relief goods). Supplies in these types of disasters were generally readily available, and there were few problems in identifying qualified local suppliers for immediate contracting, acquisition and shipping to impacted sites. The sourcing task in such cases was largely tactical in nature, and the primary activities involved how to mobilize supplies to meet a time-sensitive demand quickly. For such regionally focused disasters, organizations such as FEMA are well equipped to respond, and in some cases the Strategic National Stockpile (SNS) is called in to ship supplies from warehouses that can fill gaps in supply.

The SNS originated in the last years of the Clinton administration through an act of Congress that formalized this stockpile under its current name, and required that HHS "*maintain a stockpile or stockpiles of drugs, vaccines and other biological products, medical devices, and other supplies in such numbers, types, and amounts as are determined [...] by the Secretary to be appropriate and practicable, taking into account other available sources, to provide for the emergency health security of the United States, including the emergency health security of children and other vulnerable populations, in the event of a bioterrorist attack or other public health emergency.*"[2]

The Stockpile's funding was established at $640 million, and this value remained relatively stable over time even as the SNS has had to respond to a number of natural

[2] *See* 42 U.S. Code §247d-6b—Strategic National Stockpile and security countermeasure procurements, at https://www.law.cornell.edu/uscode/text/42/247d-6b.

disasters and biological threats (Esbitt, 2003). The SNS is described by the CDC as a repository of potentially life-saving pharmaceuticals and medical supplies for use in a public health emergency in which local supplies have been or may be depleted (CDC, 2014). The supplies which make up the SNS are known as the "formulary" and include a wide range of medical products, including both pharmaceutical interventions (such as vaccines, antimicrobials, antidotes, and antitoxins) and non-pharmaceutical interventions (such as ventilators and PPE) (Siripurapu, 2020).

In ordinary times, SNS supplies are stored in two groups (Banner, 2016). A small portion—approximately 2%—of the supplies are maintained in "12-h Push Packages" designed to be quickly distributed to locations across the United States, thus providing local authorities with a wide variety of emergency response equipment designed to obviate the need for specific requests. For example, these Push Packages were put to use in the Stockpile's first major response, assisting state and local authorities in the aftermath of the terror attacks of September 11, 2001 and the anthrax incidents of the following month (Khan, 2011). The remainder of the SNS inventory is stored as "Managed Inventory" and the CDC may provide local authorities with specific supplies from this group in accordance with particular needs. Notably, the CDC and the SNS are not the first movers of emergency response—rather, the Stockpile functions to assist state and local authorities with their emergency responses, an interconnected effort mandating cooperation and communication between federal, state, and local authorities to be successful (Redd et al., 2017).

The SNS's failures have attracted the scrutiny of the public and many government agencies and committees. For example, a *New York Times* report details that, even during the early efforts to increase the Stockpile's supplies to respond to COVID-19, $626 million was diverted to purchase anthrax vaccines from Emergent BioSolutions—money which, according to some involved in managing the SNS, should have been used to buy PPE and ventilators.[3] In April 2020, a shipment of ventilators was diminished by more than 2,000 as the result of a contracting dispute, which had prevented government contractors from properly maintaining the ventilators in storage.[4] Initiatives to expand the Stockpile's quantities of needed supplies also failed, including the Trump Administration's push to expand N95 mask supplies to 300 million within 90 days starting in May 2020. By November 2020 only 142 millions N95 masks had been delivered and were

3 *See* Chris Hamby & Sheryl Gay Stolberg, *How One Firm Put an 'Extraordinary Burden' on the U.S.'s Troubled Stockpile*, N.Y. Times (Mar. 6, 2021), https://www.nytimes.com/2021/03/06/us/emergent-biosolutions-anthrax-coronavirus.html.

4 *See* David E. Sanger, et al., *A Ventilator Stockpile, With One Hitch: Thousands Do Not Work*, N.Y. Times (Apr. 1, 2020), https://www.nytimes.com/2020/04/01/us/politics/coronavirus-ventilators.html.

being held in inventory by the SNS.[5] At the same time, state governments had to compete amongst themselves for access to goods they could not otherwise get through the Stockpile.[6]

Government criticism has primarily attributed the SNS's failings to a lack of appropriations, but we believe the problems of the SNS is a function of a much larger set of problems.

A recent study by the author of patterns of PPE distribution during COVID suggests that the federal government did not have a systematic allocation strategy during the COVID crisis for distribution of PPE, and that the patterns of doing so appeared to be ad hoc. This raises an important issue: *how are governments allocating critical medical supplies in situations such as COVID, when the demand far exceeds supply?* For instance, an interview with a large mask manufacturer revealed that the monthly demand for N95 masks the manufacturer was facing was more than 1B per month—while their typical volume of orders was never greater than 2 M per month. This disparity in supply and demand requires an innovative set of allocation and distribution measures.

The interviews held with state officials in this study confirmed that the use of the SNS from the federal did not seem to follow a structured distribution strategy, and that most of the supply shortages experienced in the pandemic has fallen on the SNS, the CDC, the Federal Emergency Management Agency (FEMA) and HHS more broadly.

Once the pandemic arrived in the United States in February 2020, there was an attempt to source supplies on an expedited basis to address rapidly growing hospitalization rates and the commensurate shortages of badly-needed supplies, including ventilators. Additionally, a number of planning and interventionist strategies were adopted with the goal of making up the difference in critical materials by foraging supplies, such as 3-D printing of face shields by universities. But these measures were insufficient to prevent supply exhaustion in the early months of the pandemic; by late March 2020 most materials in the SNS had been depleted.

Meanwhile, what resources were available were the subject of competition between different federal agencies, as well as state and local governments, which resulted in inequitable distributions of stockpiled materials as well as hoarding and gaps throughout the country benefiting better-funded areas at the expense of their worse-funded counterparts. Federal responses to this state of affairs came only after the SNS had already been drained of its key materials, with invocations of the Defense Production Act (DPA) to secure a supply of PPE in mid- to late-March, although the global supply of raw materials to produce these goods was already backlogged by April.

---

[5] *See* Stephanie Armour, et al., *U.S. Supplies of COVID-19 PPE Fall Short of Targets*, Wall Street Journal (Dec. 9, 2020), https://www.wsj.com/articles/u-s-supplies-of-covid-19-ppe-fall-short-of-targets-11607509800.

[6] *See* Monika Evstatieva, *A Revamped Strategic National Stockpile Still Can't Match The Pandemic's Latest Surge*, NPR (Nov. 23, 2020), https://www.npr.org/2020/11/23/937978556/a-revamped-strategic-national-stockpile-still-cant-match-the-pandemics-latest-su.

During the interviews, several state procurement officers operating in the Emergency Operations Centers of States that experienced high COVID-19 surge highlighted several times the lack of inappropriate PPE support from the federal governments.

> "If you think that the federal government supported us properly, you are a believer. Initially we did get some supply, but I don't think anywhere near the amount that we were hoping to."
>
> "We did receive product from FEMA, but the vibe we had on it is that we were never too sure what was coming. And when it was coming."
>
> "Emergency management got some product from FEMA, but I would say it wasn't at reliable intervals and it often wasn't necessarily what we needed."
>
> "There was no big National Stockpile to dole out and so there was anything coming on our way. I know other states experienced FEMA diverting their orders last minute."
>
> "FEMA cut off funding effective September 15th for all non-healthcare PPE [...] this decision put several States in a very difficult situation."
>
> "I heard many, many stories about product ordered and coming in, but they could not get it because FEMA stopped it at customs and took it."
>
> "As time went on, working with the vendors, they [the federal government] started putting limits on each state so that they could provide a little bit to each state, but that was further down the line."

At the same time, there were states (not necessarily the most affected by COVID-19) that received a good and timely amount of federal PPE supply.

> "They get federal distributions that the emergency management team is showing us [...] they get, from the federal government, so many of these gloves. They are shipping them and they are tracking per county, how many are coming in of N95 and gowns and gloves to our hospital agencies- directly to hospital."
>
> "We did receive shipments from FEMA [...] we did get lots of help and receive some shipment of N95 and other types of PPE."
>
> "We worked through FEMA, which is our normal partner for an emergency, and they were able to get us the large amounts of PPE that had to go into federal sites for testing [...] They send us truckloads of PPE."

This uncoordinated and ad hoc distribution approach affected states' ability to be able to react to the emergency and provide hospitals with what they needed. Our interviews reflected a diverse set of approaches, which operated relatively independently, that can be discriminated in two groups.

The COVID-19 pandemic was dramatically different, and our federal and state agencies were unprepared for a disaster of this magnitude, which impacted every state in the country, every industry, every population, every hospital, and continues into its third year. The pandemic revealed significant gaps the federal response to this crisis, and has been the subject of many new task forces and inquiries. A recent study with state procurement officials suggests that several limitations of the federal responses were evident.[7]

- *Material distribution*: The Strategic National Stockpile was grievously underfunded, and thus why it lacked the supplies needed to forcefully respond to the pandemic, and was not assigned the right level of preparatory priority, in terms of resources, staffing, and budget. It was also largely focused on anthrax and biological attacks, and was never designed to respond to a pathogen-related pandemic
- *Economic hardships*: The country was largely unprepared for the discontinuities in business operations that resulted from the pandemic. As stay-at-home orders spread across the country, many individuals working in restaurants, service industries, banks, retail stores, and other locations found themselves out of work, with few options available.
- *Production shutdowns*—As COVID cases spread, material shortages began to recur due to border crossings being shut down, increased COVID-related workforce disruptions, and executive decisions to shut down operations due to the unknowns of what a pandemic meant for the economy.
- *Trade flow disruptions*—Trade between countries, including imports into the US and to other countries, was shut down as countries went into an 'isolationist' mode. This was particularly debilitating for US companies that relied on Asian supply chains, as China, Vietnam, Singapore, Taiwan, and other regions halted all trade into and out of their countries.

3.4.1 Current Capabilities

Many of the problems that stemmed from the COVID pandemic were attributed to slow responses on the part of government to recognize the massive impact of a pandemic on all facets of the economy. Slow reaction times to enforce border closures, development of rapid testing and contact tracing, and poor communication to the public lies at the root of the problem for the spread of COVID at such a rapid rate in the US. However, there also exists a number of lessons learned around the following areas, that together provide a list of missing capabilities the current administration should consider:

---

[7] Based on a study capturing interviews with 47 state chief supply chain officers which identified three primary reasons for the ineffectiveness of a federal response. https://www.naspo.org/wp-content/uploads/2021/03/2021_COVIDReportC.pdf.

3.4.2 Missing Capabilities

- *Agile decision-making*: During a disaster such as COVID, it is essential to quickly establish strong leadership for rapid decision-making, rapid deployment of resources, and agility. With so many things happening on short notice, the government cannot wait for emails and messages to be conveyed between agencies, particularly when multiple agencies such as the SNS, FDA, CDC, FEMA, DHS, and DHHS are all involved in different aspects of the required outcomes. To enable rapid decisions, a "playbook" is required that is set into motion, that gathers the right people from different agencies into a taskforce, that can operate in a cross-agency manner. Individuals representing their agencies will need to have the right level of authorization to make decisions quickly and conveying the requirements to their teams back at the agency.
- *Lack of information access* In addition to the inability to respond with needed healthcare supplies (PPE, ventilators, pharmaceuticals, etc.), the government lacked access to the data required for advanced supply and demand planning, which is critical when preparing for future pandemics. Responding to a national pandemic is a complex task with many moving parts, including advanced demand planning, supply market intelligence, analyzing markets to assess the global availability of critical healthcare supplies and creating sourcing plans for every key need that might arise. This task became essentially unachievable given the lack of dependable and real-time data on the current status of COVID cases, demand requirements for supplies across a healthcare network, current inventories of supplies, and upstream availability from suppliers and distributors capable of shipping to different facilities across the countries.
- *Lack of supply chain management expertise.* Beyond the lack of fundamental supply and demand data in the supply chain, our federal healthcare response system also lacks the personnel and capacity to determine the content of the changes that will address its response problems. In particular, agencies lacked knowledge of demand forecasting, acquisition and contracting, strategic sourcing, and modern warehouse management practices. Such knowledge is essential in the middle of a pandemic. Our healthcare systems were completely reliant on external global sourcing networks for supply of PPE and medical equipment, yet government agencies had a dearth of skills on how to manage such complex global networks to secure required supplies during the COVID-19 pandemic.

In light of this (these) factors, the obvious question arises: *what can governments learn from this experience and prevent this situation from happening in an emerging global threat, and what can be done to improve response to new variants during the current pandemic?*

The IBM Center for The Business of Government and the Shared Services Leadership Coalition recently convened a roundtable of experts from government, industry, and academia to address these issues.

Objectives for the roundtable included:

- framing the government's supply chain challenges more specifically in the context of major performance problems in the national response to COVID;
- understanding the current state of capabilities in the national response system, and what missing or under-performing capabilities caused performance problems to occur;
- developing a model based on how world class commercial entities approach similar challenges, and attributes associated with these capabilities;
- conducting a gap analysis between government and industry models; and
- considering the crucial role of informal networks in addressing such a complex national challenge.

Some of the insights form this roundtable are summarized below.

3.5 Requirements for a Government Response

The development of a solution for improving the government's response and agility to future pandemic and healthcare emergencies is no small task. Addressing such a problem of national scope requires a "whole of government" approach, but must also tie in critical industry players who play essential roles in healthcare as well. There exist many tensions already between agencies, and the requirements for government contracting using the FAR also present many barriers for effective and timely responses from the private sector. Yet we know from the COVID experience how critical it is to have an agile government response, and one that can tie in industry knowledge, market intelligence, and distribution/sourcing capabilities in a rapidly moving crisis. Timeliness, access to real-time information and indicators, and supply chain talent are all essential in such environments, yet there does not exist a mechanism today to bring this about.

In thinking through how to develop these capabilities, our roundtable led to a number of discussions that focused on the role of shared governance models. Shared services are a helpful metaphor for development of shared governance models, and provide a useful approach for organizing diverse actors to achieve a common set of goals. Employing a shared services model involves identifying critical networks of actors, establishing data standards, role/responsibility identification, collective sources of funding, and requirements to measure success.

So what would a shared service model look like for an improved government response, when so many actors are required?? We address this issue in Table 3.1, which summarizes the outcomes from our research roundtable. We also developed a high level budget to estimate what the cost would be of establishing and running such a center of excellence. Our estimates were that it would run approximately $10M a year for 10 years. This is a drop in the bucket compared to the economic loss driven by the COVID pandemic.

Table 3.1 Summary of critical insights

	Current state	Desired future state
Supply chain issue: Global independence, Persistent and agile/market intelligence	• Reliant on external supply base primarily located in Asia • Absence of early warning signals and lack of mechanisms for what to do with the signals • Absence of coordination determining government actions when early warning indicators are present	• Prioritization of critical supply list and government policies targeted supporting a domestic "stop gap" manufacturing capability • Create a medical intelligence office to provide early warning of healthcare threats • Creation of a supply market intelligence capability to monitor supply materials
Supply chain technology: Transparency and traceability, Flexibility/agility, equitable distribution	• Lack of material visibility within the SNS and inability to detect shortages in hospitals and distribution networks • Significant expiration of many materials required for COVID, no recovery plan • Inability to replenish stocks as majority of products were manufactured in Asia, not the US	• Control tower needed to create a national healthcare monitoring system • Create "supply chain immunity" to create real-time monitoring of inventory , including barcodes, supply mapping, event monitoring, and global network trusted suppliers • Recruit ad hoc industry experts into a supply chain task force and allocate based on critical need
Government policy issue: Leverages existing capabilities and delivers effective cross-government shared services, respects constitutional roles and responsibilities, enables innovation throughout the ecosystem, political viability and sustainability	• Poor data quality and lack of data standards • Federal agencies were competing with one another over decision rights, state agencies operating independently • No formal governance structure in place to manage the crisis-informal interconnected network emerged	• Inventory visibility systems with common data standards across multiple industry tiers as a contractual requirement • Cross-agency informal network to devolop linkages between DLA, FEMA, SNS, and other critical agencies • Defined governance and playbook for creation of taskforce with a pre-defined governance apparatus from the outset • Consider using a similar structure to the "War Production Board" used in WWII

The categories that are essential for success are classified into three categories: **Supply Chain Issues, Technology Issues, and Government Policy** issues. Each of these issues is described first in terms of the current state observed during the COVID crisis, and the future state that would envisioned in a shared services environment. In the next chapters we go through each of the areas of supply chain immunity necessary to attack problems observed in the current state, the critical requirements for success, and how these elements would be carried out in a "desired future state".

Traceability and Transparency 4

This chapter discusses the issues and lessons learned during the COVID-19 response regarding transparency and traceability in supply chains. Organizations and consumers need to be able to see where their products and services are coming from and have the ability to track them as they flow through the supply chain. This is especially important when there are security assurance concerns related to the source of an item and its relative quality, as was the case with items such as PPE and vaccines during the COVID response. This chapter provides a description of the "current state" as of our time working on the COVID-19 response and the "future state" that we hope to see. We note that some future state concepts may already be underway.

R. Handfield and D. J. Finkenstadt, *Supply Chain Immunity*, Synthesis Lectures on Operations Research and Applications, https://doi.org/10.1007/978-3-031-19344-6_4

4.1 Transparency and Traceability

This characteristic refers to the ability to detect material levels in the supply chain through real-time visibility to inventory levels. Accurate visibility requires a strong focus on data hygiene with continuous quality improvement through analytics technologies, novel solutions such as distributed ledger technology (i.e., blockchain), artificial intelligence, multi-cloud environments, etc. Inventory turns can thus be managed, as material that is close to being expired can be released and sold to the healthcare system in question, and replenished on a timely basis. Another critical component of transparency and traceability is supply chain mapping. Understanding who supplies what to each of the suppliers in any particular value chain, affords the end item producer the insight necessary to both proactively and reactively respond to potential supply chain disruptions. For instance, knowing that a critical component in an end item was sourced from SE Asia during the pandemic would have allowed manufacturers, and in fact did allow some manufacturers, to secure inventories quickly or proactively design their supply chains with a more geographically diverse set of lower tier vendors. Looking forward, and backward, understanding what lower tier vendors are located in areas likely to see tropical storms or flooding allows value chains to posture for unavoidable events and create business continuity plans that can be quickly deployed instead of trying to map their chains in a time of crisis when vendors are the least likely to be responsive to the thousands of calls they may be receiving. A critical component of any supply chain mapping activity is the use and engagement of the actual vendors that are participating in your product line. Having a more networked supply chain affords the opportunity to quickly communicate problems and solutions in the face of a crisis.

4.1.1 Current State

A lack of technology for material visibility within the SNS, FEMA, and state procurement offices led to a lack of demand insights and the inability to detect shortages in hospitals and the national stockpile. There were no barcodes or similar tracking systems to monitor where inventory was in the system or to find the expiration dates of materials in storage. This lack of visibility hindered sound management of the SNS. In 2020 the SNS relied on a manual count of inventory and manual updates to its antiquated Department of Defense material system, with an antiquated inventory management system providing no visibility into materials' expiration dates, similar to recent findings reported in regard to the Veterans Affairs' COVID-19 inventory readiness.[1] For instance, an audit of the SNS stockpiles in January 2020 revealed that the stock of N95 masks, gowns, and gloves had been depleted during the H1N1 pandemic a decade earlier and never replenished,

---

[1] US Government Accountability Office. VA acquisition management: supply chain management and COVID19 response. Report no. GAO-20-638T. Washington, DC; 2020.

and many of the masks were past their expiration dates.[2] We further discovered that significant shortages of PPE were not being reported publicly by the CDC during this period. A report by the **National Healthcare Safety Network (NHSN)** recorded on June 10, 2020, which is part of the CDC, is the nation's most widely used healthcare-associated infection tracking system, and reported that 40% of hospitals could not get N-95 masks.

4.1.2 Future State

Preparing for a pandemic requires the ability to monitor many different things at once, from the dynamics of the Asian health-care market to the shifting nature of supply and demand across multiple categories such as PPE, drugs, vaccines, ventilators, and testing kits. A significant investment needs to be made in creating a shared service capability staffed with experienced supply management professionals, who are knowledgeable in developing category strategies for these materials. Many of these issues were problematic because of a lack of visibility technology, a lack of a barcoding system for track and trace of material locations, expiration dates, and consumption, and a lack of market intelligence on what was happening in the supply markets for these items.

To ensure agility and flexibility, a system for monitoring the current condition of inventories not only within the SNS, but across major healthcare distributors and hospitals will be critical. We envision having a "national healthcare monitoring system" with access to an inventory visibility system tied to a "control tower", with real-time analytics that tracks the current status of material in critical stockpiles throughout the country. During the COVID response no one knew, at an enterprise level, where products were coming from, where they were being sent, and who was receiving them. Further, national monitoring of critical materials through a modern QR or barcode inventory tracking system would be required, which would in turn require investment across a trusted network of hospitals, distributors and manufacturers. This is not expensive technology and relatively easy to deploy.

We need to move away from the idea of simply increasing the Strategic National Stockpile, and think more in terms of the "Strategic National Sourcing" framework. What is needed is a sophisticated approach for development of category strategies, combined with deep supply market intelligence around how to construct strategies to mitigate risk. "Supply market intelligence can be defined as a process for creating competitive advantage and reducing risk through increased knowledge of supply market dynamics and supply base

[2] Bender M, Ballhaus R. How Trump sowed COVID supply chaos. "Try getting it yourselves." *Wall Street Journal*. August 31, 2020. https://www.wsj.com/articles/how-trump-sowed-COVID-supply-chaos-try-getting-it-yourselves-11598893051#comments_sector. Accessed October 12, 2020.

composition." (Handfield, 2010, p. 43).[3] Market research, in a public context, is the collecting and analyzing of information about capabilities within the market to satisfy agency needs (Federal Acquisition Regulation, Subpart 2). This can consist of surveillance and investigation techniques. Surveillance is a continuous awareness process whereas investigation consists of targeted, comprehensive analysis for a direct need. We note that supply chains and markets can be viewed as having informational attributes that can be viewed in the aggregate or at discrete, finite levels. We can 'zoom in' or 'zoom out'. It can also be viewed along a temporal dimensional attribute. Any future governance framework should consider these attributes and look for useful, analogous frameworks from which to learn.

Our team of experts and advisors supporting the federal supply chain task force efforts noticed early in March 2020 that the supply of PPE necessary to combat COVID for the near, intermediate and long haul was woefully insufficient. The protective personal equipment most in demand included face protection (goggles, shields, N95 respirators, surgical masks etc.), body protection (gowns, coveralls, aprons etc.) and hand protection (gloves and sanitizer). Despite our early insights into this issue and our early warnings, most hospitals are still re-using masks and other equipment, and nursing home and professional medical associations are experiencing chronic levels of PPE scarcity.[4] Many of these agencies report having to reuse disposable masks and gloves, and with calls for reporting of shortages to be diverted to the White House, the likelihood of further shortages is becoming potentially hidden and elevated. Practitioners admit that, though the FDA has provided emergency use authorization for ultraviolet PPE decontamination, using decontamination procedures on disposable items is not ideal. It is incredible that in an advanced country like the United States, that five months into this pandemic our healthcare system still did not have availability of PPE. We sought insights into this question and came up with a metaphor to describe this situation: "blurry vision".[5]

4.1.3 A Supply Chain Spectrum for PPE

We see the issues related to supply chain traceability and transparency analogous to those of the electromagnetic spectrum. In the natural sciences the electromagnetic spectrum refers to the range of light that exists, from radio waves to gamma rays. Most of this light is actually hidden to us and we use others methods to observe it. Our Supply Chain

---

[3] Handfield, R. (2010). Supply Market Intelligence: Think Differently, Gain an Edge. Supply Chain Management Review, 14(6), pp. 42–44, 46–49.

[4] https://www.nytimes.com/2020/07/08/health/coronavirus-masks-ppe-doc.html

[5] Finkenstadt DJ, Handfield R. Blurry vision: Supply chain visibility for personal protective equipment during COVID-19. Journal of Purchasing and Supply Management. 2021 Jun; 27(3):100689. https://doi.org/10.1016/j.pursup.2021.100689. Epub 2021 Mar 28. PMCID: PMC8005248.

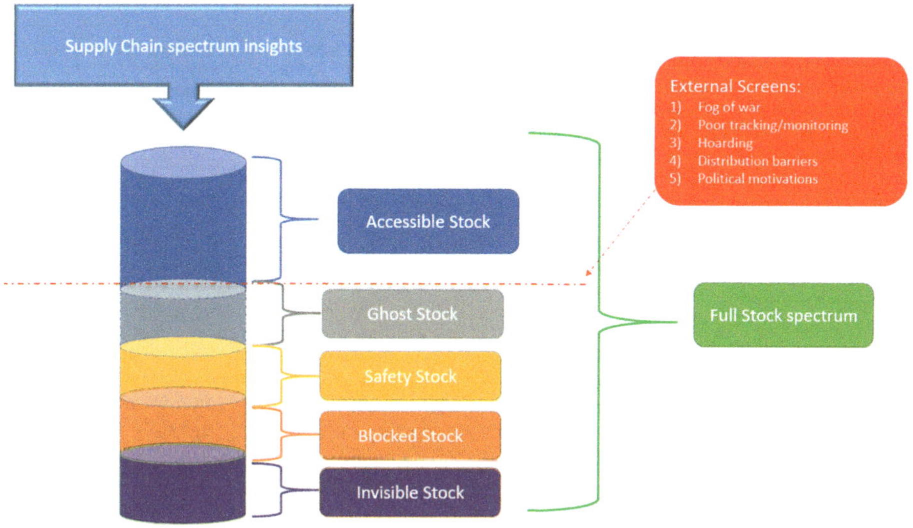

Fig. 4.1 Supply chain spectrum for PPE during COVID-19. *Source* Finkenstadt and Handfield, 2021

Spectrum for PPE (Fig. 4.1) employs the spectrum as a useful metaphor for depicting the current level of visibility into healthcare materials and PPE within the supply chain.[6]

In a perfect world all stakeholders would have full insight into system-wide inventory supply levels that match precisely to their accessible stock. A portion of this stock can be considered ***accessible***, (e.g. it can be seen, acquired and shipped to an approved buyer who needs the material in a relevant time of need). However, ***external screens*** exist that can hide the true quantity of stock in the system. The stock in the system falls into one of two ranges: the visible (e.g. accessible) and the hidden. The hidden spectrum includes ***ghost, blocked, safety*** and ***invisible*** stock. Ghost stock includes stock levels reported erroneously or fictitiously based on self-serving behaviors (political motivation etc.). Safety stock covers items kept based on hoarding behaviors as a form of irrational or inequitable risk mitigation or the inability to accurately forecast needs. Blocked stock are those items we can either not see or access due to distribution barriers (tariffs, blockades etc.). Finally, invisible stock, are items that players in the supply chain (customer, manufacturer, distributor etc.) cannot see due to mass confusion, or poor supply tracking and monitoring systems.

We observed significant challenges within the systems used to manage the Strategic National Stockpile (SNS) and hospital supply systems themselves. We observed the early depletion of the SNS in February 2020, and an inability to replenish these supplies in a timely manner. We were unable to find a clear use of barcoding systems (or other

[6] Ibid.

Table 4.1 Types of hidden stock within the supply chain spectrum

Hidden stock type	External screen	Reasons for screen	Potential root causes
Ghost stock	Political motivation	Re-election, reputation, optimistic speculations, career-building, public pressure to provide rapid updates	Bureaucrats measured on hospital availability
Safety stock	Hoarding	Risk mitigation, capitalist focus, poor demand forecasting capabilities	Medical workers and administrators
Blocked stock	Distribution barriers	Tariffs, blockades, poor logistics, lack of global independence in the supply chain, prohibitive price gouging	Customs agents, brokers and other agents
Invisible stock	Fog of war and/or poor tracking and monitoring systems	Unprecedented global need, inadequate command and control, years of failure to invest in people, systems and processes to monitor supply chain risks	Buffers, lead time settings in MRP systems, outdated hospital systems, FEMA, DHS, wholesalers motivated by profit, poor governance

Adapted from Finkenstadt and Handfield, 2021[7]

modern inventory management systems) to allow tracking and tracing of material in the supply chain. We were also informed that there was widespread use of manual processes prone to human error and an inability to detect expiration dates on key supplies and equipment. There was a singular lack of federal level market intelligence and supply chain transparency. These problems were often undetected, unannounced, and never properly audited and reviewed, so that they likely exist today (Table 4.1).

4.1.4 Improving Our Vision

Our team established a set of recommendations to the Joint Acquisition Task Force to revamp the SNS (or a future response system) to enable clarity over the full spectrum of supplies in the system. We recommended that the federal government rethink the role of the SNS, and embracing the concept of supply chain immunity. We need to take certain

[7] Finkenstadt DJ, Handfield R. Blurry vision: Supply chain visibility for personal protective equipment during COVID-19. Journal of Purchasing and Supply Management. 2021 Jun; 27(3):100689. https://doi.org/10.1016/j.pursup.2021.100689. Epub 2021 Mar 28. PMCID: PMC8005248.

Table 4.2 Recommendations for improving stock visibility

SC immunity objective	Hidden stock uncovered	Action/Key result
Traceability Establish systems for traceability of all healthcare materials through barcode readers	Ghost, safety, blocked, invisible	Legislation and/or policy to enable blockchains across the entire SNS and eventually tied to state and local healthcare and industrial supply/demand centers. Fast (real-time) and accurate stock and capacity reporting at a national level. Reduces fog of war and holds politicians accountable
Flexibility Increase stock flexibility with inventory planning	Safety, blocked	DoD and Veterans Affairs hospitals become living stockpiles[8] that hold and rotate excess inventory instead of a JIT system. They serve as the custodians of this excess inventory for maintaining currency and local distribution/allocation in the time of crisis
Responsiveness Improved sourcing and category management	Blocked, invisible	Develop and maintain strategic sourcing plans for every key PPE need that might arise in an emergency. This is accomplished through detailed category management structures and market intelligence guidelines. Work with universities to establish agile response
Globally independent Increase material distribution equity	Safety	Reduce large lot buys and move to joint-public/private purchases with phased delivery to ensure PPE is current, usable and that all parties are incentivized to be transparent and equitable in their demand for PPE

Adapted from Finkenstadt and Handfield, 2021[9]

actions as it applies to supply chain visibility in order to move to a proactive stance. Table 4.2 provides a short summary of our initial recommendations related to improving

[8] Guinto, P. (2020) Personal Correspondence.

[9] Finkenstadt DJ, Handfield R. Blurry vision: Supply chain visibility for personal protective equipment during COVID-19. Journal of Purchasing and Supply Management. 2021 Jun; 27(3):100689. https://doi.org/10.1016/j.pursup.2021.100689. Epub 2021 Mar 28. PMCID: PMC8005248.

visibility that have potential impacts cutting across other areas for improving supply chain immunity during global catastrophes:

4.1.5 Conclusion

To our knowledge, our recommendations have not been fully adopted. The primary focus still seems to be on building up new stockpiles in depleted areas for "reserve capacity" until a major event occurs. In many cases, we have learned that the federal government is appropriating orders from manufacturers of PPE and other materials going out to hospitals, and placing these in warehouses. It appears that the SNS has not fully responded to the risk that large inventories of raw material can expire at or around the same time. This can lead to a reliance on "reserve capacity" from suppliers who are already unable to meet demand.

Transparency and traceability are necessary for supply chain success, especially during a global crisis. Hospitals' age-old strategy of hoarding material cannot be allowed to occur and will require regulatory changes. There will also need to be significant changes in federal contracting guidelines, to allow for better data management and preventive actions such as establishing agreements with vetted and audited suppliers, to ensure their capability to replenish the strategic national stockpile (SNS). The SNS itself requires on-site management, to drive appropriate turning of inventory and avoid large lot buying and large concurrent expiration dates for large parts of its contents, (which occurred in January 2020 right before COVID hit). These changes will required the use of qualified supply management professionals who have experience in applying the tools of category management, strategic sourcing, contracting and supply management. While our recommendations are not exhaustive, we believe it sets a useful beginning and general practices for improving traceability and transparency in critical supply chains such as PPE.

Flexibility and Global Independence 5

5.1 Flexibility Through Improved Contracting

Once transparency is achieved and we identify risks such as hacking, counterfeiting, human disruption etc., we must then determine what to do about it, how to respond. That takes planning and it takes flexibility. We have to develop systems that can adjust rapidly to changing conditions. Again, one must first be able to see the conditions change and know the products or services at risk—so traceability and transparency are the first order, but then organizations and governments have to build up resiliency muscles and agility reflexes. That requires that entities strategically design in the ability to flex.

Future state supply chains must be able to pivot to additional sources and means of channeling requirements in the face of disruptions. This requires advanced planning, effective category intelligence, and strategic sourcing plans for every key need that might arise in an emergency. Planning Teams needs to develop war-gaming situations/simulations, and capacity requirements that span both domestic and global sources. Firms need to develop "playbooks" to enable them to have a quick 'go-to' reference in times when the "fog of

R. Handfield and D. J. Finkenstadt, *Supply Chain Immunity*, Synthesis Lectures on Operations Research and Applications, https://doi.org/10.1007/978-3-031-19344-6_5

war" impedes information gathering and requires rapid decisions. During a contingency, events pop up that drive the need for immediate decisions and actions. The faster a person can observe and orient to the need to decide and act, the more likely they are to prevent unnecessary loss or casualty.

The military uses playbooks and handbooks for contingencies in a variety of organizations, many related to contracting and acquisition management. For example, the Department of Defense's Defense Contingency Contracting Handbook has been used for over a decade to "help DoD Contingency Contracting Officers (CCOs) and other Operational Contract Support (OCS) staff meet the challenges they may face in contingency environments." The Marines develop tactical battle playbooks, based on countless after-action-reports that establish lessons learned from successes and losses in the field. DoD planning and operations that include contracted support are subject to the requirements of the Federal Acquisition Regulation and its Defense supplement (DFARS). They are also directly guided by Joint Publication 4–10, Operational Contract Support. But the DoD recognizes that, during a contingency or natural disaster, people in the thick of crises can't be expected to download, carry or process thousands of pages of rules. So, they offer an easy to follow and easy to carry handbook. The handbook answers the basic questions or who, what, where, how, and when. This also allows them to pull considerations that may not be of daily concern to the front of mind. For instance the CCO Handbook begins with a chapter on fraud, waste and abuse; covering issues like human trafficking and contract fraud that may concern defense personnel in non-emergency situations but are much more likely to occur during a crisis in some of the environments in which DoD personnel deploy to and operate. Where the regulations may tell the reader prohibitions or mandates, the handbook provides real-time support information. In the case of contract fraud, for instance, it describes potential schemes, conditions that make fraud more likely, and common fraud indicators. Governance is spelled out as well, so that it is made clear who is in charge under different circumstances. These details would have been very helpful to many firms and governments during the PPE sourcing stampede that has occurred post-COVID. The US Army Corp of Engineers (USACE) has a contingency operations playbook as well. The Center of Army Lessons Learned (CALL) Special Study No. 16-01 (Oct 2015), USACE Overseas Contingency Operations Playbook. The playbook is rather large at 412 pages but covers an enormous variety of issues. One of the most impactful features of the USACE playbook is its use of "Enduring contingency engineer mission principles captured from experiences in Operation Iraqi Freedom and Operation Enduring Freedom". This is a table of seven guiding principles that were developed from real lessons learned during OIF and OEF. It provides a descriptive table of principles that includes adapt, design sustainability, develop capacity, integrate command and effort, maximize reach back, standardize process, and train. Its use of case studies to highlight lessons learned for each principle is highly useful.

Federal requirements should embed industry standards to create maximum flexibility and increase alternatives in the event of need. Establishing standards on what types of materials are required is helpful, as this enables procurement to identify suppliers that can provide this quickly. Generally speaking, industry standards will allow procurement to source more easily, as they are more likely to be commodity items. This is the opposite of stockpiling of unique items, but rather involves contractual requirements and effective supplier development to ensure availability of supplies. Such contracts are useless if suppliers are unable to obtain supplies from global sources overseas, which means that material must be in a secure location within the borders of the U.S. Holding material in anticipation of a need will require managing inventory turns and expiration, which private sector companies are likely in a better position to do, as they have other customers who they can sell material to and replenish it.

Firms and agencies need to consider optimal contracting mechanisms for establishing governance in times of use, while not constraining availability to pivot towards other suppliers. Governance can either be codified in a contract or manifest relationally across a spectrum ranging from standardized contracts to custom and even verbal agreements. We must build in flexibility in governance, funding, and available sourcing. This means considering agreements over contracts (governance), reducing color of money constraints (funding) and developing robust category market intelligence practices (sourcing). In the case of the US government, we should shy away from mandating or pushing large indefinite-delivery, indefinite quantity (IDIQ) type contracts for items and services with just a few suppliers in favor of agreements such as Blanket Purchase Agreements, Basic Ordering Agreements, Basic Agreements and even standing price quotes under FAR 13.103. Agreements allow the parties to hash out the particulars of governance that can stimmy rapid decision-making during crisis yet provide the ability to pivot towards additional sources when current sources cannot satisfy needs. This is true for firms as well. Funding is also more fluid when it is not tied directly to a contract based on minimum order quantities and maximum ceiling amounts.

Some have argued that one needs the contracting mechanism to ensure firms are on the hook to perform. Our experience is that one of two conditions will always be the case; (1) the firm can perform an order under the agreement (i.e. a contract) and will never turn down work it can perform, (2) the firm can't perform the order under an agreement and having forced them to do so under an IDIQ mandate won't solve that issue. Now there are always concerns about developing suppliers and maintaining strong supplier relationship management. This holds true for governments and firms. We are not suggesting you have no formal contracts with such suppliers, just that you establish additional agreements with suppliers in case of emergencies. For those formal contracts we also suggest force majeure terms and conditions that allow the buyer to shift mandatory ordering and funding from said contract to others when it becomes apparent that the existing source cannot meet requirements in time. We also need to rethink our 'color of money' practices within the government in times of crisis. During a national pandemic or other calamity, we cannot have our dollars held hostage to budgetary programmatics when doing so actually puts the future of such programs at risk. For instance, the F-35 program doesn't get to continue if all the personnel supporting it can't get vaccinated or obtain necessary PPE.

Finally, we suggest increased flexibility via mutual aid networks. Mutual aid networks have popped up all over the world during the COVID pandemic. However, they generally consist of local government and non-profits sharing food and supplies for needy families or communities etc. We suggest that this model can be expanded to large agencies and firms across the country regarding supply chain items necessary to continue operations during a crisis to reduce hoarding and other negative behaviors related to sourcing common, finite goods and services. During our time supporting the COVID-19 supply chain response with the DoD we noted many instances of state and federal agencies, commercial hospitals and private firms competing for common PPE items. In many cases these items were obtained in excess of what was absolutely needed while other entities suffered dramatic shortages. This was especially alarming in the hospital systems where some were flush with PPE while others were washing gloves and reusing them. Mutual aid networks can establish ground rules for sharing resources between otherwise competing or non-associated entities. When goods and services become part of a 'global commons' they are no longer simple free market commodities. There must be an orchestrated approach to supply and demand during crises or we all shoot ourselves in the foot. Firms can establish grounds for enacting robust information sharing during emergencies. During a contingency, information can be the most valuable commodity. Firms and agencies should also establish clear prioritizations with each other regarding use of shared ordering on contracts. So, if Company X has a large contract for cleaning materials and their sub-tier supplier Company Y doesn't have any such sources, Company Y can utilize the contract vehicle once Company X has met its internal need. Further Company Z, a competitor, may have the same options during a declared contingency, but know clearly that they are lower priority than Company Y, and therefore seek to establish additional sources. It may seem counterintuitive that competing firms would ever help each other survive, but some calamities reach far beyond the market and impact the nations in which companies operate. In those cases, it is paramount that companies see the bigger picture and consider the overall future health of the system in which they operate. And please know that while COVID is the catalyst for this concept, it is by far not the only reasonable concern in which to use it. With climate change, additional diseases and cyber terrorism threats on the horizon, we need to get started baking this concern into every part of the supply chain and contracting process.

Supply chain and contractual flexibility has never been more necessary. Post-COVID agencies and firms will have to look to each other for lessons and support. Government can learn from industry with regards to flexible contractual arrangements and advanced supply chain visibility. Industry can observe the robust contingency sourcing preparedness planning and exercise protocols of the Department of Defense. We can all help each other. Supply chains have become the Achilles Heel of so many companies, organizations and governments during this pandemic. We must get better. Step one was transparency, step two is increased flexibility. Both are strong examples of why we can no longer divorce the vital intersection of supply chain and contract management. But flexibility requires the ability to have freedom of movement; it requires independence.

5.2 Global Independence

A ***Globally Independent Supply Chain*** is one that is not reliant on foreign manufacturers for production of goods. For sectors where this is not possible, the goal should be to facilitate government policies that establish a system of acquisition and supply for areas which are critical to national security, which may include development of a domestic network of trusted suppliers who are willing and capable to be part of the response system fabric. This may require additional ties to industrial policy, such as the "Make Buy American Real" proposed rulings by the Biden Administration.

5.2.1 Current State

During the COVID crisis in 2020, the pandemic differed from prior regional disasters in a number of different ways, which led to the inability to coordinate a federal supply chain response. The most fundamental and significant characteristic was a severe shortage of critical medical materials, specifically PPEs (e.g., N95 masks, nitrile gloves). This shortage was not relegated just to the US, but was global in nature, and the emergency continued to rage month after month. Many of these items required to respond were not available stockpiles at the SNS. During COVID-19, both sourcing activities and supplies emerged as the weakest link in the disaster relief operations. COVID-19 was an epic disaster on a national and global scale, which impacted every state in the union. State agencies were unable to reach out for assistance from other states (who were equally overwhelmed), and challenges existed in getting support from federal response agencies, which were equally disabled in their responses.

The federal government's reliance on health care suppliers that are primarily overseas and beholden to the export policies and priorities of other nations has led to significant shortages. As we have mentioned, even 3M in the United States was not able to produce masks because all the sources of materials (fabric, elastics, nose bands) were produced in China. Our discussions with state CPO's suggested that partnerships with private sector companies, such as the association between the state of Michigan and General Motors, facilitated access to Chinese suppliers through assets that were on the ground in Shanghai. Many distributors were unable to get supplies, and were inundated with promises from bogus suppliers in Asia that they could produce masks. They were also flooded with orders from hospitals, who were desperate and were placing orders with everyone, making it difficult to understand what the actual demand levels really were. In the words of one SNS official we interviewed, "I had no idea that almost all of healthcare supplies like PPE were manufactured overseas, and purchased from abroad by medical distributors." This lack of awareness reveals how unprepared this national institution was for the COVID onslaught that ravaged our healthcare system and our population.

In addition to the problem of most healthcare supplies being manufactured overseas, there was a lack of visibility into the current inventories being held across the country, and a pervasive problem also involved the lack of an allocation strategy, which often led to hoarding. These problems were discussed at length by participants in our workshop, and led to a number of ideas for a desired future state.

5.2.2 Future State

Creating a globally independent US-based supply chain would entail development of a prioritization of critical supplies that are required for national security—and for each one, a policy for either development of a domestic capability, or a means for managing the supply of these items from global sources. As an example, individuals pointed to the fact that during COVID, domestic manufacturers sprung up to produce N95 masks, supported by DoD grants. However, following COVID, many of these same mask manufacturers are declaring bankruptcy, as their price point is much higher than the raft of Chinese manufacturers that are selling to healthcare distributors and hospitals. This is leading us to the exact same condition we were in before.

To address this issue, a recommendation would be to create government industrial policies that are targeted at supporting a domestic "stop gap" manufacturing capability. Secondly, partnerships should be developed with distributors to enable visibility into their inventory systems, and ensure they enter contracts which set aside inventory for government allocation under different conditions of duress. This will require a set of common data standards and a common architecture to create a dashboard and control tower .

In addition, a multi-agency materials inventory portfolio based on in-depth supply market analysis is needed. At a minimum, this should include specialists in the following categories: disposable medical supplies (PPE, gowns, gloves), pharmaceuticals, plastics and resins, medical equipment, biologics, healthcare personnel, and respiratory products. This will require team of supply market analysts with special knowledge of these categories, that track the condition of critical supply markets for medical supplies, the supply risks within those markets, and acquisition strategies to manage the risks. Multi-tier supply chain mapping can provide clues as to critical points of risk that can "shut down" the US healthcare sector, based on multiple forms of risk assessment.

Because re-shoring all healthcare products is not practical or cost effective, we believe this office should also be responsible for developing a "make or buy" set of analyses that develops recommendations for investment in critical manufacturing infrastructure within the United States. This should be deployed in conditions of elevated market risk that creates extreme exposure to other countries where the bulk of manufactured supplies are produced. This might include critical pharmaceutical products and raw materials, as well as other critical materials deemed to be of strategic importance. A "cold eye" view of this capability should be conducted by such an agency to provide an impartial and objective assessment of the likelihood of success for re-shoring.

This is important, because there are infrastructural challenges with the idea of re-shoring manufacturing to the United States. Our discussions with manufacturing executives suggest that once an organization commits to outsourcing to third parties in low-cost countries, there is a minimum planning horizon of five years involved, as this requires supplier qualification, audits, start-up, quality certification, and on-going ramp-up. In many industries, sourcing executives have embedded their supply chains in Asian regions, noting that *"...these jobs will never return to Western countries."* As an example, 80% of the world's production of certain medical products are produced by four manufacturers in one province in China. To establish alternative sources that are competitive, qualified, and at-scale would cost much more than the 25% in tariffs companies are paying today in the U.S. to import from China.

Roscoe et al. (2021) developed a framework of supply chain strategies for geopolitical risk mitigation (see Fig. 5.1), which provides some guidelines to the federal supply chain on whether to adopt centralized/regionalized or localized supply chain designs according to how entrenched their suppliers are in a particular geographic location as well as how severe the geopolitical disruption is perceived to be.

The Y-axis of Fig. 5.1 shows the shifts in the external business environment, which have rendered it difficult to localize or shift the supply base, because of the entrenched nature of the supply base, or the cost-prohibitive elements for doing so. We note here that many Chinese industries were established with government investment, and the cost of capital for developing local sources is a significant barrier for investing in local supply capacity. The X-axis refers to the perceived likelihood of on-going political risk and disruption that is likely to continue, including the likelihood of on-going tariffs, customs

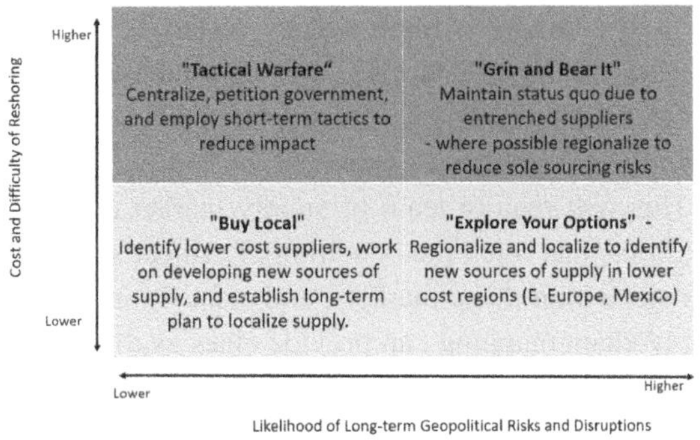

Fig. 5.1 Framework for supply chain strategies for geopolitical risk mitigation

duties, quotas, and export restrictions, resulting from a major and ongoing geopolitical event such as Brexit or the US-China trade war. The guidelines for this framework are shown in the Appendix F.

6 Persistent Market Intelligence

Another critical requirement for supply chains is the development of a ***Persistent source of Medical and Market Intelligence***. A primary requirement to enable an effective response to supply chain threats is the ability to monitor events globally and establish early warning signals that may indicate a potential threat on the horizon.

Current State: In the case of the pandemic, early warning of the COVID virus spreading was evident in medical intelligence signals in December and January, but this knowledge was not effectively translated into whole of government actions to mitigate the tidal wave that flooded our nation. Improved responsiveness requires agile networks that can quickly bring critical agencies into discussion where missions intersect. Multiple interpretations of intelligence is a requirement for effective translation into effective government policy making decisions. Today, much of the intelligence in this area is spread out across multiple agencies, including the Intelligence Community, the FBI, the State

R. Handfield and D. J. Finkenstadt, *Supply Chain Immunity*, Synthesis Lectures on Operations Research and Applications, https://doi.org/10.1007/978-3-031-19344-6_6

Department, and the Department of Defense. An important criteria for assessing the usefulness of intelligence is the extent to which early warning signals can be leveraged from all of these resources, to highlight potential threats that may impact national security.

Future State: A requirement for the future was discussed by many of the participants in the room, and consists of several components. Two critical elements of intelligence are required: medical intelligence, and supply market intelligence. In any crisis, the early part of the crisis is fraught with uncertainty, and often there is not good access for public health data within the United States healthcare system. Social media is often a complicating factor which does not always help. There is a need for an established source of truth that can provide objective information on what is happening in the global healthcare ecosystem, to provide an early warning system.

6.1 Medical Intelligence Signals

To begin with, a medical intelligence signal is required that can provide early warning indicators of problems that lie ahead. This was also noted in a recent book by Scott Gottlieb (the former head of the FDA) who notes that the US can no longer rely on the WHO and other nations like China for early detection of novel pathogens.

> ...global conventions, many implemented afters—SARS-1, were supposed to keep us fully informed of these risks and they were disregarded. This should change the future course for how we deal with these threats. We'll need to rely much more on our own tools for gathering information about novel pathogens and menag—cing outbreaks. We can no longer depend largely on global cooperation and the competency and transparency of other nations. COVID wasn't caused deliberately, but it was enabled and nurtured by the intentional quashing of information.[1]

He also notes specifically the need for a national intelligence service that can better identify and alert the country on the presence of alarming new medical threats:

> In short, COVID proved that we cannot trust as our only backstop the international conventions, cooperation, and obligations that govern our global public health order. These institutions need to be strengthened. But they cannot be our only means for being informed about new risks. We need more active efforts to gather information when there are signs of troubling outbreaks. We will have to rely more on our tools of national security, including our intelligence services.[2]

The immediate question that arises is what such a medical capability consists of. Several participants in our workshop discussed the need for a medical warning systems:

[1] Gottlieb, S. "Uncontrolled Spread: Why COVID-19 Crushed Us and How We Can Defeat the Next Pandemic", New York: Harper Collins, 2021, p. 46.

[2] Gottlieb, S., 2021, p. 54.

> "Being surprised by COVID has been a patterns, and shows the lack of health security warning failures. WE don't have a warning system nor do we have a system of training the recipients of that information to understand the indicators. We can't just pay attention to patients, without determining what the global impact will be. There does exist a system of conveying warning information—and we do understand what these patterns look like, and can alert people to the fact that there is a situation occurring that is connected to the air traffic grid, which is how pathogens travel to the U.S. We need to develop the equivalent of a "hurricane warning" system."

A comprehensive capability that relies on both global health security risk signals, as well as tacit human intelligence signals through a network of trusted experts around the world, is at the heart of a threat response system. This would include the following:

- A funded agency capability of observing direct indications of reports of human disease, coupled with indirect indications and warnings such as official acknowledgement and action, demand for medical services, local perception of threat, comprised infrastructure, and change in business practices.[3]
- Validation of medical threats with tangible multi-indicator analyses, and early warning signals communicated to others within the national threat infrastructure.
- A multi-disciplinary approach with specialists that have specific knowledge over different pathogens and biological threats.
- A system for conducting simulations and scenario analyses that can provide a range of possible scenarios, tied to contagion through air-borne travel, and a system of alerting healthcare providers of such risks using reliable and objective reporting. Today, this is done by "monitoring the FDA website", and guessing at the hidden meaning of different indicators that might be observed around the world.

A parallel idea in the defense sector is NORTHCOM, which provides command and control of Department of Defense (DOD) homeland defense efforts and to coordinate defense support of civil authorities. USNORTHCOM is responsible for the security of America's homeland—protecting people, national power, and freedom of action. They are charged to deter, detect, and defeat threats to the United States, conducting security cooperation activities with allies and partners, and supporting civil authorities. In the case of an emerging pathogen, there is no early signals monitoring system. NORTHCOM or a similar organiztion would provide early indicators of potential healthcare threats in addition to traditional threats such as ballistic missiles, both of which are equally serious (Fig. 6.1).

---

[3] Polyak, M., Black, M., Collmann, J., and Wilson, J. "Emergency of SARS in the People's Republic of China, 2002–2003: A Case Study to Define Requirements for Detection and Assessment of International Biological Threats", working paper, 2021.

Fig. 6.1 National Threat Response System

6.2 Supply Market Intelligence

A second component is a party (either internal or external to the government) tasked with monitoring conditions across supply markets with the following properties, that include both a strategic and a tactical set of components.

The tactical element to this function would involve development of acquisition contracts with critical private sector providers, to ensure that the government can procure the right materials in quantities that are aligned with the perceived level of threat identified by the Health Security Risk team. This might involve development of new suppliers, as well as potential surge sources domestically that could be alerted to possible preparation in light of new information.

This capability is reliant on the ability to create a market intelligence capability that is focused on the availability of critical medical supplies required to respond to threats. The last pandemic illustrated how our country was held hostage to export restrictions from other countries where the majority of PPE and medical supplies were manufactured. It is unlikely that the US will develop a healthcare manufacturing infrastructure that covers all possible medical threat outcomes any time in the near future.

One of the current programs discussed in our forum involved the Warstopper program. The program is designed to provide analytical support to maximize vendor capacity, provide industrial equipment, stage raw material, subcomponents, and manage vendor managed inventory systems, and establish industrial base maintenance contracts to maintain vital domestic industries. More details on Warstopper are provided in Appendix D.

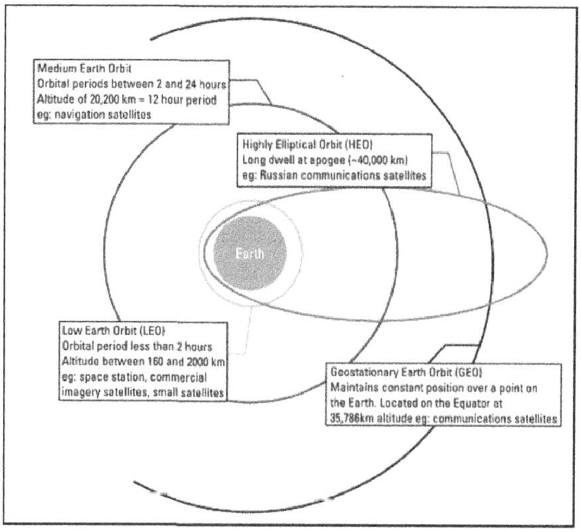

Fig. 6.2 Summary of different orbital regimes for persistent surveillance platforms[4]

6.2.1 An Orbital Regime of Market Intelligence

"Supply market intelligence can be defined as a process for creating competitive advantage and reducing risk through increased knowledge of supply market dynamics and supply base composition."[2(p43)] In a public context, market research is the collection and analysis of information about capabilities in the market to satisfy agency needs (Federal Acquisition Regulation, Subpart 2). This may consist of surveillance and investigation techniques. *Surveillance* is a continuous awareness process, whereas *investigation* is a targeted, comprehensive analysis of a direct need. We see supply chains and markets as having informational attributes that can be viewed at the macro and micro levels. Investigation can also be time dependent. We believe that there are useful analogies in intelligence collection that can be used for informing a future governance framework for managing critical supply chains.

Our team had previous experience in space acquisition and intelligence collection. We noticed a very useful analogy in space that is transferable to market intelligence gathering known as Activity-Based Intelligence Principles and Applications.[3] This concept recognizes that, teams must rely on a suite of sensors deployed at various orbital regimes in order to maximize intelligence collection and optimize the use of finite exploratory resources when the area is to broad or too deep. In the case of space, that body of observation is the Earth's entire surface, at to a certain degree so it is in the case of supply chains. This is an informative analogy for persistent supply chain awareness. Space-based persistent surveillance considers four primary orbits: (1) low Earth orbit (LEO), (2) medium

Earth orbit (MEO), (3) geostationary Earth orbit (GEO), and (4) a specialized type known as a highly elliptical orbit (HEO) (Fig. 6.2).

The further out the orbit, the longer the monitoring asset can dwell and the broader the area that can be observed. But this dwell and breadth comes at a cost of detailed depth perception. Therefore, a constellation or portfolio of orbits at various levels are managed to gain a full sight picture. We see clear parallels between space-based persistence and supply chain persistent surveillance. We imagine that a governance framework for future supply chain immunity must also be persistent (see Appendix A) (Table 6.1).

This orbital regime imagines a future of market surveillance, research, and intelligence that considers these 3 facets of orbital persistence and observation to be considered: (1) Dwell time, (2) Revisit rates, and (3) Resolution. Dwell time refers to the length of uninterrupted time spent observing a market area of concern. For example, in Market Investigation Orientation, a market investigation into PPE needs for shortages in a certain area would be required long enough to secure sourcing, and then resources would move along to the next immediate need.

In the case of a Market Intelligence Orientation (MIO), governments may need to maintain uninterrupted focus on the impacts on barriers to critical supplies and services, such as tariffs and trade agreements. Revisit rates refer to the frequency with which market investigation returns to and considers as specific areas of interest. As before, in MIO, a PPE shortfall may be addressed quickly and resources may move along, but it may be necessary to revisit the region's PPE needs periodically until they appear to have stabilized. Then it may be prudent to move this concern into a Market Surveillance Orientation regime to consider the long-term sustainability of PPE sourcing for this and other regions of interest.

The third and final consideration is resolution; referring to the depth and breadth of the market sight picture. Environmental variables determine how deep and how wide we need to see into a market for certain goods or services. Again, as an example, PPE shortfalls would require deep resolution, quick dwell, and rapid revisit (eg, MIO orientation), but the overall need to enable the long-term production and distribution of domestic PPE for future pandemics would require a longer dwell, necessarily slower revisit rates, and a broader resolution (eg, MSO orientation).

Some supplies and services reside in obscure markets and require a specialized intelligence-gathering method, what we refer to as a SMO orientation. Governments should consider an SMO orientation when supply chains require long dwell, deep resolution, *and* rapid revisit rates. In these cases, we would spend a substantial amount of time observing the market, delve deeply into the details necessary to secure uninterrupted and reliable sourcing, and then rapidly return to ensure these sources do not face new global challenges. These are specialized orientations because these market require a heavy commitment of resources to conduct such analysis.

---

[4] Biltgen P, Ryan S. *Activity-Based Intelligence: Principles and Applications*. Norwood, MA: Artech House; 2016.

Table 6.1 Market orbital regimes for persistent supply chain immunity.

Analogous space orbit	Features	Notional supply chain orbit	Features
Low earth orbit (LEO)	Short dwell, rapid revisit rates (every 90 min to 2 h). More finite details are available with lower-power sensors. Example: Imagery satellites	Market investigation orbit (MIO)	Short dwell, rapid revisit. Drills into supply chain availability, risks, and constraints on a direct need, immediate time horizon basis. Targets the need, responds, and then revisits quickly as needed for future supply chain requests. Informs MSO. Example: Finding PPE for a supply chain shortfall in Nebraska hospitals
Medium earth orbit (MEO)	Medium dwell, revisit rates every 12 h. Ability to cover entire Earth in pairs. Example: GPS	Market surveillance orbit (MSO)	Medium dwell over product and service categories of the health care supply chain (eg, PPE, ventilators, pharma). Considers supply chain availability, risks, and constraints in a given portfolio of spending for a near-term planning time horizon. Informs MIntO. Example: Defense Product Act investments in bolstering domestic production of nonwovens
Geosynchronous earth orbit (GEO)	Long dwell, revisit rate every 24 h. Follows equatorial rotation. Example: Communication satellites	Market intelligence orbit (MIntO)	Long dwell over entire marketplace. Considers supply chain availability, risks, and constraints for a complex network of global supply and service needs over a continuous time horizon. Example: Considering the trade agreements necessary to bolster domestic supply chain resilience
Highly elliptical orbit (HEO)	Long dwell, used to provide persistence over hard to see northern latitudes. Distance from Earth is variable. Example: Russian communication satellites	Specialized market orbit (SMO)	Long dwell in niche areas of the supply chain. Considers supply chain availability, risks, and constraints for a complex network of global supplies and services in areas that require special access or expertise. Example: Considering the long-term investment in capital and manpower to mine rare earth elements from domestic or intracontinental sources

Adapted from Handfield et al. 2020; Finkenstadt and Herrmann 2022 [5]

[5] Finkenstadt, D. and Herrmann, P. Orbital Market Surveillance Contract Management Magazine, June 2022.

As an example, rare earth elements (REEs) are a finite resource that are very hard to source in the continental United States. Thus we could develop strategies to find more secure sourcing of REEs, such as a Pan-American plan to use natural resources in Canada, the capital of the United States, and affordable manufacturing in Central and South America. This would reside in the SMO orientation until we observe long-term, consistent, and secure sourcing. Then these markets may move into a MIntO or MSO regime.

7 Equitable Distribution

Understanding the elements required for equitable distribution during a pandemic involve many complexities associated with government policy, allocation, inventory visibility, and a host of other issues. This chapter draws on discussions from a workshop held by the IBM Business of Government, entitled "Enabling a More Resilient and Shared Supply Chain Strategy for the Nation: Lessons Learned from COVID-19. The workshop was attended by a number of subject matter experts from industry, government, think tanks, academics, and physicians. https://www.businessofgovernment.org/report/supply-chain-strategy-covid-19

R. Handfield and D. J. Finkenstadt, *Supply Chain Immunity*, Synthesis Lectures on Operations Research and Applications, https://doi.org/10.1007/978-3-031-19344-6_7

7.1 Equitable Distribution

During a pandemic, the demand for materials can come from many different kinds of organizations, at different times, and with claims on the common goods. We have seen large integrated delivery systems, individual hospitals (inside and outside these systems), government delivery systems, including the military and Veterans Affairs, prisons, nursing and senior residential facilities, and rural hospitals and clinics, all seeking products. Importantly, all have had access or a lack of access to different sources, especially traditional distributors and group-purchasing organizations. There were hospitals in the same city experiencing vastly different levels of access to medical goods and services based on their buying power. Nowhere was inequitable access to medical items like vaccines more stark than in the marginalized populations in the US, such as the tribal nations. There is still a major inequity in the low-income countries around the world. As of the writing, only 58 of WHO's 194 Member States had reached the 70% vaccination target. In low-income countries, just 37% of healthcare workers had received all courses of their primary vaccination.[1] The "alternative markets" that emerged during COVID-19, consisting principally of pop-up "brokers" with personal contacts in Asia or Central America that were not part of the usual PPE production system, targeted many provider organizations. An equitable system will be responsive to need, as opposed to demand, and be guided by a set of ethical principles that facilitate triage and distribution and are not subject to behaviors that threaten the evolving commons. The development of national allocation strategies, based on hot spots and simulation analysis to better predict how to stem the spread of biological threats and viruses would be important, as well as development of strategies for effective pathogen protection and public communication channels.

7.1.1 Current State

As noted, the Strategic National Stockpile was never designed to be able to support a national pandemic. Prior to the 2020 COVID-19 pandemic, the SNS contained about $8 billion of medical assets. However, as we shall see, many of these assets were well beyond their useful expiration date and could not be employed during the COVID crisis.

Disparate means of communication and coordination among public agencies were apparent to everyone. Prior to and during the early pandemic the Division of the Strategic National Stockpile occupied a low level within the Office of the Assistant Secretary for Preparedness and Response (ASPR), a group of public-health experts in the Department of Health and Human Services. In this location, the SNS has little influence and national visibility and is not resourced appropriately, often with reduced budgets. In this location, managers struggle to get access to information from other agencies, and they

[1] https://www.who.int/campaigns/vaccine-equity.

have little national visibility to enable them to request such information. Ideally, the SNS would require the opinions of experts from many sectors, including epidemiology, health care, distribution, occupational safety, cyber security, drug administration, the intelligence community, the State Department, state agencies, and public health.

The COVID-19 pandemic put a sudden and significant strain on the supplies stored in the SNS. Despite the SNS's purpose to provide for emergency health security and to respond in the event of a bioterrorist attack or other public health emergency, the Stockpile had not attained public prominence prior to the COVID-19 pandemic, nor did it benefit from a level of funding commensurate with its mission. As such, the SNS did not have the tools or supplies it needed to address the pandemic.

An equitable system for distribution is especially needed during an emergency, and a federal policy is needed. During a pandemic the demand for materials can come from many different kinds of organizations. We have seen large integrated delivery systems, individual hospitals (in and outside of these systems), government delivery systems including military and VA, prisons, nursing and senior residential facilities and rural hospitals and clinics all seeking medical supplies. Importantly, all have had access or lack of access to different sources. The "alternative market" that emerged during COVID-19, consisting principally of suppliers with personal contacts in Asia that were not part of the every-day PPE production system, targeted many of these provider organizations. An equitable system will be responsive to need as opposed to demand and be guided by a set of ethical principles that facilitate triage and distribution. To have an equitable system requires input from the various provider organizations regarding demand on their systems—but also focuses on preparedness (just in case)—which, if credible, may well prevent hoarding. If we can see where things are going, we can alleviate the need to rely on distributors and vendors to allocate material to the right places, whereas a demand sensing capability at the SNS level can drive allocation to the right states and counties most in need. Resource availability is key, but we note that information availability may be just as important, if not more so. We note that current COVID19 supply strategies have become a zero-sum game given asymmetric information, and new forms of governance are required to address these shortfalls.

7.1.2 Future State

Asking distributors to warehouse goods and then be responsible for distribution is not feasible. Prior research has highlighted a number of structural issues with healthcare distribution that make it problematic for distributors to house finished goods inventory buffers or at point of care. One of the biggest challenges is the allocation of goods, which historically has been not equitable. During the COVID crisis, the SNS failed to serve a large number of healthcare institutions, namely smaller hospitals in less populated states. This was clear during many of the conversations we had with the National Association of

State Procurement Officers as well as with the National Governors Association. Further, private distributors and GPO's will always first serve their primary customers based on who has the greatest buying power and based on prior existing relationships, and there is thus a need for increased visibility and fair allocation mechanisms that are transparent to all. (Note that the CDC NHSN has the data to demonstrate that major shortages of PPE and masks were not equitably distributed.).

While there are no "magic bullets" that emerge as a solution to the problem of equitable allocation in a crisis, one possible solutions suggested is the "Dollar-a-Year Men" solution. This is discussed in Appendix G. The notion of using industry experts from the private sector lead a recovery effort would serve to support improved government insight and agility. The expert advisors could pool their knowledge, drawing on their private-sector (and likely international) knowledge and connections to identify what goods are needed and where, the present supply of those goods, and any competing demands. This would allow for the integration of medical, production and supply chain management expertise in an area of the government presently dominated by only medical knowledge, with the potential benefit of the SNS being much better equip to assess the health of its supply lines, in order that HHS more broadly can protect the health of the nation. In this sense, the recent national supply chain directive highlights that the SNS should contribute to medical supply chain resiliency (and not a static inventory level), and the prescribed responsibilities of this resource should change.[2]

7.2 Government Policy Issues

Equitable distribution of materials involves advance planning, that often spans multiple government agencies. Such agencies often fail to coordinate properly - one of the vehicles suggested for coordination and governance of allocation in a pandemic involved the development of a shared services capability. Government policy issues must be redesigned to enable all of the above elements discussed in this paper. We need to establish a system of governance that ensures all of the agencies in the government are effectively leveraged, such that the expertise resident in the different agencies can be combined to address the complex issues that underlie the response to a global pandemic. Specifically, we require a system that captures the following elements: (1) leverages existing capabilities and delivers effective shared services outcomes, (2) respects constitutional roles and responsibilities and enables innovation and networks of innovators at all levels of government, and (3) is acceptable to all, and is thus politically and operationally viable and sustainable.

[2] *See* https://www.whitehouse.gov/briefing-room/presidential-actions/2021/02/24/executive-order-on-americas-supply-chains/.

7.2.1 Leverages Existing Capabilities and Delivers Effective Outcomes

State, local and industry capabilities already in place and working effectively should be leveraged as fully as possible before building new Federal capabilities for effective allocation of scarce resources. To what extent could new national capabilities be leveraged by multiple Federal, state or local government entities as a shared service capable of responding to a variety of emergency response scenarios? This capability is required to ensure that the right level of expertise is being leveraged across agencies, to bring a coordinated and intelligent equitable response that builds on the strengths, knowledge, and network of experts that can work together in a unified forum to solve a problem.

7.2.1.1 Current State

A component discussed at length in one of our workshops was the issue of poor or missing data quality and data standards. For an effective shared service organization to function, shared data standards across agencies will be essential for rendering decisions about allocation of materials. Participants also noted that "we have to be realistic vs. inspirational when thinking about a national threat model. It will take years to break down siloes between agencies to establish a common architecture, but what we require is a system for interoperability, one that is not too federalized and which "screams" for agility. Having a public–private commercial model would lend itself to that, and there is a need to focus on information that is critical, and to establish who is doing what, and communicate that throughout the enterprise.

Agility in government requires trusted networks to build interoperability. It is no good to simply order disparate groups to team up together in an emergency to figure out a solution of how to deploy materials needed for healthcare response across the entire country. It is critical to establish the guard rails for each agency, and determine who is in charge of what activities, based on capabilities, not based on charters. Rather than setting up a massive monolith organization, we need to establish national networks.

7.2.1.2 Future State

Asking Manufacturers to reserve capacity/quantities of material to supplement the SNS is not going to happen. We now know we cannot rely on this strategy—foreign manufacturers will voluntarily or be forced to serve their country's needs first. Analysis we conducted with S&P Market Intelligence shows how exports into the US were restricted during this period. Our manufacturers most often rely upon foreign supply chains, and this is not going to change overnight. This is simply not a workable proposition. Reserving manufacturing capacity is simply not possible, as most of the time this requires significant advance notice to scale up, and manufacturers do not have control over the capacity of their tier two suppliers in foreign countries. We are not going to be able to control manufacturing capacity which even if contractually reserved, and many of these products unavailable to tap into during the COVID crisis (and remain so today) During any major global crisis, this will similarly be the case.

Instead, contractual requirements must be supplemented by inventory visibility systems tied to a control tower, as well as blockchain (or other distributed ledger) transaction channels. A blockchain creates a trusted network of suppliers, through a private and secure technology network, that allows instantaneous ordering, payment, and notification of receipt. Data and key metrics that are renewed and updated on a daily basis is essential for equitable distribution. A missing component of the COVID response was the inability to track where products were coming from, where they were being sent, and who was receiving them. The hoarding that is occurring can be prevented by inventory visibility systems, that employ barcode and QR code tracking of material through the supply chain, through a trusted network of distributors and manufacturers. Consumption of supplies should also be tracked, so that supply allocation decisions can be made in real-time based on daily or even hourly updates on what is happening vs. self-reporting demand that can contribute heavily to the tragedy of the commons scenario. This technology is not overly expensive to create but requires a centralized mandate and infrastructure to pull required data into a data lake that can serve as the single source of truth. This data lake must be curated carefully by a centralized group of information technology (IT) professionals, to ensure that data quality, reliability, and timeliness is not compromised. Traceability and transparency can reduce the risk of profiteering, counterfeiting and quality degradation in critical supply chains as well. We mandate that blockchain and visibility are critical features not a nice to have for the future strategic national stockpile (SNS) and should be used by all healthcare logistics material allocation solutions.

7.2.2 Respects Constitutional Roles and Responsibilities

The Federal role should focus on setting standards and enabling interoperability for information sharing and coordinating response efforts, rather than orchestrating "command and control" responses from Washington, DC. Alignment of federal, state, and local actions should be clearly defined through a "playbook" that designates the roles of each party in the response. The response system should also enable entrepreneurial responses from individual innovators and networks of innovators at all levels of government and industry.

7.2.2.1 Current State

During the peak of the COVID crisis, federal agencies were competing with one another over their decision rights and ownership of issues. An equitable and fair means of deploying materials in the stockpile was missing. Today, no such policy exists, as there has never been a situation comparable to COVID whereby every state in the country required emergency medical supplies. Further, state procurement agencies were operating independently, which led to hoarding and gaps throughout the country, often with the bigger and more populous states getting priority and the less populated or lower-funded states being left out. A system for tracking inventory across state lines and creating a commons-based system of supply that shows the nationwide demand and supply requirements was needed for the equitable distribution and allocation of materials.

The Defense Production Act was invoked for PPE essentially after the fact, as the global supply of raw materials to produce these goods was already backlogged by April. Government edicts to control production did not function in a global supply chain ecosystem that did not have raw materials available domestically. This situation revealed a lack of adequacy, capability, and governance to create and manage a commons to respond to a national pandemic situation. We attribute this to a number of inherent problems in both the national pandemic response and the general lack of integration across the entire US health care system.

7.2.2.2 Future State

Our workshop panelists from government noted that developing a cross-agency network to allocate equitably across the country as politically sensitive and dire as a national pandemic requires "baby steps", and cannot be evolved overnight. As one government expert noted:

> "Grand strategies sound great, and setting up some broad frameworks such as the DPA can be helpful, but we need to address the problems in smaller bit-sized pieces. If ASPR is the wrong place to host the SNS, we should adjust the federal approach to managing the problem. We should rely more on commercial partners to optimize our response, and partner with our other Allies around the globe. We also have to be cognizant of the political realities faced, which often entail multiple objective that are sometimes at odds with one another. For instance, contracts and acquisition strategies can be formed within the agenda of a national industrial strategy. To the extent that we promote an industrial strategy and relate it to the support of domestic manufacturers in the context of more strategic agile supply chains, we can be consistent across multiple agendas."

Another executive noted the following:

> "Across the pandemic it is not just about material movement, but it also involves working across a patchwork of government issues and federal benefit programs. There were clear gaps between agencies like the DLA, FEMA, and the American Red Cross, and very little interoperability between their systems. We need to know what is in our own warehouses, but also build partnerships with suppliers through our acquisition channels."

One of the biggest challenges that lies ahead for government involves ensuring improved data sharing between agencies. During COVID, members of the SNS emphasized that they were not even able to get basic data from FEMA or FDA on the number of cases, and where they were occurring. As we understand it even today, the DLA and FEMA don't communicate well with one another. Our proposed model requires a willingness of agencies to communicate, which in turn is a legal issue and a trust issue. We need to put in measures to address the legal issues, and then begin to work on the trust issue through communication, to be less reactive and more proactive.

There are other activities in the government today to improve communications across agencies, but the concern is that if there are multiple parallel activities going on, all of them will get bogged down. Government systems are often "crickety", but when national emergencies occur they are able to mobilize if the right governance is established. The future state needs to find ways in which government and industry can reach agreement on setting priorities for production and robust stockpile management approaches, and to agree on specific actions with specific agencies responsible. For instance, there is already a lot of activity on rare earths and microelectronics, both areas where the US is exposed internationally. Perhaps we should focus on a few of these areas, or even a few medical supply areas, and focus on these.

The current administration is putting a lot of emphasis on on-shoring activities, but the reality is we will only likely get to 10% domestically produced medical goods. But we also need to think about how to incentivize investments in domestic supply chains, and in so doing may also over-emphasize on-shoring by making it a "jobs" issue, and not a "risk" issue. We are seeing a lot of industries start to restructure their global supply chains and moving to the West to reduce supplier risk, and also perhaps moving towards a new north–south axis with Latin America and Mexico. These are promising areas where government policies may intersect well with these trends.

Any future state needs to clearly define who is responsible, based on the type of emergency the country is dealing with. An excellent example of such collaboration was the "swarm mentality" that took place during the Boston marathon bombing, when dozens of different police forces worked together but stayed in their own lane. During COVID, it was never clear who was in charge, and there was no master plan for determining which issues would be handled by which agencies. Clearly defining the "swim lanes" will be an important component of the future state.

An important component of a future state model might consider using an "Other Transaction Authority" (OTA) as a contract vehicle for acting rapidly in an emergency. OTA's are a special contractual vehicle used for specific federal agencies to expedite and simplify access to mission-critical technologies, that by-pass the complicated and burdensome requirements of the Federal Acquisition Regulations (FAR). If OTA is used, it will be critical to define and understand what third parties providers will supply to the government and the public. Many of the OTA concepts were built for non-traditional, small firms. Unfortunately, many OTA contracts have gone to large firms who seek to avoid cost accounting and IP restriction requirements. OTA is typically used to avoid using the Federal Acquisition Regulations.

7.2.3 Political and Operational Viability/Sustainability

The proposed solution for equitable allocation must be acceptable to a broad, bipartisan community of political decisionmakers at the national level and operationally viable and sustainable at the state and local levels in a deeply divided America.

7.2.3.1 Current State

Unlike prior emergencies like 9/11, when there were a lot of exercises going on across agencies, the COVID crisis did not have a lot of supply chain governance mechanisms. Executives in our forum asked if there was a way we could enable interconnected preparedness and have protocols in place across agencies to manage such a crisis. One noted that "OMB performs in such cases, because it has a built-in structure to frame the policy, identifying where it is coming from, and a format for presentation. OMB may not be the right agency to run the show, as they are not operational in nature, but they do have a good system of governance." The critical element that was needed was a framework that was anticipatory, so that the agencies could begin to manage it immediately when it happens, and have all the resources in place.

Executives at our workshop noted how there was no structure in place to deal during the crisis. The government developed an interconnected network overnight, which involves having to talk to agencies twice a week, to determine where the money was going and how it was being spent. The team also reached out to the business community, to create a larger structure of external interconnected networks, to enable a common mission.

7.2.3.2 Future State

A desired future state would entail developing a "Board" or "Taskforce" that has a predefined governance apparatus that is focused on developing a structure approach for equitable distribution. This Board should have an initial meeting with representatives from the right agencies to define the common mission of the board. The goal would be to rapidly enable an interconnected network, which would require having some pre-defined sourcing agreements in place. Because it is impossible to determine the exact nature of the problem at hand and its characteristics, given that the taskforce would not have any level of perfect foresight, it will be critical to enable a structure that will perform in as agile a manner as possible.

Having the previously mentioned early medical intelligence warning signals would trigger the formation of such a task force, and allow it to begin due diligence in terms of supply market intelligence, identification of current state conditions for critical inventory, sourcing channels, and hospital preparedness levels. These details can provide the basis for sound decision-making based on evidence and information, not on assumptions, suppositions, or wild planning. Ideally there should also be some form of industrial policy protocol in place that could be mobilized, and intelligence networks reaching into the private sector, especially in medical distribution, would need to be in place prior to the emergency. For instance, during COVID no one on the COVID task force had any idea about the lack of N95 masks or the poor condition of existing ventilators in the SNS, until it became too late to be able to do anything about it.

A good model for developing such a task force is the case of the War Production Board. The WPB was envisioned as a new administrative agency vested with full power and responsibility for the complete mobilization of industrial resources, with the objective to assure the most effective prosecution of war procurement and production (Civilian Production Administration, 1947).

When the WWII came to the United States with the attack by the Japanese at Pearl Harbor, production efforts, which until that point had been preparatory—directed towards anticipating future needs and addressing them with ample supplies—now had to pivot to a reactionary posture, filling needs as they arose—which necessarily included developing additional industry to meet new demand. To meet this more difficult task, the government created a correspondingly more powerful agency than had existed previously: the War Production Board. Additional details on the War Production Board are found in the Appendix.

7.3 Conclusions

The IBM workshop in 2021 produced significant insights into the challenges of whole of government supply chain and shared service challenges using the COVID-19 experience as a proxy and drawing conclusions for larger process, technology, and policy strategies. Among positive observations was that embedded in the United States is tremendous problem-solving capacity which comes from networked individuals who step outside assigned roles and designations and join with others to bring their collective talent to bear on urgent problems facing the country. This capacity was on display in the roundtable discussions as participants representing a wide spectrum of views assembled to share experiences, address what has been learned and how to prepare for the next pandemic or other whole of government management challenges.

The following recommendations emerged from this roundtable on how to create an equitable distribution policy before and during a pandemic event:

- Establish a new set of government industrial policies that are targeted at supporting a domestic "stop gap" manufacturing capability, that can meet the immediate emergency needs for material.
- Government should be developed with distributors that enable visibility into their inventory systems, and with fixed "set aside" inventory for government allocation under different conditions of duress. This will require a set of common data standards and a common architecture to create a dashboard and control tower.

- Establish a multi-agency materials inventory portfolio stockpile based on in-depth supply market analysis. At a minimum, this should include acquisition and material specialists in the following categories: disposable medical supplies (PPE, gowns, gloves), pharmaceuticals, plastics and resins, medical equipment, biologics, healthcare personnel, and respiratory products. This could be a virtual stockpile that utilizes existing government storage facilities, such as the VA.
- Establish an independent office responsible for developing a "make or buy" set of analyses that develops recommendations for investment in critical manufacturing infrastructure within the United States. This should be deployed in conditions of elevated market risk that creates extreme exposure to other countries where the bulk of manufactured supplies are produced.
- Develop an external party that can develop and sustain a medical intelligence signal to provide early warning indicators of healthcare pathogens and other healthcare crises that are imminent. This would result in a capability of observing direct indications of reports of human disease, coupled with indirect indications and warnings such as official acknowledgement and action, demand for medical services, local perception of threat, comprised infrastructure, and change in business practices
- A party (either internal or external to the government) tasked with monitoring conditions across supply markets with the following properties, that include both a strategic and a tactical set of components. The tactical element to this function would involve development of acquisition contracts with critical private sector providers, to ensure that the government can procure the right materials in quantities that are aligned with the perceived level of threat identified by the Health Security Risk team.
- Creating a shared service capability staffed with experienced supply management professionals, who are knowledgeable in developing category strategies for these materials. This team would be tasked with development of category strategies, combined with deep supply market intelligence around how to construct strategies to mitigate risk.
- Creation of a "national healthcare monitoring system" with access to an inventory visibility system tied to a "control tower", with real-time analytics that tracks the current status of material in critical stockpiles throughout the country.
- Establishing a pandemic planning team to develop demand sensing capabilities, war-gaming situations/simulations to inform category strategies, and capacity requirements that span both domestic and global sources. Requirements should embed industry standards to create maximum flexibility and increase alternatives in the event of need.
- Using industry experts from the private sector to improve the integration of medical, production and supply chain management expertise in an area of the government presently dominated by only medical knowledge, with the potential benefit of improved preparedness of healthcare supply lines.

- Development of contracts that require manufacturers to reserve capacity/quantities of material to supplement the SNS, along with contractual requirements requiring development of inventory visibility systems tied to a control tower, as well as blockchain (or other distributed ledger) transaction channels.
- Consider using an "Other Transaction Authority" (OTA) as a contract vehicle for acting rapidly in an emergency. OTA's are a special contractual vehicle used for specific federal agencies to expedite and simplify access to mission-critical technologies, that by-pass the complicated and burdensome requirements of the Federal Acquisition Regulations (FAR).
- Development of a "Board" or "Taskforce" that has a pre-defined governance apparatus from the outset. This Board should have an initial meeting with representatives from the right agencies to define the common mission of the board. The goal would be to rapidly enable an interconnected network, which would require having some pre-defined sourcing agreements in place. Because it is impossible to determine the exact nature of the problem at hand and its characteristics, given that the taskforce would not have any level of perfect foresight, it will be critical to enable a structure that will perform in as agile a manner as possible.

Part III
Immunity

8 Major Supply Chain Events of 2020 Beyond COVID

We have covered various forms of symptoms and treatments for sourcing sickness that aid in supply chain immunity in the previous chapters. Much of this was developed based on our experience responding to the global pandemic. However, since 2020, we have witnessed a host of other supply chain events that point to the need for greater supply chain immunity and stronger decision-making in the future. We don't just see pandemics impacting our supply chains. We see cyber-attacks, weather disruptions and other man-made disruptions that continue to impact our ability to meet demand and need across the globe with minimal friction. This chapter provides a small summary of such disruptions and foot stomps the need for better means of decision-making and supply chain management in a post-pandemic world.

8.1 Supply Chain Cyber Attacks

One of the first threats to our national security is the massive growth in cybersecurity attacks on both the public and private sector. The volume of cyberattacks is escalating at an incredible and unsustainable rate. One estimate provided by experts we interviewed is that cybersecurity costs will escalate to more than $5 trillion by 2025. The Identity Theft Resource Center found that cyberattacks on supply chains rose by 42% in 2021.[1] Companies and governments have been held hostage by ransomware attacks on supply chain nodes. In March 2021, the National Counterintelligence and Security Center published a summary of major supply chain software attacks since 2017[2] (Fig. 8.1). This

[1] https://www.cips.org/supply-management/news/2021/april/troubling-rise-in-supply-chain-cyber-attacks/.

[2] https://www.dni.gov/files/NCSC/documents/supplychain/Software_Supply_Chain_Attacks.pdf.

R. Handfield and D. J. Finkenstadt, *Supply Chain Immunity*, Synthesis Lectures on Operations Research and Applications, https://doi.org/10.1007/978-3-031-19344-6_8

figure shows ten major supply chain software attacks from 2017 to 2021. Software provides persistent access opportunities for sly intruders, one in which the hostage is the supply chain itself. Strong supply chains rely on highly interconnected networks. This level of interconnectedness also creates dependencies that serve as a fertile opportunity for massive cyberattacks without strong security measures.

Solarwinds Summer 2021 brought the SolarWinds attack. SolarWinds Orion Russian hackers entered the SolarWinds Corporation's systems as early as January 2019, based on recent reporting from the company.[3] SolarWinds Corp. is headquartered in Austin, Texas and provide IT infrastructure management software and services within the US and abroad.[4] The breach expanded and malware spread to thousands of SolarWinds customers, including the U.S. Justice, Commerce, Homeland Security, and State departments and the U.S. Treasury. In late June 2021, the U.S. Security and Exchange Commission opened investigations into whether or not companies affected by the SolarWinds hack tried to mitigate the financial impact by purposely hiding the attacks from investors and customers.[5]

JBS Holdings In June 2021, the largest meat company in the world (by sales) JBS USA Holdings, a Brazilian firm, was forced to pay $11 million to Russian-based REvil ransomware hackers[6,7] to regain access to critical operating systems across their networks. REvil is the same group that shut down Colonial Pipeline Co. operations on May 7th 2021, disrupting oil distribution in much of the eastern United States. Colonial was forced to pay $4.4 million to regain control of their systems, though the FBI recovered more than half the ransom.[8]

Kaseya Software On the eve of July fourth, REvil hackers infiltrated Kaseya VSA software. VSA software allows IT departments to remotely manage and monitor computers. The attack had spread to more than 1500 companies as of July 10. Kaseya was forced to ask 40,000 customers to immediately disconnect its software.[9,10,11] Many of

---

[3] https://seekingalpha.com/pr/18325545-hackers-targeted-solarwinds-earlier-previously-known.

[4] https://finance.yahoo.com/quote/SWI/profile/?guccounter=1&guce_referrer=aHR0cHM6Ly93d3cuZ29vZ2xlLmNvbS8&guce_referrer_sig=AQAAAFKCL7Eye1PTSmiu5Pv4Ooc_2F5UMxQNqf8o4Y5TOSoCML5AMP9JMM20AhMFdgBfEemniyZs6I2jpTtB8dygOJgCFC3znkiQ2YH8T36sUeQpnz-PbRcLzFC418KSRR7VifjfiK1h5o1P1gtG0FKbG6TJ52RN6jqJ-4P1lNIB9Knj.

[5] https://www.reuters.com/technology/us-sec-official-says-agency-has-begun-probe-cyber-breach-by-solarwinds-2021-06-21/.

[6] REvil operates as a ransomware-as-a-service provider, leasing the malware to outside parties for use in crippling networks for a ransom.

[7] Pronounced "Are Evil".

[8] https://www.cnn.com/2021/06/07/politics/colonial-pipeline-ransomware-recovered/index.html.

[9] https://www.washingtonpost.com/technology/2021/07/02/kaseya-ransomware-attack/.

[10] https://apnews.com/article/joe-biden-europe-government-and-politics-technology-business-fc0df4c42f8cd6148bf936ca24bb5cbe.

[11] https://www.washingtonpost.com/politics/2021/07/08/cybersecurity-202-kaseya-attack-is-revolution-sophistication-ransomware-hackers/.

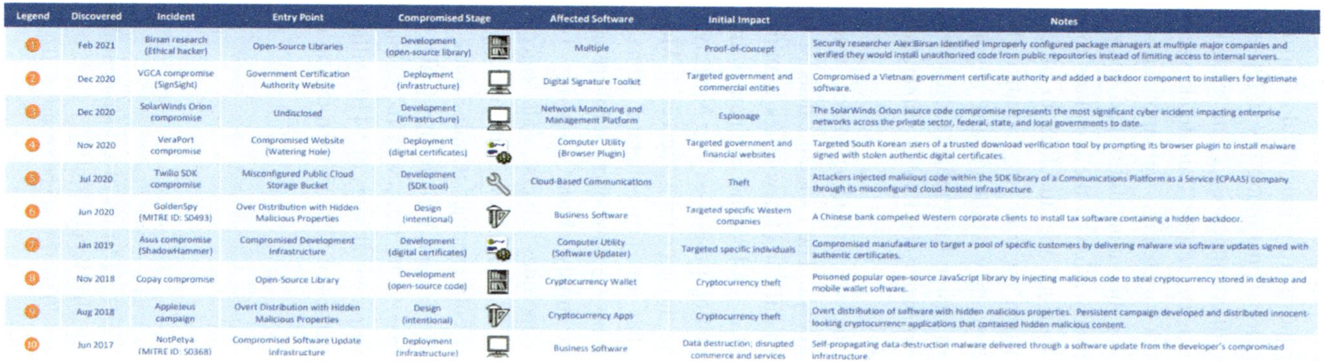

Legend	Discovered	Incident	Entry Point	Compromised Stage	Affected Software	Initial Impact	Notes
1	Feb 2021	Birsan research (Ethical hacker)	Open-Source Libraries	Development (open-source library)	Multiple	Proof-of-concept	Security researcher Alex Birsan identified improperly configured package managers at multiple major companies and verified they would install unauthorized code from public repositories instead of limiting access to internal servers.
2	Dec 2020	VGCA compromise (SignSight)	Government Certification Authority Website	Deployment (infrastructure)	Digital Signature Toolkit	Targeted government and commercial entities	Compromised a Vietnam government certificate authority and added a backdoor component to installers for legitimate software.
3	Dec 2020	SolarWinds Orion compromise	Undisclosed	Development (infrastructure)	Network Monitoring and Management Platform	Espionage	The SolarWinds Orion source code compromise represents the most significant cyber incident impacting enterprise networks across the private sector, federal, state, and local governments to date.
4	Nov 2020	VeraPort compromise	Compromised Website (Watering Hole)	Deployment (digital certificates)	Computer Utility (Browser Plugin)	Targeted government and financial websites	Targeted South Korean users of a trusted download verification tool by prompting its browser plugin to install malware signed with stolen authentic digital certificates.
5	Jul 2020	Twilio SDK compromise	Misconfigured Public Cloud Storage Bucket	Development (SDK tool)	Cloud-Based Communications	Theft	Attackers injected malicious code within the SDK library of a Communications Platform as a Service (CPAAS) company through its misconfigured cloud-hosted infrastructure.
6	Jun 2020	GoldenSpy (MITRE ID: S0493)	Over Distribution with Hidden Malicious Properties	Design (intentional)	Business Software	Targeted specific Western companies	A Chinese bank compelled Western corporate clients to install tax software containing a hidden backdoor.
7	Jan 2019	Asus compromise (ShadowHammer)	Compromised Development Infrastructure	Development (digital certificates)	Computer Utility (Software Updater)	Targeted specific individuals	Compromised manufacturer to target a pool of specific customers by delivering malware via software updates signed with authentic certificates.
8	Nov 2018	Copay compromise	Open-Source Library	Development (open-source code)	Cryptocurrency Wallet	Cryptocurrency theft	Poisoned popular open-source JavaScript library by injecting malicious code to steal cryptocurrency stored in desktop and mobile wallet software.
9	Aug 2018	AppleJeus campaign	Overt Distribution with Hidden Malicious Properties	Design (intentional)	Cryptocurrency Apps	Cryptocurrency theft	Overt distribution of software with hidden malicious properties. Persistent campaign developed and distributed innocent-looking cryptocurrency applications that contained hidden malicious content.
10	Jun 2017	NotPetya (MITRE ID: S0368)	Compromised Software Update Infrastructure	Deployment (infrastructure)	Business Software	Data destruction; disrupted commerce and services	Self-propagating data-destruction malware delivered through a software update from the developer's compromised infrastructure.

Fig. 8.1 NCSC recorded major supply chain software attacks since 2017[12]

[12] Ibid.

Kaseya's customers manage IT services for smaller firms, causing a downstream effect, for example, forcing Swedish grocery store chain, COOP, to shut down 800 stores for two days. REvil adjusted initial ransom requests from individual companies in lieu of a mass payout of $70 million in ransom for a global encryption key to release those affected.[13,14]

Equifax Breach (2017) In 2017, Equifax announced a data breach that impacted almost 150 million people. The data breach was the result of an attack on an Apache Struts vulnerability. Apache Struts is a framework for creating web applications and is widely used by Fortune 100 companies. A patch for the vulnerability had been available at the time of the attack. The unpatched vulnerability allowed the attackers to gain access to login credentials for three servers, which then led the attackers to other servers where they continually accessed personal information for over two months. The attackers stole the data piece by piece from 51 databases so they wouldn't raise any alarms. The investigation exposed numerous security lapses inside the organization's network, but the initial attack occurred because Equifax did not patch the known vulnerability in the third party software.

Merck (2017) In 2017, several businesses, including Merck, were affected by a massive ransomware attack known as NotPetya. The incident caused a worldwide disruption of Merck's operations and halted the production of new drugs, significantly impacting revenues for 2017. The attack was intended for Ukraine companies and infected a Ukraine software company that provided accounting programs. The NotPetya ransomware spread to Merck through a software update to the accounting server at Merck's Ukraine office. Merck was not the intended victim, but was just collateral damage. Other companies, including Maersk, were also impacted by the third-party software update.

Merck has an incredibly complex supply chain, supplying 140 countries, 60,000 SKU's, and over 200 products. This includes tablets, vials, injectable', implants, and inhalers, to name a few. The company is also involved in cold chain under difficult circumstances, including distribution of Ebola vaccines in sub-Sahara Africa which goes to small villages with no refrigeration. Ken Frazier, their former CEO, provided an overview of the cyberattack, and noted that "our SCM team was incredible during this period! Despite the quick response, Merck lost potential sales of $410 million in 2017 and 2018, and had to pay other hack-related expenses of $285 million on top of this loss.[15]

---

13 https://www.forbes.com/sites/daveywinder/2021/07/05/70-million-demanded-as-revil-ransomware-attackers-claim-1-million-systems-hit/?sh=79cccfae57c0.

14 https://www.washingtonpost.com/politics/2021/07/08/cybersecurity-202-kaseya-attack-is-revolution-sophistication-ransomware-hackers/.

15 https://www.fiercepharma.com/pharma/merck-insurers-fight-over-1-3-billion-damages-from-cyberattack-bloomberg.

8.1.1 The Critical Elements of Cybersecurity[16]

These cybersecurity breaches mentioned above all have one element in common: the lack of a supplier cybersecurity program and an ability respond to such threats in a timely manner. In the best case, an organization can identify the threat and respond quickly. The worst case? Not recognizing the threat, allowing significant damage to be wrought, or even worse shutting your supply chain or leaking your customers' data! Today many organizations have established cybersecurity programs, but only a few are expanding their scope to include supplier cybersecurity. Organizations cannot establish supplier cybersecurity programs without a strong and resilient internal cybersecurity organization in place. Let's first establish the critical elements of an effective internal cybersecurity program.

The CIA triad is a well-known information security model used to guide organizations creating security processes, procedures and policies. For over 20 years, it has embodied the fundamental security objectives for data and systems. It represents the three primary objectives around information security: Confidentiality, Integrity and Availability. Effective cybersecurity begins with the CIA triad, which requires organizations to consider three principles of cybersecurity risk. Confidentiality ensures only authorized users access or modify data. By addressing this principal, system access is controlled to prevent unauthorized disclosure. A common approach to support confidentiality is the use of encryption. Integrity ensures data is maintained in its correct state. It essentially refers to the trustworthiness of the data. File permissions and access control are among the methods to ensure integrity is maintained. Availability means authorized users are able to access data/systems when needed. While this part of CIA may not seem security-focused, it is a key part of securing the data. Even if confidentiality and integrity is maintained, the data (or systems) will be unusable if it is not available to authorized users. Attacks against availability include distributed denial of service (DDoS) attacks and ransomware. Protecting availability typically involves having backup systems and redundant systems. Most cybersecurity incidents can be traced to a problem with one of those three; therefore, the CIA triad is the recommended starting point for any cybersecurity program. It is an essential playbook for managing what amounts to a complete shutdown of an organizations' supply chain.

When considering the CIA triad in a supply chain environment, there are four basic elements of supplier cybersecurity to address. First **threat prevention** is about establishing an infrastructure that blocks intruders before they are able to penetrate the environment or cause any damage. It includes policies and tools such as firewalls, VPNs, network segmentation, email filtering and other solutions that prevent threats from getting to end users. Second, **network protection** helps ensure that if something gets onto a network, the accounts are protected, the data is protected and suppliers/contractors have awareness about how to look for threats and avoid inadvertently downloading malware. It also involves having controls and restrictions around sensitive data, and ensuring files are

[16] Handfield, R., and Earp, J., Supplier Cybersecurity, CAPS Report, 2022, forthcoming.

encrypted in transit. Third, **detection controls** establish guidelines on how to monitor the environment, systems, employees, and signal an alert if a suspicious event arises. Finally, the last bucket is **incident response**, which refers to the ability to identify, contain and eliminate an attack while minimizing any resulting damage. It addresses how to escalate and ramp up, how to ensure backup and recovery measures are established ahead of time, and establishing an annual disaster recovery testing protocol to ensure an organization can respond to a crisis. These four buckets form the basis for our playbook described later in this research.

8.2 Supply Chain Weather Disruptions

8.2.1 Natural Rubber

North American manufacturers that use natural rubber in their production processes can expect higher prices and over the next six months. Automotive, tape, and defense companies will suffer most. They require unique properties of certain grades of natural rubber. Natural rubber and latex have few substitutes. Roughly $350 million in natural rubber is brought to North America by distributors and sold to first- and second-tier automotive manufacturers and the entire tape industry.

While the economic value of these transactions is not large, they are of strategic importance to the North American economy. Without engine mounts, automotive assembly lines can't run. Without tape, Amazon can't ship packages. As shown in Fig. 8.2, rubber futures went to an all-time low of $135 in April 2020 and have since risen to about $221. While this increase is not insignificant, based on our forecast, this market is poised to experience a real shock in prices as the delayed impact of market forces come to bear later in Q3.

Low prices resulting from the pandemic and the global economic slowdown have reduced the incentive for growers, tappers, and producers to manufacture natural rubber. As a result, production was estimated to be down 6.5% for the quarter ending December 2020 versus the same quarter in 2019.

As North American manufacturing came back online in Q1 2021, there was a supply deficit. The shortage is exacerbated by a La Niña weather event that is building momentum and expected to persist. Heavy rainfall in Southeast Asia interferes with natural rubber harvesting and hurts productivity among growers, tappers, and producers.

In January, in four provinces of Southern Thailand, two million acres of natural rubber plantation were flooded by heavy rainfall. Production in the region could be reduced by 100,000 metric tons.

Suffering losses in the palm oil and coal markets, Singapore's banks have withdrawn funding from all commodities, affecting companies critical to the natural rubber supply

Fig. 8.2 Global Natural Rubber Prices—Investing.com

chain. Without funding for raw material purchases and to sustain factory operations, producers are buying less from the tappers, producing less for inventory, and withdrawing from the market.

These combined forces have significantly curtailed the availability of natural rubber and there is no quick or convenient way to make up for the production shortfalls.

It is likely that the supply deficit will be felt well into Q3 and Q4 2021.

Prices today remain below the inflation-adjusted trend over the last several decades, but the last time this perfect storm of conditions existed, the price increased five-fold and took several years to settle.

These conditions have created a strategic trade issue. The economic recovery in China began before that of Europe and North America, so while prices have been artificially low, China has been buying natural rubber beyond its immediate demand. The Chinese likely now exercise soft control over more than 70% of the world's production of natural rubber.

Right now, tire manufacturers are unconcerned because they purchase directly from the producers. This is done with the tacit cooperation of the Chinese. Should trade relations further degrade, this consent cannot be assumed.

The withdrawal of funding by the Singapore banks is also giving rise to a strategic vulnerability in the distribution of natural rubber to manufacturers in North America.

All but one of the distributors serving the North American manufacturing base are Chinese-owned or controlled. Absent bank funding to support the purchase and transport of natural rubber along the fifteen-thousand-mile supply line from Southeast Asia to North America, the single distributor not under the influence or control of China will be forced to withdraw from the market.

China's possession of most of the current natural rubber inventory and the continuing trade tension between the U.S. and China should concern all U.S. stakeholders. While there has been no indication that the Chinese have influenced U.S. distribution as yet, the risk to the U.S. economy cannot and should not be ignored.

The brewing rubber crisis offers a rare view of a supply chain crisis as it develops. It is a reminder of the importance of maintaining visibility of and protecting a supply chain all the way to the source of strategic parts. Mitigation of the crisis before it fully develops is imperative.

U.S. banks, legislatures, and corporations must act immediately to protect and maintain diversity in the global distribution of natural rubber. Doing so will require strategic funding for distributors whose interests align with those of the North American manufacturing base. Now is the time to act to prevent Chinese government interests from attaining monopoly control.

8.2.2 Texas Freeze

In mid-February 2021, a polar vortex hit Texas, driving temperatures down to levels not seen since the late-1980s.[17] The freezing conditions shut off water to over 12 million people and power to over 4 million causing the largest energy disruption in U.S. history.[18,19] Supply chain disruptions hurt grocery distribution in New Mexico, fuel distribution, energy and even the already threatened semi-conductor supply chain.[20]

The world's largest petrochemical production cluster was brought to a complete halt.[21] U.S. prices for polyethylene, polypropylene and other chemical compounds used to make auto parts, computers, and a vast array of plastic products, reached their highest levels in years.

However, Houston-based, Kinder Morgan, Inc., an energy infrastructure company and gas pipeline owner, managed to make record earnings in excess of $1 billion during the first quarter of 2021[22] as other Texas power producers and utilities lost billions. Kinder

[17] https://news.un.org/en/story/2021/03/1086752.

[18] https://www.nbcnews.com/news/us-news/texas-contending-water-nightmare-top-power-crisis-n1258208.

[19] https://abc13.com/2021-texas-winter-storm-weather-how-many-people-lost-power-boil-water-advisory/10356914/.

[20] https://www.ismworld.org/supply-management-news-and-reports/news-publications/inside-supply-management-magazine/blog/2021/2021-03/the-texas-freeze-repercussions-and-risk-mitigation/.

[21] https://gov.texas.gov/uploads/files/business/petroleum-snapshot.pdf.

[22] https://www.dallasnews.com/business/energy/2021/04/22/pipeline-operator-kinder-morgan-posts-1-billion-windfall-from-texas-winter-storm/.

Morgan attributes their positive position to strategic natural gas logistics networks and storage assets that kept operations alive during the Texas freeze.[23]

Prices for polyethylene, polypropylene and other chemical compounds used to make auto parts, computers, and a vast array of plastic products, have reached their highest levels in years in the U.S. as supplies tighten.[24] For example, prices for polyvinyl chloride, or PVC, have more than doubled since last summer, according to S&P Global Platts. The shortages in PVC and other materials are even shutting down automotive production lines, as well as other manufacturers that I've been speaking with. As plants slowly restart, it will take months for inventories to recover, with prices likely only returning to normal levels in June," said Joel Morales, analyst at IHS Markit.

Discussions with experts on the Gulf Coast have also shared the ongoing challenges faced in this region. The delayed impacts of the petrochemical industry were urgent after the freeze. It is also testament to the fragility of this supply chain. This region endured two spikes of COVID and two tropical storms, all of which disrupted operations. This time, it took a whole week for recovery after the storm went through on a Sunday. This caused disruptions to multiple industries. Almost all packaging today is petroleum based, and in the lower 48 states, the bulk of refineries (which feed petrochemical plants) are in the Texas and Louisiana regions, (see map). This means that the majority of refining and petrochemical capacity is in a region susceptible to not only hurricanes, but also freezes like the one we had. This episode was a humbling experience, and, if variants in weather due to global warming continue to occur, companies need to look at alternative ways to manage their plastics and materials supply chains (Fig. 8.3).

One of the big bottlenecks is nitrogen and refrigerants. There is only one manufacturer in North America capable of producing the high-grade ethylene required: Eastman Chemical (located in Texas). It is a major input into LNG and other refining facilities, and is only produced in three locations worldwide: Texas, China, and Korea. If it shuts down, most of the plants in this region would be impacted.

In another example, to get access to nitrogen, one company interviewed relied on shipping it in from Oklahoma, as the pipelines were non-functional. The truck took 6 days to make the normal 1-day travel time, due to road conditions. This required having 3 drivers on board to ensure that working hours were not violated!

As a result of this exposure, more companies are beginning to explore using biobased chemicals as alternative sources of supply. A recent study by the Green Chemistry and Commerce Council (GC3) shows that this sector is growing rapidly. Using biobased feedstocks not only promotes a sustainability platform, but also provides alternatives to a supply chain that is susceptible to disruption. In the past, the cost of these biobased feedstocks has been greater than petrochemicals, which has slowed their adaption by industry.

---

[23] https://www.forbes.com/sites/greatspeculations/2021/04/23/kinder-morgan-the-texas-freeze-winner/?sh=19c90263bac0.

[24] https://scm.ncsu.edu/scm-articles/article/the-big-texas-chemical-freeze-raises-issues-on-resiliency-of-the-petrochemical-supply-chain.

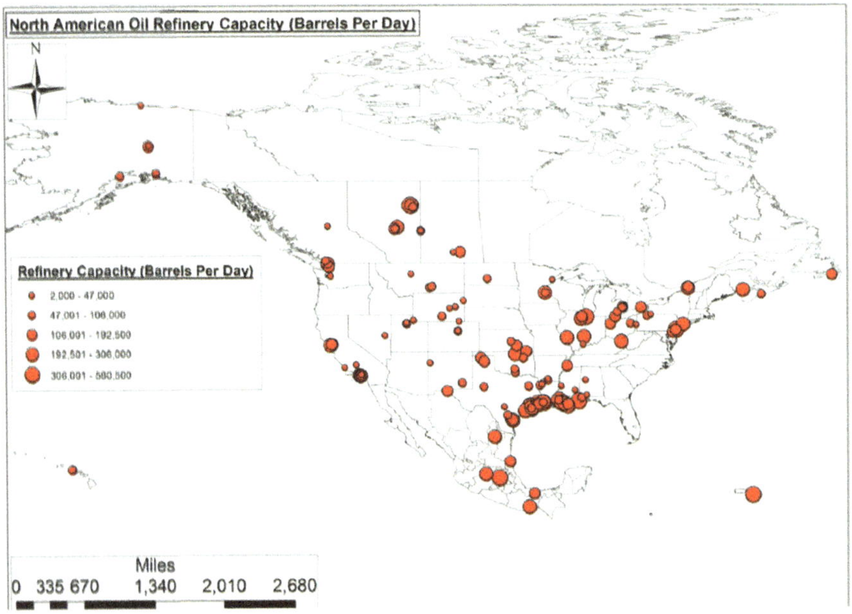

Fig. 8.3 North American Refinery Capacity

However, as prices have spiked with petrochemicals, and are unlikely to recover until the fourth quarter, this may be an ideal time to go green when it comes to plastics and chemicals.

8.2.3 Factory Fires

The incidence of factory fires increased by approximately two-fold in 2021. This is thought to be an unfortunate byproduct of COVID as typical inspection procedures were delayed or waived altogether across the globe as a result of COVID and the re-initiation of such activities constrained by a tight labor market. Since most fire prevention efforts requiring at least annual inspection and effectivity checks, for things like fire extinguishers, sprinkler systems, fire hydrants, and actual planning for new and modified structures, one year of eased requirements proved to cause significant fragility in the supply chain.

8.3 Manmade Disruptions

8.3.1 Suez Canal

On March 23, 2021, Ever Given, Evergreen Marine's super container ship, ran aground in the Suez Canal, which moves more than 13% of the world's freight. The main passage between Asia and Europe was blocked.[25,26] The ship remained stuck for six days, creating a traffic jam of more than 370 ships.[27] Ships were forced to take an alternate route around Africa, adding nine days to traditional shipping times.

8.3.2 Saudi Oil Fields

In March, Yemen's Iran-aligned Houthi rebels launched six drones at a facility belonging to Saudi Aramco. The attack caused more than half the world's oil production to cease, momentarily driving oil prices to their highest level in over a year.[28] In 2019, the same rebels attacked another Saudi Aramco processing facility and oilfields causing a 5.7-million-barrels-per-dayloss in production.[29]

8.3.3 3M Coolant Factory Shutdown

In April, a 3M factory was temporarily shut down after Belgian authorities found high levels of toxicity in soil samples. Perfluoroalkyl and Polyfluoroalkyl substance (PFAS) is a coolant required in chip manufacturing and the Belgian plant provides 80% of the world's supply of this coolant. This created another shockwave in the supply chain that forced other manufacturers of the coolant to shift work to make sure that this shortage did not make the 2020 semiconductor bad dream a nightmare.[30] Even so, there was an impact to semiconductor manufacturing and an even more pronounced impact to resins,

---

[25] https://www.forbes.com/sites/loracecere/2021/03/29/the-ever-given-is-moving-but-your-supply-chain-will-not/?sh=37c68af6744f.

[26] https://www.popularmechanics.com/technology/infrastructure/a36073961/ever-given-ship-stuck-in-suez-canal-timeline/.

[27] https://www.forbes.com/sites/loracecere/2021/03/29/the-ever-given-is-moving-but-your-supply-chain-will-not/?sh=37c68af6744f.

[28] https://www.cnbc.com/2021/03/08/brent-breaks-70-after-houthis-attack-saudi-arabias-oil-facilities.html.

[29] https://www.spglobal.com/en/research-insights/featured/saudi-oil-attacks.

[30] https://www.siliconexpert.com/blog/3m-shutdown-impacts-chipmakers/#:~:text=May%2011%2C%202022%20%E2%80%93%20Global%20chipmakers,a%20single%20factory%20in%20Belgium.

like those used in carbon fiber manufacturing, as lesser producers of the coolant had to absorb impacts in other areas in a labor constrained economic environment.

8.4 Climate Disruptions

Climate disruptions represent a segment of manmade disruptions caused by rapid weather pattern changes driven by global warming. A significant increase in the number of supply chain disruptions caused by chronic and extreme weather changes has increased dramatically and the supply chain is generally not ready for it. In a recent study and later publication in the Harvard Business Review, Resilinc and the University of Maryland found that "Eighty percent of all sites in the United States and 48% of all sites in China and Taiwan have either no business continuity plans or no alternative sites."[31] Recent events, including sea level rise across the globe, extreme heat throughout Europe and India threatening health and food production, changing precipitation patterns leading to droughts in areas like the Western US and India and record setting floods in other parts of India and in Nebraska in 2019, contributed to an estimated 49% increase in the number of manufacturing sites that have experienced an increase in climate variability induced supply chain disruptions in the last ~ 20 years.[32]

Major supply chain disruptions due to weather-related incidents are on the rise, and one can't dismiss these as a "once-in-a-century" event any more. The Texas freeze and Hurricane Ida severely disrupted petrochemical production, which had a ripple effect on the supply of resins and plastics, which is used in many different types of resins and plastics (think of impact on pallet shrink wrap), as well as toys (which use both plastic AND semiconductor chips), consumer goods packaging, and many, many countless other areas of industrial production that rely on PVC, citric acid and carbon dioxide, (used in soda drinks), nitrogen, and other process chemicals.

8.5 Labor Shortages

Labor shortages have wreaked havoc on supply chains. Again, there has been a confluence of factors starting with a decrease in population growth with an aging population in many nations across the globe, a corresponding crackdown on immigration which has been a primary source of new labor for decades, and pandemic related changes to worker expectations for work life balance. These impacts have hit nearly every industry, across agriculture, manufacturing, hotels, travel, service industries across both white and blue collar professions. It has also been a spark for labor disputes across the globe, further impeding both supply and movements of supplies creating poor financial returns during

[31] https://hbr.org/2022/05/how-exposed-is-your-supply-chain-to-climate-risks.

[32] https://hbr.org/2022/05/how-exposed-is-your-supply-chain-to-climate-risks.

a time of high inflation. The great resignation or great reshuffle has hampered operations globally and has not yet appeared to subside. The status of both imigration regulations and general workforce trends must be closely monitored throughout the supply chain as it poses perhaps the most significant threat to the supply chain as strikes and lockouts cause companies to cease virtually all operations, including the distribution of inventory.

8.6 Russian Attack of Ukraine

Apart from being an unnecessary humanitarian disaster perpetrated by a dictatorial leader and terrorist, the attack on Ukraine has wreaked havoc on already critically ill supply chains. Although Russia and Ukraine are two relatively minor economic players, their contribution of raw materials, food, and fuel to the global supply chain is significant enough to be the primary driver of inflation, hunger, and suffering unseen in the last 3–4 decades. While morally just and militarily savvy, the coalition of international partners banning Russian goods has led to large price increases for nearly all fuels and great concern over Europe's coming winter. Ukraine, while a lesser economic player than Russia, is a major provider of grain and inert gasses (neon, krypton, and xenon) for the rest of the world. The inability to access Ukrainian goods as a result of the conflict coupled with drought and extreme head in other food producing nations and has led to disastrous food shortages in many African nations as well as developing nations across the globe. The neon shortage has also hit the semiconductor industry hard during an already painful time as Ukrainian neon represents 90% of the gas used in semiconductor manufacturing in the US. In its July 26, 2022 earnings call, Raytheon identified supply chain issues as a driving factor and risk in its financial performance. One specific challenge noted is the lack of titanium castings available internationally as a result of sanctions against Russia. As damaging as this unjust attack has been, it leaves the authors wondering about the potential magnitude of a Chinese attack on Taiwan. Two major economic players could potentially be removed from the global supply chain as economic sanctions against China would be very likely and access to goods from Taiwan would almost certainly be denied.

8.6.1 Shortfalls in Capital Expenditures and Production Capacity

One of the unfortunate circumstances supply chains are facing is a shortfall in overall production capacity for targeted products, largely due to a shortfall and slowdown in capital expenditures (CapEx). This is partially the result of COVID, but is primarily due to a lack of capital investment over several years, not just in 2020–2021. Companies across sectors and the globe announced capital expenditure cuts ranging from 10 to 80%, with the largest cuts in oil and gas, and transportation. Despite the impact of capital expenditure reallocation on overall ROI, research shows that prior to COVID a third of companies reallocated only 1% of their capital from year to year. During COVID, as a result of the lack of capital expenditures, production cycles are at a low point. This is likely due to across-the-board cuts by senior leaders without a structured approach for reallocation that gives priority to strategic growth objectives. The problem is that capacity expansion cannot occur quickly once it is halted. In the semiconductor and pharmaceutical industries, factories don't pop up overnight. It can take years for a new pharmaceutical production facility to be built and to pass FDA inspection. Computer chip fabrication plants require painstaking care to produce and certify clean room conditions. These production cycles grow only when investors are sure there is demand for them. The same is true for beef and pork herds, and poultry farming. It takes almost two years for cattle herds to expand their calf production.

Cattle. The cattle cycle is at a low point globally, with production down in Australia and New Zealand. This situation does not have anything to do with processors in the United States. Protein imports of beef are at an all-time low. However, exports have been strong; China has bought a record amount of beef this year, and all of the substitutes for protein are also tight. Prices are elevated globally and there is not a lot of supply—Brazil, Canada, and Australia have limited production as well. The cycle will eventually turn around. It is expected that in three years there will be an oversupply of protein, demand will be scaled back, and prices will drop.

Semiconductors. The demand for semiconductors continues to surge, but capital expansion over the past three years has not kept up with demand levels. When the pandemic hit last year, many companies scaled back orders for computer chips amid a global economic slowdown. Revving up production has proven challenging as continued virus transmission in countries like Malaysia, Vietnam, and the Philippines leads to new complications. The consequences have been far-reaching, resulting in widespread shortages of electronic goods and automobiles.

Almost 50% of chip production can be traced to the Taiwan Semiconductor Manufacturing Company (TSMC), and another 25% to Samsung. Samsung plans to expand its capital expenditures by $205 billion over the next three years, an expansion of 30%. The company's second half 2020 expenditures of 18.3 trillion (South Korean) won were up 24% over the 14.7 trillion won spent in the first half of 2020, and they were up 33% over the 13.8 trillion won spent in the second half of 2019. The company's CapEx increased 46% in 2020. The obvious outlier is TSMC, which is forecasting a 54.1% increase in capital spending from $17.2 billion in 2020 to a range of $25 billion to $28 billion in 2021. About 80% of the $25–28 billion will be spent on advanced process technologies: 3 nm (nanometer), 5 nm and 7 nm. The remainder will be split roughly equally between packaging/mask making and specialty technologies. A small portion of the total will be spent on construction of the new fab in Phoenix, Arizona, which began this year.

Over the past 10 years, TSMC has underestimated how much the company will spend on CapEx, TSMC's capital spending is related to expected conditions over the next one or two years, and it often changes over the course of the year. The 2021 CapEx plan will result in a capital intensity higher than the long-term mid-30 percent target range. This reflects the fact that capital expenditures for semiconductors of lags demand. Semiconductor companies are cautious because a new fabrication plant can cost more than $1 billion, and represents a significant outlay of capital. For that reason, manufacturers prefer to extend lead times rather than build ahead of demand. In 2021, however, manufacturers severely underestimated the demand for semiconductors, and investment is strongly lagging the actual demand. This cycle hopefully will improve in 2023.

8.6.2 Summary

Indeed, it seems that supply chain problems are impacting virtually every industry: commodities, manufactured goods, energy, oil and gas, pharmaceuticals, health care, to name a few. The Federal Reserve claims that such shortages are a short-term, temporary issue. Federal Reserve Board Chair Jerome Powell argues that price inflation will soon pass once **supply constraints ease**. We remain dubious, and believe it is far more likely the global supply chain community will continue to see high prices and constraints for some time to come, likely into 2022 and possibly 2023. Indeed, the International Monetary Fund (IMF) downgraded its 2021 U.S. growth forecast on October 12 by one percentage point, the most for any G7 economy. The IMF cited supply chain disruptions and weakening consumption—which have been partially driven by supply chain bottlenecks. The problem can be seen in the shortage of new cars amid the computer chip shortage.

The current set of supply chain disruptions is precipitated by a combination of COVID-19 cases and energy disruptions in manufacturing hubs around the world, labor shortages, lack of capital infrastructure investment, misaligned transportation resources (a lack of containers, ships, and trucks,) and surges in consumer demand. This has led to massive supply chain shortages, including a scarcity of microchips, massive bottlenecks in US. and Chinese ports brought on by a combination of pandemic shutdowns and bottlenecks caused by a surge in demand for consumer goods by a world forced to sit at home. In all of these situations, there has been a disconnect in the ability for free market price-levels and significant shortages. There is widespread speculation that the Christmas season will be marred by massive waves of product shortages. Consumers continue to order and manufacturers continue to ship, but the global system is not prepared to logistically process a flow of irrational buying-selling behaviors brought on by the chaos of a pandemic. Even grocery store shelves are sometimes empty of certain products. The supply chain is gummed up, and the cascading set of bottlenecks can only be understood in the context of "systems thinking," which views the end-to-end supply chain as a series of nodes that must flow seamlessly. Each node, however, is also a bottleneck, which creates a series of problems that hinder the flow of materials from source to consumer. A question that is on everyone's mind, including the popular press, is whether consumer demand and suppliers are being pressured to purchase and produce more in a world of point-and-click ordering combined with perceived scarcities.

8.6.3 A Global Supply Chain Crisis Model

In short, pandemics are not the only risk to global supply chains. Many events happen every year that put global health and safety at risk, many of them through direct or indirect supply chain disruption. The cacophony of events since COVID-19, shown in the Figure above, highlights the need for a better way to observe and manage supply chains for global security. These events are also occurring in an age of data saturation. We end our book by offering a new model towards a better way of planning and acting in a data saturated world full of disruptions (Fig. 8.4).

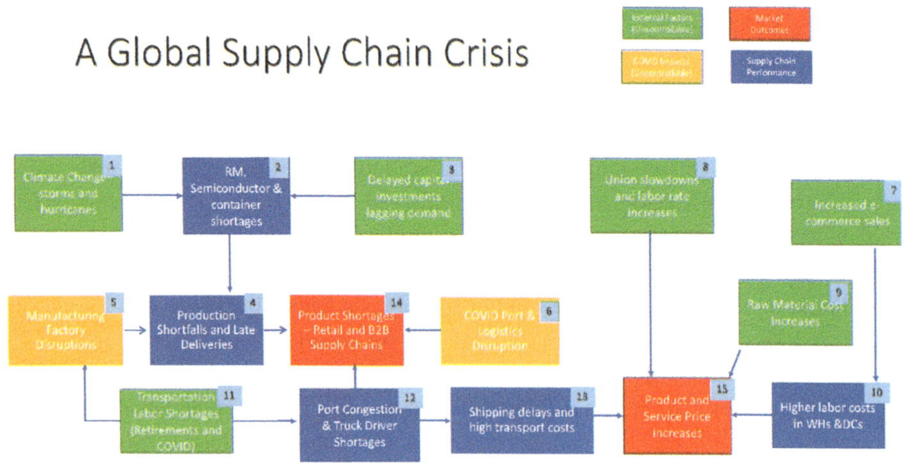

Fig. 8.4 A global supply chain crisis

9 A New Model for Emergency Decision-Making and Planning in a Data Saturated World: The Goals, Decisions, Signals, Data Model

COVID-19 exposed significant weaknesses in our national security, especially as it relates to supply chain management. Lack of supply chain traceability, flexibility, global independence and monitoring and equitability are just some of the weaknesses that have caused massive inefficiencies and performance issues across public and private sourcing for vital items such as PPE and sanitation materials since the pandemic hit US soil at the start of 2020.[1] Data is a primary observation for any planning activity, but can become a 'holy grail' of sorts in a contingency environment. We observed two major data management conundrums during the federal pandemic response that fundamentally altered our perception of supply chain risk management: Issue 1—Too many competing goals, with not enough data to meet them; Issue 2—Too much data and not enough direction to leverage it.[2] These issues don't just crop up during global contingencies like a pandemic, they exist in most public and private organizations, especially in our age of what Spivey coined as "Data Saturation" and Debra Bass has named "InfoObesity".

---

[1] Handfield R, Finkenstadt DJ, Schneller ES, Godfrey AB, Guinto P. A Commons for a Supply Chain in the Post-COVID-19 Era: The Case for a Reformed Strategic National Stockpile. Milbank Q. November 2, 2020. https://doi.org/10.1111/1468-0009.12485.

[2] Finkenstadt, DJ, Handfield, R. and Guinto, P. How Firms can Plan for Risk in a Data Saturated World: The Goals, Decisions, Signals, Data (GDSD) Model. California Management Review - Insights, October 10, 2022: https://cmr.berkeley.edu/2022/10/how-firms-can-plan-for-risk-in-a-data-saturated-world-the-goals-decisions-signals-data-gdsd-model/

R. Handfield and D. J. Finkenstadt, *Supply Chain Immunity*, Synthesis Lectures on Operations Research and Applications, https://doi.org/10.1007/978-3-031-19344-6_9

9.1 Issue 1—Too Many Competing Goals

The first type of data management issue we observed during the pandemic involved the over abundance of competing goals. Some agencies were primarily focused on increasing the supply of ventilators, while others were focused on increasing the fidelity of epi models for predicting the virus' spread. Hospitals were competing for resources such as personal protective gear and medical staff without a shared understanding of the national need. There was no centralized mechanism to aggregate disparate hospital need data in order to prioritize and allocate resources efficiently and equitably, resulting in the haves and the have nots during a major medical crisis. Many response priorities were being driven by the news media, and not be actual need. In fact, our research suggested that most hospitals either did not have the data within their systems, or were loath to share it with others. The government was also not fully open with the public about the nature of shortages, as we found in reviewing hospital shortage data developed by the NHSN.

9.2 Issue 2—Too Much Data and not Enough Direction to Leverage It

At the same time we faced a dearth of centralized data goals we faced an explosion of disparate data related to virus spread, market intelligence on vendors claiming to provide product X of service Y or federal agency spreadsheets on stated need for medical equipment and PPE. The result, chaos. For example, state CPO's were inundated by sales pitches for PPE from multiple distributors, many of which were not vetted or who did not even have PPE in their warehouses. This led to a number of instances of fraud, which were investigated in detail by the FBI. Similarly, many organizations are overwhelmed by purchasing transaction data, which often leads them unable to discern potential fraudulent activities that may be occurring, such as multiple vendors with the same address, or having the same business owner, or over-charging over contracted prices previously negotiated.

We are not claiming that we could have managed this event any differently, we are offering our observation surrounding how we have continued to plan for and respond to disruptive events using pre-data saturation models. Prior to the age of data saturation we could rely on traditional management decision models such as Boyd's OODA Loop (Fig. 9.1).

The late Air Force Colonel John Boyd tell us, observation is the first step to adapting and responding to the unexpected, followed by orientation, decisions and actions (the OODA loop).[3] This model is taught in federal personnel development courses such as professional military education and many of the government's leaders, including the authors,

[3] Boyd, J. R. (1996). The essence of winning and losing. (Unpublished briefing). Edited by Chet Richards and Chuck Spinney (2010). Retrieved from http://dnipogo.org/john-r-boyd/.

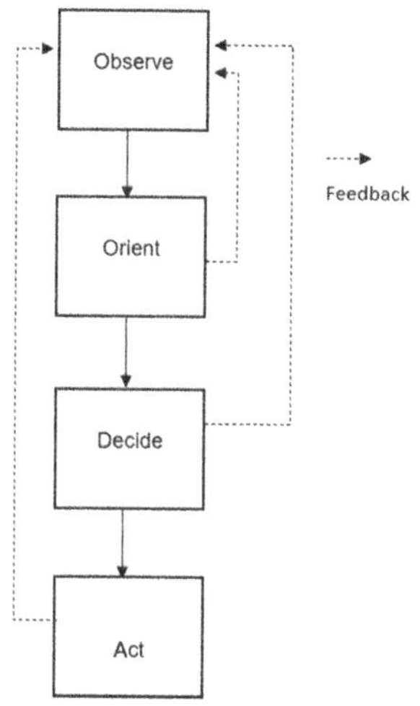

Fig. 9.1 Boyd's OODA Loop

have managed their decisions according to this model. The weakness in the OODA loop in a data saturated world is that it tends to focus users on the observation first…i.e. data, vs. the outcome goal, or intended effect. This can be overwhelming in a world of overwhelming volumes of data, leading to decision swirl. We admire the data instead of strategically leveraging it. It may seem counterintuitive to first ignore the data in a world of endless data, but that is precisely the point, data saturation has occurred because our ability to collect data has outpaced our ability to support, filter and manage it.[4] Trying to do this during an emergency event with full force fog-of-war underway is exponentially worse. We have to recenter our planning and response management processes to focus on outcomes as the step 1 of the process, and see data as a resource to strategically tap into with intent to cause kinetic effects.

Our team of experts working with the federal response to COVID-19 supply chain issues found that data was the largest hurdle in terms of observation and orientation. This, in turn, impacted state, local and federal agencies, firms and non-profit's ability to decide and act in an efficient, interoperable and strategic manner. Availability of data, data sharing across institutional stovepipes and the ability to synthesize and interpret data intelligently were all woefully lacking. They still are. The common thread in all this data failure was a lack of corporate strategy at every level. This is not necessarily organizational negligence, more of a new paradigm that has been on the horizon for years but has had limited understanding in terms of how to shift towards it. There are very few strategies on how data will be gathered, what will be gathered, how it will be visualized for decision-making and what types of decisions are even necessary. We see an immediate need to develop a strategy methodology for value chain data that leverages the benefits of value stream mapping adapted for this particular purpose.

[4] Why Too Much Data Is a Problem and How to Prevent It (forbes.com).

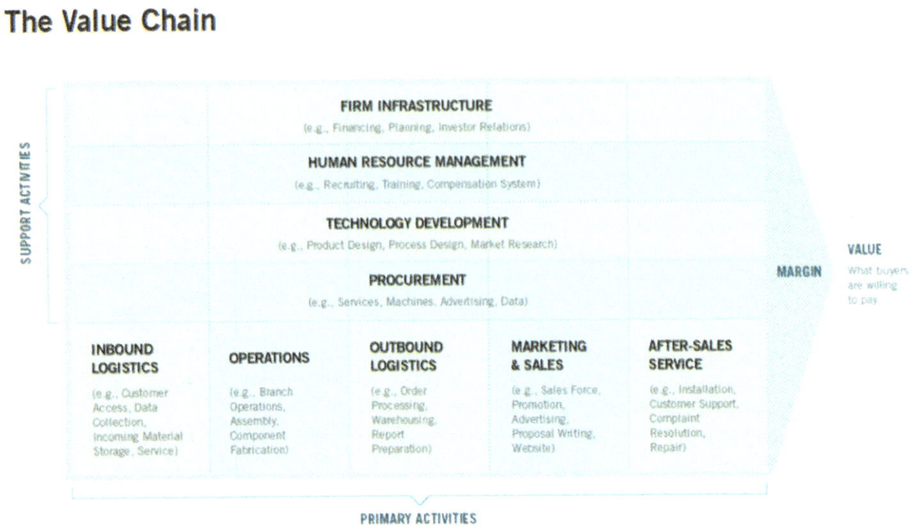

Fig. 9.2 The value chain taken from Harvard Business School. https://www.isc.hbs.edu/strategy/business-strategy/Pages/the-value-chain.aspx

The concept of value chains and value stream mapping are not new. Michael Porter introduced value chains in 1985.[5] Value chains expand the traditional supply chain to include both primary activities of inbound logistics, operations, outbound logistics, marketing and sales, and service as well as secondary activities such as procurement, human resources management, technology development and infrastructure (Fig. 9.2). Value stream mapping was introduced by Toyota's Taiichi Ohno and Shigeo Shingo in 1980. It has been used by organizations in most sectors including automotive, aerospace and construction. There are seven primary tools for use in value stream mapping[6] (Table 9.1).

These tools have varying purposes, but all demonstrate similar features. Each seeks to summarily visualize a complex process, quantify key indicators of performance, and provide decision support for a firm in day to day operations. Additionally, we see value stream mapping as a tool that can be leveraged to tune our focus on data necessary to make critical, emergency decisions in a value chain. Value stream mapping can help organizations strategize the best data gathering, monitoring and reporting solutions to support their objectives and key results. Public, private and non-profit organizations can better manage their contracts and supply chains with stronger market intelligence, supplier

[5] Porter, Michael E., "Competitive Advantage". 1985, Chap. 1, pp. 11–15. The Free Press. New York.

[6] Hines, Peter; Rich, Nick (1997). The seven value stream mapping tools. International Journal of Operations & Production Management; Bradford 17(1), pp. 46–64. http://dx.doi.org/10.1108/01443579710157989.

Table 9.1 Seven value stream mapping tools

Tool	Description
Process activity mapping	Detailed, line by line list of the process including distances moved, time taken, and number of people involved to inform a flow chart of activities from the industrial engineering discipline
Supply chain response matrix	Diagram of critical lead time constraints for a particular purpose
Production variety funnel	A visual display of complexity along the supply chain from the operations management discipline. The funnel shape occurs as the items move from production plants that deal in less complex materials (i.e. raw) to more complex end items
Quality filter mapping	Graphical display of product, service and scrap defects. Scrap defects are essentially product defects caught in production before they are delivered
Demand amplification mapping	A systems dynamics approach to displaying varying demand over time within the chain
Decision point analysis	Visually mapping the supply chain to show points where production based on actual demand is overtaken by that based on forecasted demand
Physical structure mapping	A visual approach to demonstrate the value chain from raw materials to distributers in terms of number of firms and cost contribution at each level

monitoring and contingency planning. To do this they require stronger signals for rapid decision-making. These signals must flow up from the value chain.

In a world of endless data, you must be able to 'tune' the signal to eliminate noise.

COVID-19 came at a time in our history when the global chain was not in America's favor. In the last 20 years the global value chain has shifted to see China as the primary trade mediator from most of Europe and Asia to America.[7] Some OEM supply chains can expand to as many as 1.25 M suppliers at the various tiers, yet many have no idea who is in their tier 2 and below supply chain. Many of the products needed to respond, such as N95 masks, other forms of PPE and pharmaceutical APIs all come directly from suppliers in China. We are also witnessing exponential growth in supply chain management concerns stemming from COVID-19 impacts on traditional supply chains and emerging supply chains popping up in response to COVID-19. Such concerns include:

- Lack of tracking and tracing beyond the prime contract level
- Lack of demand or need forecasting
- Lack of disruption prediction, alerts and responses

[7] Global Value Chain Development Report 2019, World Bank, World Trade Organization.

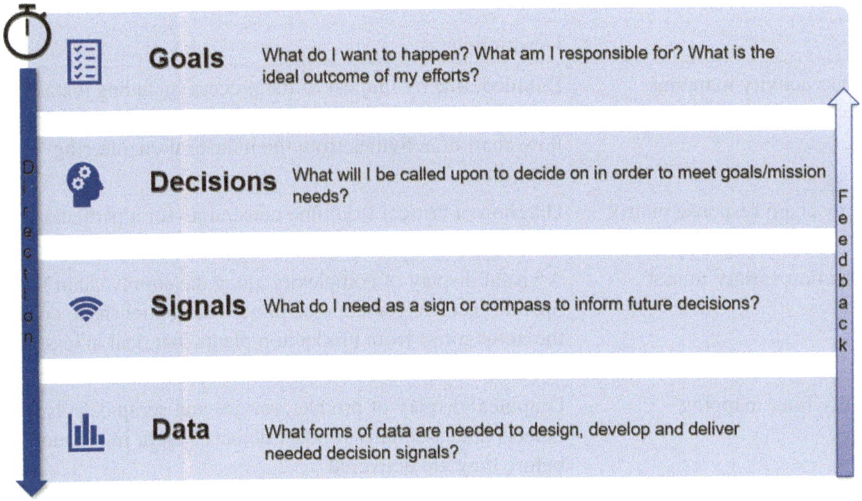

Fig. 9.3 Finkenstadt et al.'s Goals decisions signals data model

- Growing fraud and counterfeit products and sources
- Lack of supply chain human capital
- Lack of channel distributer capacity
- Lack of data fidelity and integrity for digitizing the SCM process
- Ill-formed contracts as a result of ill-informed supply chains.

We are all aware that we need 'better data' for 'better decisions'. But we can't just keep grabbing at more data. We must have an intentional strategy to capture and use this data for supply chain and contract decisions. That starts with value stream mapping and replaces the OODA Loop with what we call the Goal-Decision-Signal-Data model (in the case of a healthcare value chain we begin with VSM) (Fig. 9.3):

Step (1) Value Stream Mapping based on Goals → Lay out the entire value chain process in 'stream' form, from end customer (medical provider receiving PPE or patient receiving vaccine) all the way back to raw materials based on delivering to stated goals.
Step (2) Decisions → Ask yourself what critical decisions you will face in managing this value chain. What will you need better intelligence on for making decisions? (Allocation of supplies, sourcing of critical human resources or technology, process or policy changes, etc.). How will this impact sourcing instruments such as contracts? What flexibilities do we need?
Step (3) Signals → Ask yourself what would serve as a signal for decision-making? Do you need to know things like delivery delays, vendor vetting results, supply chain

disruptions anticipated from weather or other disasters, market impacts from global policies, contract performance and delivery term issues etc.?

Step (4) Data → Ask yourself what data would lead to informative signals that can lead to intelligent decision-making within your focal value chain. Obtaining needed data will require collection. How do you collect it easier and with less threat of human or system error? Storing data takes capacity. How will you collect what you need and store it? Data hygiene is important. How will you ensure 'clean' data or clean data if it becomes corrupted? Data needs to be verifiable and protected. How will you ensure what you are collecting is real and immutable?

Continuous Feedback → Provide feedback back up through the GDSD model, between all levels, to ensure that the data collected can provide the needed signals to inform the decisions necessary to meet the intended goals of the value chain. If it does not, then that data is not useful for this purpose. Also consider if the data collected, signals developed, or decisions made could impact other goals that may be of interest to the organization. In other words—don't throw the baby out with the bathwater (Fig. 9.4).

Why it matters

As mentioned before, failure to follow a GDSD model in a data saturated world can lead to misguided focus. The main goal of our response to COVID19 has always been to stop the spread while delivering necessary supplies to save critically ill patients. When we started with COVID19 response everyone was staring at epi model for the University of Washington. Why? Well mainly because that's the initial focus of pandemic response, how many are sick or are expected to be sick, when and where? But we quickly realized that that would not help us respond with action. It was just a clean visual of a nasty problem. We needed to make decisions about where to acquire needed materials and where to send them. Next, we started to receive data and interactive maps showing social distancing behavior and requested PPE and medical device needs from healthcare facilities. This was a step closer, we could see where spread was likely to occur, availability of hospital beds and the stated needs of these locations. This was much more useful, yet is still missed vital data to provide a clear signal for decision-making. Material stock visibility issues were not accounted for. Such behaviors included hoarding behaviors and political motivations that misrepresented the required need for things like PPE, poor tracking and inventory management that led to misreported stock levels, and additional distribution information that would allow us to see potential barriers from sources and too delivery locations. Finally, after months of pandemic response, the CDC issued burn rate calculators that allowed medical facilities and support activities to estimate the needed levels of material based on usage rates. This should still be improved to rely on scanner technology to remove human error and misrepresentation, however, it is a step closer to the true data needed to meet the original goal…stop the spread of the virus. This is just a brief example of the problem with taking an observe, orient, decide and act approach to a data saturated world. Instead, we must use strategic foresight to ask ourselves what we want to happen in good times and in bad, what decisions will support those goals, what signals will support those decisions and what data is pivotal to developing such signals. Then we need to prioritize the curation and management of that data to enable more efficient and effective responses in the future. Firms and public agencies should shift their focus towards the GDSD model for managing data.

The GDSD model should also be used to manage supply chain risk as a whole, to assure not only goods are distributed effectively in times of crisis, but also that goods will be available to distribute. The goal of supply chain risk management is to assure availability of goods and services from responsible and ethical sources in the face of any potential disruption through mobilization of the entire value chain. The decisions range from utilization of different sources of supply, maintaining dual or multiple sources, increasing inventory levels, seeking new entrants or different product designs, selecting different vendors to diversify geographic and geopolitical risk or to seek other novel solutions. The signals include any potential event that could create a supply chain disruption ranging from the solvency of suppliers, exposure to environmental disasters, weather disasters,

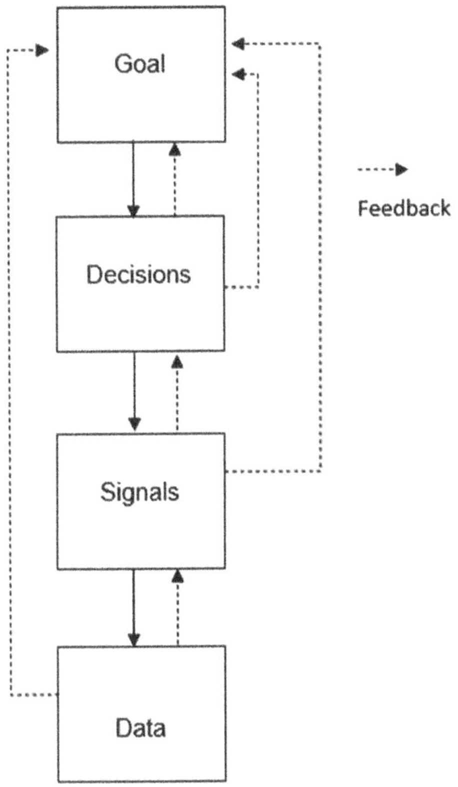

Fig. 9.4 The GDSD model

factory fires, cyber-attacks, geopolitical instability, criminal activity, and more. The data begins with knowing who is in the value chain, from raw materials all the way to the buyers of end items and how to rapidly connect with those parties with the speed necessary to overcome any potential obstacle. In reverse, it is clear that supply chain risk management is a team sport. The first step is to know the roster of the entire value chain and how to reach all tiers of the value chain rapidly. Ideally, this transcends data alone and includes designing a framework for the entire value chain to rapidly receive the signals of potential supply chain disruption as quickly as possible. During a time of crisis, building networked communication channels down supply chains that could extend out a dozen tiers or more is a sign of poor planning that will manifest in failure to position for timely decisions that will allow the team, the entire value chain, to achieve the overall goal. The dynamic nature of supply chains, with record levels of M&A, new sourcing decisions every day, a record level of supply chain disruptions, global instability and procurement regulations rapidly changing across the globe necessitate this framework to be dynamically applied with revisited or continuous monitoring at every level of the framework.

9.3 Conclusion

To be honest none of the authors ever anticipated being thrown into the scenarios we faced early on in the pandemic. We all feel both blessed and haunted by our experiences on the various supply chain task forces. Sometimes ignorance is bliss. But at this point we know too much to keep it all to ourselves and pretend like a post-COVID world can be operated like a pre-COVID one. COVID didn't actually create many of the issues discussed in this book or the flood of events in our news every day. But it did create a moment where enough eyes and ears were trained on issues that have been bubbling under the surface for some time in our world. It raised the awareness and threat level for supply chain risk. We hope our book can serve to provide a consolidated (though never exhaustive) list of lessons learned and tools to help business and government leaders, civilians and public servants navigate and solve the complexities and uncertainties of global supply chains, logistics and geopolitical market forces. The threat is real, but there is still much we can do, if we will commit ourselves to the cause. September 11, 2001 created a watershed "never forget" moment for anti-terrorism and national security. We believe that COVID-19 is that same watershed moment for global markets, supply chains and economic warfare concerns. We hope it too can become a "never forget" moment and help the next generation safeguard freedom and justice for all by finding a cure and building immunity from years and years of sourcing sickness.

Appendix A: National Contingency Supply Chain Cell (NCSCC) Notional Construct

Practically we should establish a national supply chain contingency team that operates cells centering around specific category portfolios and constantly communicates with each other. This allows for persistent market intelligence and supply chain awareness that is necessary to move us from the "light switch" linear contingency model mentioned earlier to an orbital regime of market awareness and response. We offer the NCSCC notional construct based on our experiences in the COVID 19 response. The concept is a hub and spoke system that has Team 1, Prevention, centralized and Response, Relief and Recovery teams operating across the nation in hubs that report to these centralized cells. Data Management operates at all levels. The Prevention team supports the supply chain awareness and early mitigation efforts necessary to prevent national emergencies as well as manage them should they occur. They aggregate to a central command post/tower that has situational awareness at a national level. The Response teams concerns themselves with supplies and services necessary to trace, deliver and secure crisis response material. The Relief team orients itself around life support and treatment concerns. The Recovery team maintains and responds with supplies and services necessary to repair and return to order. Finally, the Data Management Team has embedded support across all cells with a centralized nerve center that assists Prevention in control tower functions.

NCSCC organizational example (Fig. A.1).

Team 1—PREVENTION—This team should vet all materials entering the system and pay attention to external market factors and clinical testing signals that may indicate early warning of a potential episode on the horizon. The team should also be monitoring development of vaccines, and Team 1 should provide key market intelligence to the community on episodes or factors that may impact shortages or shifts in the national stockpile or other response mechanisms. This team should work closely with the control tower team to update metrics and intelligence. One of the missing links quickly discovered during the current crisis was the capability and capacity to adequately and quickly test PPE and other critical components entering the supply chains.

R. Handfield and D. J. Finkenstadt, *Supply Chain Immunity*, Synthesis Lectures on Operations Research and Applications, https://doi.org/10.1007/978-3-031-19344-6

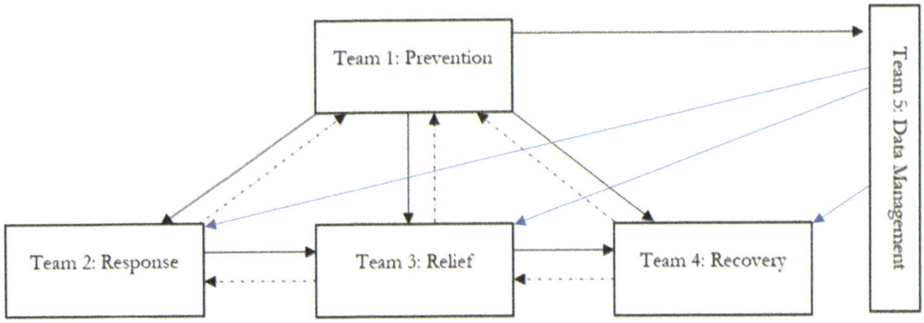

Fig. A.1 NCSCC structure (— Direct line; - - - Feedback)

Team 2—RESPONSE—This team is tasked with rapid logistic distribution of hospital supplies, medicines, and medical devices once an episode has occurred. The team will also deploy security services and products to states or hospitals based on immediate need, and track chain of custody and payment to key suppliers delivering to various points in the system. The response team should also have access to the "burn rate" of states and hospital systems who are going through supplies, and update forecasts and acquisition requirements on a timely basis through a Team 1 and Team 5 interface.

Team 3—RELIEF—Humanitarian relief efforts are needed to provide at-risk communities with food, water, shelter, first aid, PPE, durable medical supplies, treatment medications, and emergency medical services. This team will need to also track where material is distributed and have a deployed field force for ensuring it is arriving at locations securely and safely.

Team 4—RECOVERY—This team is tasked with recovery efforts during an emergency and post-episode. They should specialize and be involved in construction and supplies of destroyed or impacted field locations, assessment and assignment of economic restoration analysis, counseling services for those experiencing severe trauma and death, and other services that can help individual return to normal life planning activities.

Team 5—DATA MANAGEMENT—this team will work across all the teams to provide mobile data analytics to support activities in the field or wherever required. The team will develop a system to scrape information from systems into a control-tower environment to provide real-time visibility to material, including expiration date, usage rates, minimum order quantities (MOQ's), current suppliers, and other risks in each category. A critical need this Team will address is maintaining a list of approved, trusted suppliers that can quickly provide needed materials and products. Data quality is also critical in any crisis and the incomplete reports, missing data, multiple estimates, and other data quality problems during the current crisis has hampered efforts all across the country. Many systems (and even states) are still years behind other industries in data transfer relying on faxes and unsecure telephone calls to transmit critical data. It would be critical that this team be statutorily empowered to collect necessary information from all entities who may need assistance. The team would leverage the legal requirement to motivate need agencies (e.g. local and state governments, private healthcare organizations etc.) to provide data and data access. The teams task would be to work with all entities (specifically any accredited healthcare providers) to create either a standard interface for automated data pulls or a means by which the entity could quickly provide inventory and usage data. Ideally, this would occur before a time of crisis and when not automated be subject to random auditing (structured auditing during crises) to assure that accuracy/diligence is ingrained in the organizations before it is too late. Should organizations not comply, their SNS aid eligibility would be suspended until compliance is achieved.

The following example can provide an illustration of how these teams could work together in a center-led pandemic organization. The Prevention team (1), while conducting prevention surveys of major hospital risks, the team notes that PPE is a high-risk area and then puts the other teams on alert and into action based on specific risks identified. A PPE shortfall could in this manner be better handled if a cell structure is in place, whereby the Prevention team notifies the Relief team (3) of a need to start strategically sourcing the PPE and planning a smart allocation model with the Response team (2). Strong communication, command and control is key. Any one team could identify a potential risk in the process that should be flowed to other teams depending on their role (e.g., if Recovery notes an issue that could have helped them that flows back to the Response effort then it should become situational awareness for all teams).

Strategic human capital is a vital element of a successful acquisition function.[1] NCSCC Teams need data experts to interface with Team 5, functional experts, disaster response experts, security experts, acquisition experts (specifically contracting and program management), a financial point of contact, a centralized legal team that supports them all under Prevention, academic/research interface liaisons, and supply chain and operations research experts. In our experience working with the private sector, we observed that a control tower hub should be in a location that: (1) enables the government to recruit and retain top talent based on quality of life, (2) includes a large existing and robust acquisition workforce for rotational development and (3) enables proximity to major enablers such as medical supply firms, research labs and universities, national labs and federal agencies. The two top areas that come to mind for our team are Dayton, Ohio and San Antonio, Texas. In addition, we believe that control towers work best when they are connected to mobile networks of field personnel who are kept abreast of key metrics and data wherever they are working. We could also consider two hub locations with a central hub function (i.e., remote centralization). National team cells need to be rotationally infused with key members who have experience at the control tower, Prevention team level. The use of mobile federal civilians and military would support this strategic human capital infusion effort. Regional cells would be established and staffed with teams 2–4 based on the current FEMA ten region construct or in other forms such as time zones or modified

---

[1] Government Accountability Office (GAO). (2005, September). Framework for assessing the acquisition function at federal agencies (GAO-05-218G). Washington, DC: Author.

over time as critical distribution hubs evolve to support a globally independent supply chain model. Of course, cyber-security measures would need to be in place to prevent hacking and fraud.

We have established a set of high-level objectives and key results that these teams should initiate and categorized them into the four main issue areas noted by Patrucco et al. (2020).[2] Patrucco et al. noted common international supply chain issues for public agencies and healthcare teams in responding to COVID19 that directly parallel with our observations. The four issues that stood out were supply shortages, governance issues, competence, skill and system issues, and regulatory issues. These recommendations on the team structure are consistent with the ideas of adaptive co-management for complex systems, that seek to relate science to policy for effective resource management.[3]

Our team has also begun to think through the initiatives and priorities that a NCSCC and the new SNS should focus on in preparation for the next major national or global event:

Issue	Objective	Key result
Supply shortages	Complete traceability of SNS Stock	Full DLT (Blockchain) enablement across SNS
Supply shortages	Increase SNS flexibility	DOD and VA becoming living stockpiles
Supply shortages	Increase SNS flexibility	Develop/maintain strategic sourcing plans for every key need that might arise during an emergency. Wargame and exercise these plans with supply chain stakeholders
Supply shortages	Increase SNS resiliency	Reduce large lot buys and move to joint-purchased phased delivery
Supply shortages	Use SNS volume to enhance DPA	Multi-year purchases with all DPA/DPA-like vendors
Supply shortages	Increase SNS global independence	Universities and labs funded for JIT development of critical needs/fallback source material/items for SNS (example: NC State's textile school during early phases of COVID response)
Governance issues	Increase SNS transparency, responsiveness and flexibility	Stand-up a National Crisis Supply Chain Cell (NCSCC) structure

[2] Patrucco, A, Harland, C., Telgen, J., Ferk, P., Knight, L., Lynch, J., Peters, E., & Tatrai, T. (2020) Procurement in a global crisis: a call for extraordinary action. Working Paper.

[3] Armitage, D., Plummer, R., Berkes, F., Arthur, R., Charles, A. Davidson-Hunt, I., Diduck, A., Doubleday, N., Johnson, D., Marschke, M. McConney, P., Pinkerton, E., and Wollenberg, E. "Adaptive co-management for social-ecological complexity", *Front Ecol Environ,* 2009, 7(2) 95–102.

Issue	Objective	Key result
Governance issues	Increase SNS persistent awareness	Work with industry and universities to develop an architecture for a National Health Response System using orbital market intelligence to complement and support the NCSCC
Competence, skills and systems	Increase SNS persistent awareness	Ensure all SNS system APIs are connected to federal, university and 3rd party sources for situational awareness of available stock considering the various screens
Competence, skills and systems	Increase SNS responsiveness	Consider locating SNS and NCSCC personnel in locations that can attract top talent or allow for remote work. Consider areas that allow for greater quality of life or co locating with agencies that can allow for cross training
Competence, skills and systems	Increase SNS responsiveness	Commit to training all NCSCC personnel in areas of strategic sourcing and contingency contracting (i.e. expertise resides within Air Force contracting and at the Naval Postgraduate School)
Regulatory issues	Increase SNS awareness and responsiveness	Legislation to mandate all DEA level 1 and II pharma daily reporting be accessible to NCSCC and SNS
Regulatory issues	Increase SNS responsiveness and flexibility	Policy to enable mass ordering agreements with suppliers and alternate sources with reduced or automated public solicitation requirements
Regulatory issues	Increase global independence	Create tax incentives for domestic suppliers or SNS supplies that are US made

A.1 Further Details on Key Recommendations

A.1.1 Government-Academic Partnerships

Universities and labs should be funded for agile development of critical need and fall-back source materials for SNS, especially vital PPE. During the initial COVID response observed this working simply by chance. Despite the move of most nonwoven manufacturing offshore, universities such as North Carolina State have maintained a robust infrastructure and continued investment in textile manufacturing. Other universities have established additive manufacturing capabilities that served our country during the early days of the pandemic. Still others schools like Arizona State have completed work on PPE alternatives (i.e. nondisposable active protection materials).

A.1.2 Category Management

Category management involves the development of strategic sourcing plan based on sound market intelligence and monitoring, and could have reduced the likelihood of blocked or invisible PPE stock. Government agencies would develop and maintain strategic sourcing plans for every key PPE need that might arise in an emergency. This is accomplished through detailed category management structures and orbital market intelligence updates and planning, along with a quarterly review to a council of stakeholders from multiple agencies.

A.1.3 Legislation and Policy

Legislation and/or policy is needed to mandate that blockchain technology be enabled across the entire SNS and eventually tied to state and local healthcare and industrial supply/demand centers. Fast and accurate stock and capacity reporting that is provided in real-time across the national level is sorely needed. This can the reduce fog of war and holds politicians accountable for selfish motivations. This could increase visibility for all forms of hidden PPE stock.

A.1.4 Living Stockpiles

A revitalized SNS should establish living stockpiles of PPE and other needed medical response items at government points of care across the country. Instead of housing, maintaining and reselling SNS stock on a rotational basis that is void of contemporary tracing methods, the DoD and VA facilities could carry excess capacity and act as stockpile that is more rapidly refreshed. The most significant failures of the SNS that were observed during the response were expired goods, damaged goods, and obsolete goods. Each category of failure comes with a significant negative impact to responsiveness in addition to carrying a significant cost to the taxpayer. The SNS requires enhanced data management to provide real-time view of material and first-in-first out methods to minimize waste. These government points of care are also more likely to utilize current sourcing research practices to ensure goods are state of the art. Finally, demand allocation rules should be established a priori, based on key metrics, that determine how supply will be allocated, to avoid political favors being awarded in the event of crises.

A.1.5 Global Independence

Outsourcing of manufacturing capabilities in North America has been on-going for more than 20 years. President Trump issued an executive order in May 2020 for development of domestic sources for medicines, but did not recognize the full scope of what it would take to achieve the localization of an entire industry.[4] The Biden administration has attempted to increase our federal buying focus on American made goods. However, telling the government to buy American doesn't do much if the commercial market won't make in America. Localization of supply will take time,[5] but will also need to consider whether it makes sense, if it will truly help support national security, and will suppliers support the inclusion requirements that ask them to become part of a blockchain/visibility network.

[4] Combating Public Health Emergencies and Strengthening the National Defense by Ensuring Made-in-America Essential Medicines and Medical Countermeasures, Executive Order, Donald Trump. Issued on MAY 18, 2020.https://www.whitehouse.gov/presidential-actions/executive-order-prioritizing-allocating-health-medical-resources-respond-spread-covid-19/.

[5] Handfield, R., Graham G., and Burns, L. Coronavirus, trade wars, tariffs, and supply chain evolutionary design." *International Journal of Operations and Production Management, July 6, 2020.* https://www.emerald.com/insight/content/doi/10.1108/IJOPM-03-2020-0171/full/html.

Appendix B: Workshop Protocol

Objectives for the roundtable focused on defining the current state of the government's capability in three specific areas, as well as the ideal future state capability. Participants were asked to "stretch" their imagination on critical areas for the future state and what a world class commercial set of capabilities would look like. The participants were from a number of different areas, including DoD, Congressional Offices, DHS, academia, and ......

B.1 Questions for Discussion

1. What were the major performance problems in the national response to COVID?
2. What is the current state of capabilities in the national response system, and what missing or under-performing capabilities caused or enabled the performance problems to occur?
3. What would a world class commercial set of capabilities look like, and what attributes would these capabilities possess?
4. How do the current state and a world class future state compare relative to the following criteria (What would be the enabling elements required to establish and activate this capability? What is the timeframe for development? Who should lead the initiative? We will form breakout groups that will deal with the different types of issues for a current and future state, which includes the broad categories of Supply Chain Issues, Supply Chain Technology, and Government Policy Issues, using Table B.1.

R. Handfield and D. J. Finkenstadt, *Supply Chain Immunity*, Synthesis Lectures on Operations Research and Applications, https://doi.org/10.1007/978-3-031-19344-6

Table B.1 Shared services framework for breakout group discussion

Breakout groups	Current state	Future state	Critical success factors
Supply Chain Issues: Global Independence, Persistent and Agile/Market Intelligence			
Supply Chain Technology: Transparency and Traceability, Flexibility/Agility, Equitable Distribution			
Government Policy Issues: Leverages existing capabilities and delivers effective cross-government shared services, Respects constitutional roles and responsibilities, Enables innovation throughout the ecosystem, Political and operational viability and sustainability			

Appendix C: Supply Chain Act

There has been discussion by Representative Carolyn Boudreaux from Georgia around creating a Supply Chain Act which would create an Office of Supply Chain Resiliency and Crisis Response at the Department of Commerce, headed by an Under Secretary of Commerce. Some of the details of this office are shown below.

- **Goals of the office:**
  - Promote the leadership of the United State in critical industries and supply chains
  - Encourage partnerships and collaboration between the Federal Government and industry, labor organizations, and state and local governments; and
  - Monitor the resilience, diversity, security, and strength of supply chains and critical industries
- **The office could advance these goals by, among other things:**
  - Identifying opportunities to reduce supply chain gaps and vulnerabilities
  - Establishing a coordination group (and appropriate subgroups) with private sector partners to plan for and respond to supply chain shocks and support the resiliency, diversity, security, and strength of supply chains
  - Working with existing authorities at the Department of Commerce and in coordination with the Department of State and the United States Trade Representative to promote diversified and resilient supply chains
- **The office would be responsible for issuing a quadrennial National Strategic Plan for Manufacturing and Industrial Innovation which would include:**
  - Insights and recommendations to improve the security and resiliency of supply chains and to support the creation of jobs through the growth of American manufacturing
  - Guidance to relevant agencies with respect to critical goods and services, equipment, and technologies which should be prioritized
  - Reviews and recommendations for expanding the sourcing of critical goods and services, equipment, and technology from allies or key partners of the United States.

R. Handfield and D. J. Finkenstadt, *Supply Chain Immunity*, Synthesis Lectures on Operations Research and Applications, https://doi.org/10.1007/978-3-031-19344-6

Appendix D: Warstopper

The Warstopper program was originally targeting defense manufacturing industries, but the ultimate goal is to provide government investments when readiness demand is higher than the commercial industry is willing to invest. The program began as an after-action item from Operation Desert Storm, when the manufacturer of the auto-injector used to deliver nerve-agent antidote was on the edge of not being able to meet the go-to-war requirements, This framework would be ideal for positioning an office to address shortfalls in domestic production of critical materials.

During COVID the Defense Logistics Agency supported the nation's response with direly needed supplies like ventilators and face masks through medical readiness contracts that were part of the Warstopper Program. The amount and type of items = requested fluctuated daily as needs emerged and industry ramped up production through the Defense Production Act. While the 1950 law gives government agencies more control of manufacturing during emergencies, it also gives DOD authority to partner with industry and make investments to boost readiness on a continual basis. One advantage is that industry has to accept orders from Warstopper if they normally produce the parts and must move them to the front of their production queue for commercial work.

The Warstopper Program complements those protection measures by shielding over 14,000 medical items identified by the services as being critical from becoming scarce due to resources or industrial limitations. Long before COVID-19 spread across the globe, DLA's contracting team established Warstopper contracts with suppliers and purchased vital material to ensure essential go-to-war items could be rapidly produced. Warstopper is also authorized to purchase components needed for production of critical goods like ventilators, and pay manufacturers to keep it available for use so that when a surge occurs, the contract can be executed.

As a result, DLA's Warstopper-funded contracts facilitate "guaranteed access" and delivery of material much faster than if orders were submitting on an as-needed basis, he continued. All Warstopper contracts are sized to support DOD needs but have been used to support the public through humanitarian assistance.

R. Handfield and D. J. Finkenstadt, *Supply Chain Immunity*, Synthesis Lectures on Operations Research and Applications, https://doi.org/10.1007/978-3-031-19344-6

Appendix E: Reinstituting the War Production Board

The powers entrusted in the WPB were vast. Shortly after its creation, it would subsume into itself the functions of the other rearmament agencies, such as the Office of Production Management, consistent with the WPB's purpose as a centralized procurement management agency to handle the production of war materiel.
The WPB was authorized to:

- Exercise general direction over the war procurement and production program.
- Determine the policies, plans, procedures, and methods of the several Federal departments, establishments, and agencies in respect to war procurement and production,

The WPB was also given further authority to direct all federal agencies' procurement and production operations. In sum, the WPB was authorized to take over the procurement activities of any and all government agencies, including the armed services and put them to use to meet the needs of the war effort.

It could be reasonably argued that the powers of the WPB make it a difficult model to emulate. After all, the centralization of procurement authority in one agency comes with the attendant possibility that such authority will be deployed coercively. Therefore, the argument goes, an agency such as the WPB is proper only in such dire times as the nation faced in the onset of the WWII, much as the British government marshalled its country's industrial resources to stave off a German invasion at the beginning of the war.

Indeed, it is precisely such considerations which justify the implementation of a WPB-modeled procurement board to oversee production and distribution in the event of a public health emergency which threatens the security of medical supply chains (Maier and Kumekawa, 2020). The powers of the Defense Production Act were clearly not enough, and in many ways exacerbated federal-state divisions by giving vast and misdirected authority to the federal government, but leaving constitutional and practical responsibility for their citizens' health and welfare to the states. Nor is the SNS enough. While the purposes for which the SNS operates are not themselves war, they are real emergencies in their own right, as the SNS is charged with ensuring proper supplies to respond to grave disease outbreaks and biological terrorist attacks. Further, while there is no guarantee that

R. Handfield and D. J. Finkenstadt, *Supply Chain Immunity*, Synthesis Lectures on
Operations Research and Applications, https://doi.org/10.1007/978-3-031-19344-6

any given public health emergency will rise to the same level of crisis as a global conflict, it is worth noting that the number of Americans who have died because of the COVID-19 pandemic—more than 650,000 people as of September 2021[6]—exceeds the total number of Americans who died fighting in WWII—405,399 between 1941 and 1945 (Department of Veterans Affairs, 2020). If nothing else, this difference reveals that the seriousness of the threat of public health emergencies cannot be discounted in comparison to that of war.

Just as the WPB did not outlast the end of the WWII, eventually giving way to the Civilian Production Administration as the nation transitioned out of a war economy and returned to civilian life, a WPB-modeled production board for the SNS would not need to be a permanent body. Indeed, such a body should be viewed as an emergency measure and nothing more. But the authority to create such a body must exist in the law before it is needed, or else the difficulty of responding to crises will compound. Any attempt to establish such a body in response to a future pandemic would need legislative authorization, and while Congress may act and has acted quickly to address the present crises, the experience of the COVID-19 pandemic demonstrates that even momentary delays in response can lead to crippling shortages in needed supplies.

Therefore, providing for the creation of a WPB-style board in law ahead of time is preferable, so that in the event of a major public health emergency, especially the kind that poses an existential threat to medical supply chains and national security, such a board can spring into action at the command of the president and direct the procurement activity of the federal government towards the production, acquisition and distribution of needed supplies.

[6] *See* https://covid.cdc.gov/covid-data-tracker/#datatracker-home.

Appendix F: Global Sourcing Framework

Roscoe et al. (2021) developed a framework of supply chain strategies for geopolitical risk mitigation (see Fig. F.1), which provides some guidelines to the federal supply chain on whether to adopt centralized/regionalized or localized supply chain designs according to how entrenched their suppliers are in a particular geographic location as well as how severe the geopolitical disruption is perceived to be.

The Y-axis of Fig. F.1 shows the shifts in the external business environment, which have rendered it difficult to localize or shift the supply base, because of the entrenched nature of the supply base, or the cost-prohibitive elements for doing so. We note here that many Chinese industries were established with government investment, and the cost of capital for developing local sources is a significant barrier for investing in local supply capacity.

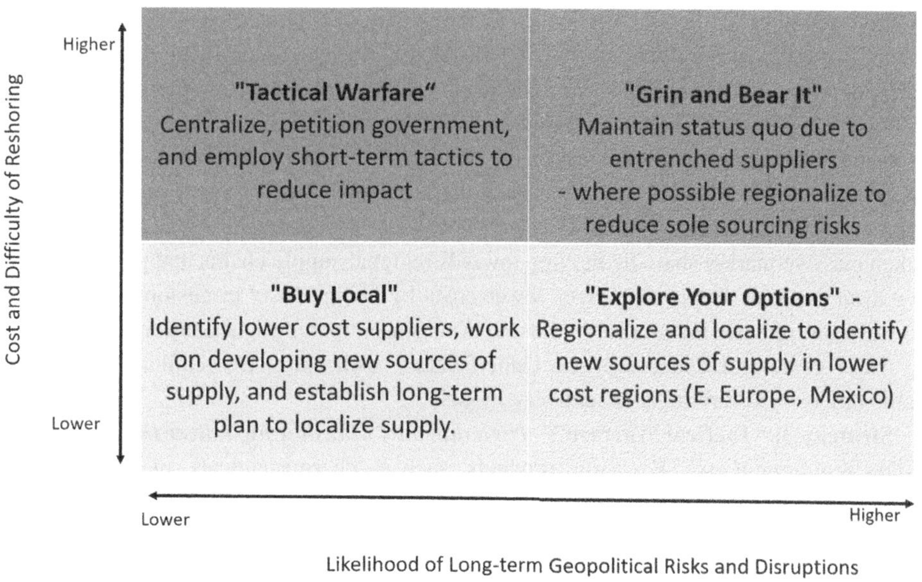

Fig. F.1 Framework for supply chain strategies for geopolitical risk mitigation

R. Handfield and D. J. Finkenstadt, *Supply Chain Immunity*, Synthesis Lectures on Operations Research and Applications, https://doi.org/10.1007/978-3-031-19344-6

The X-axis refers to the perceived likelihood of on-going political risk and disruption that is likely to continue, including the likelihood of on-going tariffs, customs duties, quotas, and export restrictions, resulting from a major and ongoing geopolitical event such as Brexit or the US-China trade war. In general, there are four strategies that emerge.

Strategy 1: "Grin and Bear It"—*High difficulty of reshoring, High likelihood of on-going geopolitical risk.* The increasing cost of moving products from an overseas supplier has been escalating, not just because of labor costs but also because of transportation costs, tariffs, duties, and supply discontinuity have dramatically increased the discussions around localization. This perception has escalated following the COVID-19 crisis, as borders were suddenly shut down for critical materials like PPE and ventilators. However, there are some economic factors that simply cannot be overcome, where entrenched supply bases produce a "Grin-and-Bear-It" approach. This approach recognizes that in some industries, supply chain redesign is difficult, if not impossible, such as in the electronics industry where the epicenter of component manufacturing and final assembly is in Asia. Under the "Grin-and-Bear-It" approach, we suggest companies will prioritize short-term tactical efforts such as building redundancies and holding inventory at different points in the supply chain. Other tactical strategies may include moving production to nearby locations (such as Vietnam) or transshipping through nearby locations to allow for a change in the country of origin customs label and the avoidance of tariffs.

Strategy 2: "Explore Your Options"—*Low difficulty of reshoring, High likelihood of on-going geopolitical risk.* The movement towards localization strategies is a function of the supply chain logics that prioritize the avoidance of uncertainty and risk, and an acknowledgment of the importance of lowering the total landed costs of goods, which occurs naturally as suppliers are located closer to customers. Localization is particularly relevant due to the size/cost ratio of goods with large, bulky, and low-margin items (such as food and beverages, vehicles, fabrications) being manufactured closer to the point of consumption because they are expensive to transport. There is also an opportunity to increase domestic sourcing to exploit local market knowledge and drive growth. Consider the case of mobile phone technology and how local producers in India and China have taken massive market share by moving towards regional supply chains that produce locally for local markets. Proximity drives lower costs by being closer to customers and closer to the point of sale. For the same reason, Amazon is opening Distribution Centers close to major centers of demand in the United States, with many US retailers moving to a same-day or next-day logistics delivery model.

Strategy 3: "Tactical Warfare"—*High difficulty of reshoring, Lower likelihood of on-going geopolitical risk.* For some products, such as pharmaceuticals, medical supplies, healthcare products, and complex tooling, we may see reduced tariff barriers as access to these products is deemed critical following the COVID crisis. For instance, we are unlikely to see a sudden surge of local production of high volume, low-cost medical products in Western economies. For products within this quadrant, the expectation is that

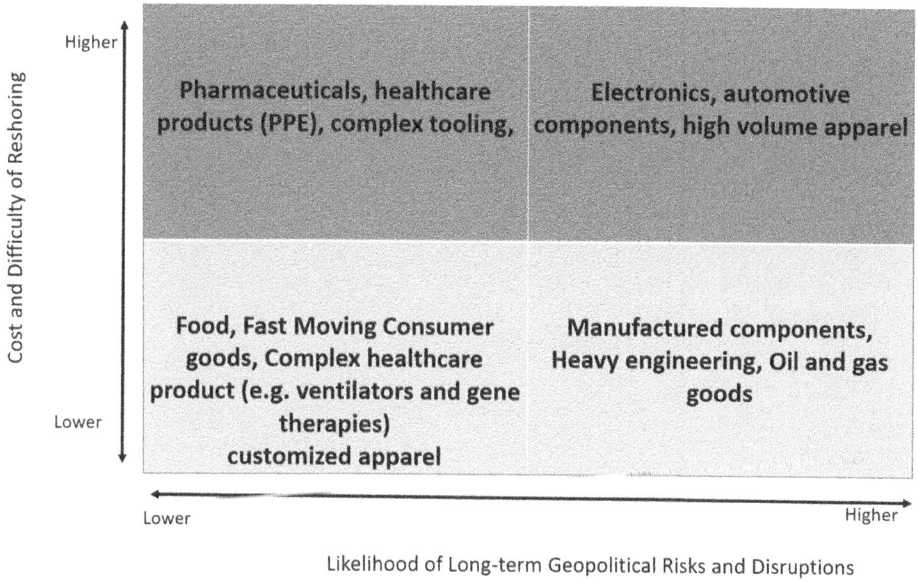

Fig. F.2 Indicative mapping of supply chain risk mitigation strategies by industry

geopolitical risks will not be ongoing, with such risks not perceived as being not substantial enough to justify the cost of relocating production. Companies in this quadrant will adopt short-term tactical measures such as tariff avoidance, regional distribution centers with inventory, and national stockpiles of goods. However, if there is maintained political pressure for these types of goods to be produced locally, such as pharmaceuticals to treat the symptoms of COVID-19 or PPE, these industries will consider shifting production onshore. In addition, we may see manufacturing, heavy engineering, and oil and gas seek to develop local suppliers of engineered products, to ensure business continuity and develop secondary sources of supply, even though costs may increase.

Strategy 4: "Buy Local"—*Low difficulty of reshoring, Low likelihood of on-going geopolitical risk*. For those products that are subject to local cultural differentiation and local sources of supply, we will see localized supply chain designs dominate, with this category expected to grow further as consumer demand for local products increases. Industries in this sector include food, especially fresh fruit, vegetables, and meat, as well as complex health products such as customized pharmaceutical products (gene therapies) and ventilators. For example, we are beginning to observe new start-up companies in areas such as customized apparel, who are seeking to develop digital apparel production capabilities in response to consumers who are seeking customized clothing and want it delivered within 48 h, and localized capabilities will become important for this sector. We now map the industries in our study to the four strategies shown in Fig. F.1 to provide an indicative framework for supply chain designs (see Fig. F.2).

Appendix G: "Dollar a Year Men" Policy Used in WWII

In the rearmament period leading up to and during the WWII, one of the most notable—and criticized—aspects of the military build-up was the presence of the so-called "Dollar-a-Year Men" (Fleischmann, 1952). These men were not government administrators by training, and yet, especially during the period before the Japanese attack at Pearl Harbor, they played a major role in the planning and procuring of the war effort. The heart of their success, and their notoriety, was that they were industry men. The colloquial term "Dollar-a-Year Man" refers to outside experts who are brought in to the government in an advisory capacity but who are paid no, or only nominal, compensation for the service they provide, such as a one dollar annual salary.

The Dollar-a-Year Men were, by and large, executives from some of the biggest companies in America, including men from corporations in Detroit, at that time a manufacturing powerhouse and home to production titans such as Ford and General Motors. The men were experts, not because of any academic background or theoretical knowledge about production and supply chains, but because they had practical knowledge gained from their careers, so they were brought on to advise and assist the government of Franklin Roosevelt in expanding the government's supply of, and the nation's capacity to produce, badly needed war materiel.

One of the recurrent criticisms of the use of Dollar-a-Year Men is that the practice of inviting industry executives into the government's decision-making process presents the potential for conflicts of interest and favoritism. While this is a valid concern, and would certainly be cause for worrying if the practice were employed government-wide on a continual basis, the practice is very useful in emergency circumstances and its deployment is also contemplated by the Defense Production Act as presently enacted: "*[t]he President is ... authorized, to the extent he deems it necessary and appropriate ... to employ persons of outstanding experience and ability without compensation.*"[7] This practice, then, is already blessed by law—provided, of course, that the outside experts are not making all the decisions for the government and they serve only in an advisory capacity consistent with their

[7] *See* The Defense Production Act of 1950, as amended, 50 U.S.C. § 4560(b)(1) (2018) https://www.fema.gov/sites/default/files/2020-03/Defense_Production_Act_2018.pdf.

R. Handfield and D. J. Finkenstadt, *Supply Chain Immunity*, Synthesis Lectures on Operations Research and Applications, https://doi.org/10.1007/978-3-031-19344-6

role—which indicates an understanding in the law that circumstances in which the government would need to invoke the DPA, such as a pandemic in which supply chains are strained to the point of breaking, are also those which would justify the employment of outside experts in spite of any potential conflict that might arise.

For any Dollar-a-Year Man scheme for the SNS to be successful, it must incorporate executives and experts from every link in the supply chain, including representatives of the health care facilities which constitute the points at which supplies are predominantly used.

The manufacturer's authorised representative in the EU is Springer Nature Customer Service Centre GmbH, Europaplatz 3, 69115 Heidelberg, Germany. If you have any concerns regarding our products, please contact ProductSafety@springernature.com

Printed and bound by CPI Group (UK) Ltd, Croydon, CR0 4YY
01/10/2026
02225875-0020

Margaret MacDougall
Editor

A Medical Educator's Guide to Thinking Critically about Randomised Controlled Trials: Deconstructing the "Gold Standard"

Springer

Editor
Margaret MacDougall
Usher Institute
Edinburgh Medical School
Edinburgh, UK

ISBN 978-3-031-25858-9 ISBN 978-3-031-25859-6 (eBook)
https://doi.org/10.1007/978-3-031-25859-6

Mathematics Subject Classification: 97B30, 62-02, 92C50, 62K05, 62A01, 62P10, 97-02

This Springer imprint is published by the registered company Springer Nature Switzerland AG
The registered company address is: Gewerbestrasse 11, 6330 Cham, Switzerland

Paper in this product is recyclable.

Preface

Background and Rationale

In 2012, I had the privilege of organising an interdisciplinary Royal Statistical Society debate involving the statisticians Prof. John H. McColl (University of Glasgow) and Gillian Raab (Professor Emeritus, Edinburgh Napier University) and the philosophers Prof. David Papineau (King's College London) and Prof. Nancy Cartwright (University of California at San Diego and the University of Durham). The theme for the debate was *Randomised controlled trials in medical research—gold standard or unhealthy fixation?* At this event, a clear dichotomy was evident between the viewpoints of the philosophers and statisticians. The statisticians argued in favour of randomised controlled trials (RCTs) as the gold standard design for primary studies in seeking an evidence base for causal relationships in clinical research, while the philosophers argued to the contrary that this perception of RCTs amounts to an unhealthy fixation. The audience vote was then cast in favour of the gold standard position. Doubtless, debate is a flawed mechanism as a source of truth and generally speaking, the fairness of the outcome is influenced by audience bias. Equally so, the lack of rigour involved in casting the vote in the above, or any similar, debate should not serve as a measure of the credibility of the status of the RCT as a gold standard.

With a view to improving understanding and use of RCTs among medical, and other health professional, students, there is, however, a need to take a more nuanced approach to the consideration of RCTs in clinical research. Drawing on the statistical and philosophical expertise of a range of authors, this book addresses the above need in two main senses. Firstly, considerable attention is given to the limitations and challenges commonly encountered in the conduct, application and reporting of RCTs. This includes exploring the many ways in which bias can infiltrate an RCT and how therefore we should seek to guard against this, where possible. Secondly, recommendations are provided as to how we can better utilise RCTs, including through reviewing findings from a basic Bayesian perspective, through judicious use of subgroup analyses and through acknowledging the contribution that findings from observational studies can make to the development of well-designed RCTs. It is also acknowledged that there are special instances where use of an RCT cannot be justified.

Arguably, any RCT inherits manifold traits and behaviours, including fallibility, from its investigators and participants. Correspondingly, associated concerns, such as integrity and adherence, and their impact on the validity of reported outcomes, can never be fully resolved. Thus, while this edited work is supportive of recognition of the RCT as an indispensable source of evidence for causal relationships, it also encourages medical educators and their students to recognise the many limitations that can arise *in pratice* in the conduct, reporting and application of RCTs.

As such, this book is not intended to serve as a compendium of RCT designs. There is a wide range of specialist journal literature in this rapidly expanding field, including in relation to the use of Machine Learning both to enhance efficiency in the conduct of RCTs and to offer novel interventions in the field of telecare. Instead, focusing mainly on the conventional two-armed, double-blinded, parallel group, individually randomised controlled trial, its primary aims are:

- To provide a comprehensive and honest consideration of the strengths and limitations of RCTs and thus a balanced representation of their role in clinical research;
- Correspondingly, with the support of topical examples, to highlight potential areas of weakness in the design, conduct and statistical reporting of RCTs; and
- In turn, to inspire and equip educators to train their students to take a more critical approach to individual RCTs as a source of evidence for clinical effectiveness, without skipping the statistical elements.

Who is This Book For and How Should They Use It?

On conceiving the proposal for this book, I had envisaged a publication primarily for use by medical educators involved in the teaching of statistics and Evidence-Based Medicine (EBM) to undergraduate medical students and junior doctors.

The rationale for this remains clear. Constraints can often be placed on teachers of medical statistics in early years of medical curricula to deliver only very basic-level statistics, using a broad-brush approach. This is to accommodate both the extensive range of subject backgrounds and aptitudes of the students on entry to medical school and to allow adequate time to cover a wide range of curricular topics to satisfy the expectations of medical regulating bodies. It is then assumed that those students who want to pursue research careers will pick up the pieces somehow through research-based self-selected programmes and that the remaining majority of students are destined for careers for which a broader and more in-depth understanding of medical statistics is superfluous. This assumption is, not only naïve but also, inappropriate. As research on the statistical learning needs of undergraduate medical students makes clear, improved theoretical and practical understanding of statistics is important for *clinical practice.*[1] The idea that, for example, aspiring primary care physicians should stick to the basics bypasses the increasing pressure that medical graduates are faced with to publish research in their chosen fields, especially at the early stages of their careers, frequently using statistical procedures not typically included in introductory or honours-level courses for medical students (see Footnote 1). The journey ahead in improving statistical learning opportunities in medicine and importantly, other health professions, is not without its opponents. Nevertheless, through provision of a range of clinical examples, this book is intended to support medical educators in pursuing opportunities throughout curricula to ensure that students are better informed about clinical trials and in particular, the reporting thereof. Medical and other students need exposure to such opportunities to enhance their future appraisal of published RCTs of potential relevance to their own patients and to enhance the statistical competence with which they will go on to report their own study findings. Indeed, given the unresolved concerns raised over several decades of statistical blunders in high-ranking medical journals (see Footnote 1), there is an urgent need for medical governing bodies to recognise adequate training in such competencies as a duty of care to patients and to ensure that provision of such training is a required standard for all medical schools.

As this content of this volume evolved it also became evident, however, that each of the chapters could additionally serve a wider range of interests, including the practical needs of physicians in interpreting research evidence and the teaching needs of philosophers of medicine who lack the statistical expertise to explain how RCTs work in practice and provide engaging examples for their students, both at undergraduate and postgraduate levels.

Practicing clinicians in general should also find some of the content in this book educational. For example, the content on diagnostic statistics in Chap. 3 highlights misconceptions about the generalisability of diagnostic findings that have been a traditional feature of medical textbooks and continue to influence the mindsets and teaching practices of physicians. Challenging such traditionalism is extremely important in ensuring progress towards better understanding of the role of prevalence and generalisability in the interpretation and use of published diagnostic statistics. The need to create corresponding learning opportunities for senior medical students is evidenced by the time constraints that typify the learning culture in early years of medical curricula, wherein medical students may have only a brief encounter with diagnostic tests.

[1] MacDougall, M, Cameron, HS, Maxwell, SRJ. Medical graduate views on statistical learning needs for clinical practice: a comprehensive survey. BMC Medical Education. 2019;20(1).

While cross-referencing is used across chapters to support further reading, many of the chapters are also designed to be read in their own right, an exception being Chap. 4, which is reliant on the foundational learning provided in Chap. 3. In its current form, the book should therefore serve as a reference for readers to dip into in order to glean ideas for their own teaching and gain a deeper understanding of both the virtues and the limitations of RCTs. The content should hopefully also make a contribution to honours courses for health scientists and master's courses for healthcare professionals.

Edinburgh, UK Margaret MacDougall

Acknowledgements

I extend my thanks to the following colleagues who, as part of the review process, provided constructive comments in relation to earlier versions of individual chapters published in this volume: Robert Newcombe (Professor Emeritus of Biostatistics at Cardiff University), Tom W. Woolley (Professor Emeritus of Quantitative Analysis and Biostatistics at Samford University), Dr. Simon Day (Director at Clinical Trials Consulting and Training), Dr. Huw Llewelyn (semi-retired hospital physician), Prof. Stephen Senn (Statistical Consultant), Julius Sim (Professor of Health Care Research, Keele University), Paul D. Baxter (Professor of Biostatistics and Education, University of Leeds), Gillian M. Raab (Professor Emeritus of Applied Statistics, Edinburgh Napier University), Dr. Rani Lill Anjum (Research Fellow in Philosophy, Norwegian University of Life Sciences), Elena Rocca (Associate Professor at Oslo Metropolitan University), Prof. Miriam Solomon (Temple University College of Liberal Arts), Prof. Tim Errington (Senior Director of Research, Center for Open Science) and Dr. Michael Wilde (Lecturer in Philosophy, University of Kent).

Appreciation is also due to Peter Sandercock (Professor Emeritus of Neurology, University of Edinburgh) for making the International Stroke Trial (IST) data available for use in Chap. 1 to support illustration of the influence of sampling variation on the results of hypothesis testing and to the Marie Curie programme of the European Commission and the Beatriu de Pinós programme of the Government of Catalonia for financially supporting research that contributed to the development of Chap. 7.

I am also grateful to the Springer Nature editorial team, including Mr. Rammohan Krishnamurthy and Ms. Jayanthi Narayanaswamy, for their courteous and supportive input. A special thanks is due to Dr. Eva Hiripi, Springer Nature Senior Editor, for her long-term commitment in serving as an informative contact during the publication process and for proving to be such a pleasant colleague to work with. I also thank my fellow-authors who have contributed their expertise to this volume, a few of whom also served as reviewers.

Above all, I unreservedly acknowledge the God who guides me and who has sustained me throughout my writing journey in preparing this edited work.

Introduction

Two recurring themes are evident throughout this book. Firstly, RCTs are not immune to many of the limitations arising from other study designs. A typically overlooked example in this respect is that of participants not having been selected randomly prior to treatment allocation—in short, non-random recruitment of participants. Secondly, there is a potential plethora of challenges to be faced in ensuring that traditional RCT designs meet their own definitive standards in relation to minimisation of bias through procedures such as blinding and random treatment allocation. These limitations and challenges cannot be resolved, however, by resorting to observational studies as a safer alternative. Those who suggest to the contrary appear, in many cases, to have misapprehended what we can reasonably expect from an RCT in referring to it as a gold standard and to have simultaneously failed to subject their *preferred* choice of study design to the same level of scrutiny.

In adopting the term 'gold standard', it bears mentioning, therefore, that conventionally, this term can be understood as referring to the best available standard of reference in terms of accuracy and reliability. This definitive criterion does not *ipso facto* assume that in any given context, a gold standard is a perfect source of evidence for assessing the truth of a given hypothesis. As intimated above, this book explicitly recognises the latter constraint as applying specifically to the RCT as a gold standard for evaluating evidence for causal relationships in clinical research. Indeed, as is particularly evident from the content on null hypothesis significance testing (NHST), a single RCT may not *always* give the 'correct' answer, but this is entirely in keeping with what we would expect more generally from a gold standard, including in the familiar context of diagnostic test accuracy.[2] Instances are also highlighted in different chapters where use of RCTs has proved inappropriate on ethical grounds or at least, observational studies have proved instrumental as an adjunct to RCTs in the discovery of causal relationships. Such transparency does not weaken the status of the RCT as a *methodological* gold standard; rather, together with the contributions on mitigating levels of bias and rigorous reporting of statistical findings, it should

[2] Peacock, JL, Peacock, PJ. Oxford Handbook of Medical Statistics. Oxford: Oxford University Press; 2011.

serve to support best practice in design and application of RCTs to ensure that the RCT meets its own design criteria. In balance, as is highlighted in Chap. 6, this does not address the need for definitive empirical evidence based on comparison of randomised and non-randomised studies that the results from RCTs are consistently of a superior nature, and there is still considerable work to be done in bringing clarity to this aspect of the 'gold standard' debate.

Chapter Synopses

In Chap. 1, I use a published RCT involving a controversial retroactive intervention as the basis for highlighting statistical flaws that can creep into the reporting and interpretation of RCTs. A recurring problem with traditional training in critical appraisal skills has been the neglect in early years of a solid understanding of the more formal aspects of statistics commonly encountered by senior medical students in their curricular research projects. In this chapter, I seek to make a contribution to addressing this deficit through critically reviewing the statistical reporting in the above RCT. Using real data from a large stroke trial to support visualisation, I then focus on the important property—well known among statisticians but requiring a greater emphasis in mainstream EBM curricula—that even in the presence of flawless experimentation, statistical significance can arise for an RCT when the null hypothesis is true. These considerations set the precedent for future chapters through demonstrating that RCT findings inherit limitations from the human agents involved and in the case of simple randomisation, by the very nature of the randomisation process. In the last section of the chapter, I challenge the popular extemporaneous use of 'by chance' in medical research, which is frequently linked to treating randomness as unmanaged chaos. In this more theoretical section, I encourage readers to consider the orderliness in simple randomisation together with the fact that our lack of access to a mechanism to explain a given chance event does not preclude the existence of a cause.

In Chap. 2, Raab provides a brief historical overview of the development of RCTs using topical illustrations involving real-life clinical trials, an approach which she carries forward on answering popular criticisms of RCTs. Her response to these criticisms includes coverage of philosophical viewpoints opposing the preferential use of RCTs in clinical research. This includes the concern expressed by Cartwright and others that the journey from establishing a causal relationship for a particular study context to that of assuming that the same relationship exists more broadly speaking, and in turn, here and now, is fraught with difficulty. Raab also answers Urbach's concern about the ethical nature of random treatment allocation, involving a placebo, in a context where prior evidence favours the intervention. Arguably, in so doing, she challenges Papineau's stance in the 2012 debate that before they can properly enter patients in a trial, doctors ought to have no view on the balance of the probabilities in favour of an intervention or control—equivalently they should have genuine prior equipoise. With the support of the clinical examples provided, these contributions in support of RCTs involve a more in-depth approach than was feasible in the original debate.

There is evidence to suggest that in some clinical fields, the probability of there being a true treatment effect conditional on the achievement of statistical significance in favour of an RCT intervention is low, that is, there is a low positive predictive value (PPV). The PPV sheds important light on the likelihood of a study hypothesis being true, given the data, which, as advocates of Bayesian statistics appreciate, makes a great deal of sense in interpreting research findings. Also of importance, therefore, in informing future use of RCTs and the interpretation of their findings is management of the PPV, including as an adjunct to traditional approaches arising from NHST.

In Chap. 3, I explain the analogies between concepts associated with NHST, including the PPV, as defined above, and concepts encountered by clinicians in the quality assessment of diagnostic tests. This includes introducing Bayes' rule. I in turn use this theoretical groundwork for discussing the challenge of managing the PPV in more depth in Chap. 4. From a teaching perspective, an important aspect of this contribution is that of steering medical students away from an infatuation with p-values towards appreciating that the probability of the null hypothesis being true conditional on the discovery of statistical significance could in some clinical contexts be rather high. As with content in other chapters, the latter emphasis ought to provide a strong rationale for dissuading trainee doctors and other future health professionals from relying on a positive finding from a single RCT—albeit a large and well-designed one—as a basis for clinical decision making. Drawing from the work of Leeflang et al.[3] and Hlatky et al.,[4] including examples involving diagnosis of coronary artery disease, I challenge traditional behaviours in the application of Bayes' rule. This includes reviewing the use of this rule in explaining relationships between diagnostic statistics and highlighting the place of multivariable modelling and covariance relationships with *prevalence* in understanding the true behaviour of *sensitivity* and *specificity* across different study populations. While this gentle diversion into diagnostic statistics is not about RCTs per se, it is important in protecting more senior health professional students, and indeed, graduates, from being deceived by heuristics at the expense of accurate clinical decision making. I also make clear, however, that, while analogies formed between the use of Bayes' rule in NHST and the more familiar context of diagnostic test accuracy can improve conceptual understanding of statistical hypothesis testing, some of these analogies rely on naïve text-book models of the parameterisation of diagnostic statistics that ought to be recognised as such in current medical curricula.

In Chap. 5, Sim et al. regularly endorse viewpoints from respected philosophers, including those of Cartwright, Deaton and Howick, *en route* to a better understanding of what we can legitimately conclude from individual RCTs. They also highlight the

---

[3] Leeflang, MMG, Bossuyt, PMM, Irwig, L. Diagnostic test accuracy may vary with prevalence: Implications for evidence-based diagnosis. Journal of Clinical Epidemiology. 2009;62(1):5–12.

[4] Hlatky, MA, Pryor, DB, Harrell, FE, Califf, RM, Mark, DB, Rosati, RA. Factors affecting sensitivity and specificity of exercise electrocardiography. The American Journal of Medicine. 1984;77.

potential value added to RCTs through assessing interaction effects across subgroups, including in addressing the limitation posed by use of the RCT outcome for the study sample as a whole in addressing individual patient needs. Their contribution is important in dispelling the popular misunderstanding that subgroup analysis is, *generally speaking*, a paradigm for data dredging and as such an inevitable threat to the credibility of published research. Through appeal to previously published RCT findings, they offer concrete examples to substantiate Feinstein's distinction between post hoc subgroups as the true source of the above paradigm and subgroups "established a priori from pathophysiologic principles" as "respectable and worthwhile".[5]

In discussing the challenges of achieving the competing ideals of internal and external validity of RCT findings, they identify the pervasive issue across RCTs of non-random recruitment as a constraint on internal validity. They also provide a useful checklist to support current and future health professionals in assessing the generalisability of their own research findings and those presented in the available clinical research literature.

In terms of striving for the truth about RCTs, an important but previously neglected area is that of identifying a credible evidence base for supporting the methodological superiority of RCTs. If, as has been a prevailing assumption in traditional EBM teaching, RCTs should rightfully own this place, what should we expect to observe *empirically* on comparing RCTs and non-RCTs? In Chap. 6, Steel and Jones make a novel researched-based contribution to addressing the above hiatus in the educational and philosophical literature on RCTs. Through conducting a literature review dating back to 1977, they review historical trends from findings comparing results from randomised and non-randomised studies. These trends relate to the meta-analytic methods used to draw comparisons, including through pooling results across studies for the same study outcome or for different study outcomes. They, in turn, relate to the direction in which the resultant aggregate effect sizes differ across study designs. This is of particular interest given the popular assumption that non-randomised studies have a tendency to overestimate the effectiveness of treatments, which is possibly influenced by concerns about study design, including lack of blinding. This in-depth literature review demonstrates a lack of consistency in the results of such comparisons, pointing to uncertainty about what can be expected of an RCT to demonstrate its superiority. Steel and Jones report evidence against the idea of there being a consistent directional bias in the above sense. They also expose the need for future research to explore whether study results are more likely to be inconsistent across non-randomised studies than across RCTs. They therefore demonstrate a sense in which the court is still out on the status of the RCT as the gold standard in clinical research. Through their literature review, Steel and Jones demonstrate that these areas remain in need of further investigation with the support of meta-analyses comparing findings from randomised and non-randomised studies.

In Chap. 7, Krauss examines the problem of bias in RCTs in some depth and proposes an optimal study design for collectively mitigating levels of bias across a

[5] Feinstein, A. The problem of cogent subgroups: A clinicostatistical tragedy. J Clin Epidemiol. 1998;51(4):297–9.

wide range of sources with the goal of addressing replication issues in RCTs. His novel, although idealistic, study design involves a conglomerate of studies which is intended to mitigate the influence of multiple sources of bias collectively. The discussion around this study design, the multi-study, involves an in-depth exposition of potential sources of bias in the design of standard RCTs, a product of which is the provision of a checklist for improving the design of future RCTs. Through factoring the potential replicability of RCT findings into the design process, Krauss's multi-study research design is intended to serve as a more reliable arbiter of truth in clinical research than those of more conventional RCTs. In presenting a response to the problem of non-replicability from a philosopher's perspective, Krauss also provides a basis for future debate involving clinical trialists and philosophers in relation to both the place of replicability in RCTs and the feasibility in practice of designing a virtually bias-proof trial with replicable outcomes.

Contents

Editor and Contributors

About the Editor

Margaret MacDougall is a senior lecturer in medical statistics and researcher in education at the Usher Institute, Edinburgh Medical School. She developed an interest in philosophy early on in her career through her Ph.D. research in mathematical logic, constructive set theory and the philosophy of mathematics. She has conceived and led 11 funded projects in medical education and statistics and supervised research staff and students at undergraduate and postgraduate levels through various spin-off projects. Margaret seeks to use her research to promote curriculum innovation and enhance student learning in statistics at undergraduate and postgraduate levels and has published on an international scale in the fields of education, medicine, philosophy and statistics. She has served as an invited speaker, chair and conference organiser at various local, national and international events relating to statistics, and learning and teaching in higher education. Currently, she serves as Vice Chair of the Teaching Statistics section of the Royal Statistical Society.

Contributors

Andrew Jones Institute for the History and Philosophy of Science and Technology, University of Toronto, Toronto, Canada

Alexander Krauss London School of Economics, Institute for Economic Analysis, Spanish National Research Council, Barcelona, Spain

Gillian Lancaster School of Medicine, Keele University, Keele, UK

Martyn Lewis School of Medicine, Keele University, Keele, UK

Margaret MacDougall Usher Institute, Edinburgh Medical School, Edinburgh, UK

Gillian Raab Administrative Data Research Centre – Scotland, Edinburgh, UK

Julius Sim School of Medicine, Keele University, Keele, UK

Daniel Steel School of Population and Public Health, University of British Columbia, Vancouver, Canada

Chapter 1
RCTs from Within: The Influence of Scientific Rigour, Randomisation and Chance on the Credibility of Reported Findings

Margaret MacDougall

Abstract In this chapter, we shall consider a published randomised controlled trial (RCT) involving a controversial retroactive intervention as the basis for highlighting statistical flaws that can creep into the reporting and interpretation of RCTs more generally. These contributions support the standpoint that even for an RCT, the credibility of the reported findings relies on the correct implementation and interpretation of statistical procedures. Within medicine and other health professions, this ought to complement existing training in critical appraisal and data skills by encouraging an interrogation of the more formal aspects of applying and reporting statistics, even at a relatively basic level. We shall also consider data from a large international stroke trial to assist in visualising the influence of random allocation error (as a specific form of random error) on RCT results. This will serve as a concrete example illustrating that for a well-designed RCT, by the very nature of *simple* randomisation, statistical significance can arise in the form of a false positive favouring the effectiveness of the intervention when the null hypothesis of no effect is true. From a philosophical perspective, we shall also consider an alternative approach to that of merely equating chance events with accidental events. This will include both recognising that such events are an inevitable product of simple randomisation and rejecting the chance-cause dichotomy.

1.1 Overview

In Sect. 1.2, a rather controversial randomised controlled trial (RCT) from the British Medical Journal (BMJ) involving intercessory prayer as a retroactive intervention will be introduced. In Sect. 1.3, the content of the relevant BMJ paper will be reviewed as a means of identifying fundamental flaws in choice of methodology and in reporting and interpretation of findings, within the context of basic hypothesis testing. This

M. MacDougall (✉)
Usher Institute, Edinburgh Medical School, Edinburgh, UK
e-mail: margaret.macdougall@ed.ac.uk

M. MacDougall (ed.), *A Medical Educator's Guide to Thinking Critically about Randomised Controlled Trials: Deconstructing the "Gold Standard"*,
https://doi.org/10.1007/978-3-031-25859-6_1

will allow us to consider good practice in the use of effect size estimates for non-parametric data and the inappropriateness of focusing on statistical significance as an end in itself. Relatedly, with the support of data forthcoming from an international stroke trial, we shall observe in Sect. 1.4 that even for a large RCT, speculating on the effectiveness of an intervention based on this trial alone is an inadequate approach to evidence-based practice and that the possibility of Type I errors needs to be kept in view. The latter observation will provide the basis in Sect. 1.5 for us to go on a short philosophical excursion involving interpretations of the concept of chance in statistics other than that afforded by reducing a chance event to merely an accident. The above sections will therefore present ideas for encouraging students to begin to take a more reflective and critical approach to the interpretation and communication of the statistical results of RCTs and other studies.

1.2 The Leibovici Trial: A Motivational Study

In 2001, Leibovici provoked substantial debate and discussion in response to his paper *Effects of remote, retroactive intercessory prayer on outcomes in patients with bloodstream infection: randomised controlled trial* [1]. In this BMJ paper, he reported findings from a two-armed, double-blinded, parallel group, individually randomised controlled trial of an intercessory prayer intervention. While a range of alternative RCT designs are currently in use, some of which are assumed in other chapters of this book, the above RCT design is well known and widely used among researchers. Thus, for the remainder of this chapter, unless otherwise stated, we shall take the above design to be the paradigm intended on using the term 'RCT'. The participant sample in the Leibovici study comprised all 3393 adult patients for whom bloodstream infection was detected at a specific university hospital during 1990–1996. The intervention was referred to as "retroactive prayer" to reflect the intention that it should take effect in the past, if at all. Prior to its implementation, Leibovici hypothesised that the intervention reduces mortality and shortens the length of stay in hospital and duration of fever. For the intervention group, intercessory prayer was implemented 4–10 years after infection, while for the control group, no intervention was implemented.

It has been suggested that the fact that this paper was published in a Christmas issue pointed to the author's intention to engage in a light-hearted experiment [2]. From this viewpoint, we are free to assume that the omission of a sound scientific rationale for the study hypothesis is in keeping with the 'Christmas spirit' and that the study intervention will never see the light of day in routine medical practice. However, noting that the BMJ is a respected peer-viewed journal involving a wide readership, including physicians, their patients and a range of alternative consumers and practitioners, the need to set the records straight for 'humour-challenged' readers at the stage of publication, rather than in response to an influx of rapid responses, ought to have been clear.

The published electronic rapid responses to the editor regarding this paper are not only plentiful (the count to date exceeding 90) but they are also almost without exception of a serious tone. Indeed, they raise important issues concerning sound ethical and scientific practice and the absence of a Biblical warrant to support the use of prayer as scientific method through treating a higher being as subject to human manipulation.

The serious tone should not come as a surprise given that Leibovici builds the rationale for this study on the current evidence in favour of remote treatment interventions in improving patient outcomes [3–5]. Indeed, Olshansky and Dossey view Leibovici's work as a call to redefine our expectations in the assessment of scientific hypotheses by appreciating that "a study violating common sense or an accepted model" need not "be summarily dismissed" [6]. In arguing that the case remains open on the efficacy of retroactive prayer as a health intervention, irrespective of the underlying assumption of distance from the patient in terms of time and space, they claim that "Retroactive prayer may be less absurd than [Leibovici] supposes, in the light of non-local phenomena". This they do in part through appealing to Einstein's initial opposition to quantum theory when, in the absence of an awareness of the phenomenon of "non-local events", it emerged as an opposing theory to relativity theory. Thus, for Olshansky and Dossey, the case remains open on the efficacy of retroactive prayer as a health intervention, irrespective of the underlying assumption of distance from the patient in terms of time and space.

Note, however, that the idea of retroactive prayer as a health intervention is not new. In 1954, Dummett presented this idea specifically within the context where the outcome, though a prior outcome, is unknown by the intercessor. He attributed philosophical credence to this idea by connecting the notion of cause with that of sufficient condition [7, 8], noting that it is not required that a sufficient condition for an event should occur before the event. Thus, he postulated a type of relationship, "quasi-causality", according to which the occurrence of an event of one kind may be sufficient for the prior occurrence of an event of another kind. Quasi-causality is characterised by absence of discernment of, or access to, the mechanism of causality and through a present action, such as prayer, as having acted in the past, not acting presently, to bring about change. It should be noted here that quasi-causality, so defined, is not a denial of divine intervention. Indeed, faith in divine intervention, by its very essence, does not require the mechanism to be revealed.

The idea of quasi-causality, and the underlying belief in backwards causation, have met with some resistance [9–11]. Nevertheless, it is also important to note that the idea of perceiving a meaningful causal nexus without discerning a mechanism has been defended as commonplace in how everyday events are interpreted rather than unique to the explanation of retroactive interventions. Examples include a ball smashing a window and the use of purely empirical evidence in medicine to link smoking with diseases such as cancer and emphysema [12]. Leibovici's experiment is unexceptional in illustrating researcher emphasis on statistical evidence that an intervention is effective [12] without requiring a replicable mechanistic explanation that can be empirically verified. This perspective is consistent with that of Howick who on the one hand refuses to support "the EBM view that mechanistic evidence is not evidence at all" but on the other hand argues that "mechanistic reasoning is not necessary to establish causal claims" [13].

Thus, differing reactions to the Leibovici trial can be expected. There are those who may wish to relegate this trial to the realm of joke science. By contrast, others may have concerns that there is a more sinister agenda, led by, let's say, *the remote intercessory prayer movement*, that poses a serious threat to the reputation of clinical practice. However, neither of these positions help to explain the apparent success of Leibovici's intervention as reported in the study findings. For example, in the conclusions section of the paper, he notes, "Remote, retroactive intercessory prayer can improve outcomes in patients with a bloodstream infection" [1].

As we shall see, regardless of the true, or originally intended, status of the Leibovici trial—cutting edge or fantasy—it has a role to play as a novel motivational study in identifying a range of statistical flaws or misconceptions associated with the reporting of published RCTs. Medical students and students of other health professions need to be equipped to look out for these flaws, both in the literature and in their future clinical practice, if they are to avoid being duped by the lens of the traditional Evidence-Based Medicine (EBM) pyramid [14]. In fact, RCTs should be just as open to scrutiny as trials involving alternative study designs. It is my intention that the topics addressed in Sect. 1.3 and the corresponding key learning points (KLPs) will serve as a useful complement to more traditional critical appraisal training in undergraduate medicine programmes and improve the statistical practices of medical graduates, among others, when conducting their own research.

1.3 Statistical Lessons Arising from the Leibovici Trial

Let's now consider some key statistical features of the Leibovici trial that ought to awaken concern.

1.3.1 Lack of Transparency in Selection of Outcome Measures

One distinguishing feature of the Leibovici trial relative to standard RCTs is that arguably, individual patient outcome data were available prior to the implementation of the intervention. Relatedly, we are not informed that the researchers were blinded to these outcome data prior to implementing the retroactive intervention. This is of philosophical interest insofar as we cannot assume that the prayer intervention used in the above trial conformed to the requirements of Dummett's notion of retroactive intervention. The absence of blinding in the above sense is also of scientific interest even for those sceptics who believe that the retroactive intervention was inert and that therefore, for all outcome measures considered, the baseline data and outcome data were indistinguishable. In Leibovici's experiment, exposure to such data could have undermined both the integrity of the randomisation process and the choice of reported outcomes. In a more conventional RCT setting, where cause precedes effect, eyeballing of outcome data prior to randomisation is theoretically and practically impossible. In this sense, even where double blinding is reported, RCTs based on retroactive interventions are more subject to scientific fraud than conventional RCTs. These two types of RCT do, however, share a common ground in terms of the need to guard against the post-hoc selection of outcomes for reporting of study findings, and the Leibovici trial is no exception.

This trial, if taken seriously, is also no different from any other RCT in so far as there is an inherent responsibility from the outset to declare the data source(s), such as specific types of patient record, for the proposed measurement outcomes and to provide the clinical rationale, as informed by the literature, for selecting them. It is through neglect of transparency in this sense that researchers circumvent the possibility of integrity checks being carried out by a nominated individual. These checks ought to confirm that the reporting of measurement outcomes is not influenced by cherry picking from the results of exploratory hypothesis testing. In the Leibovici trial, the absence of the rationale for the choice of "Mortality in hospital, length of stay in hospital, and duration of fever" as the "primary outcome measures" for analysis and reporting purposes is a case in point.

As is well known among trialists, the World Medical Association require that all clinical research involving human subjects has received prior ethical approval by an independent research ethics committee and that in turn, the research trial has been registered in a "publicly accessible database before recruitment of the first subject" [15]. It is also expected by the scientific community that this registry should be one approved by the World Health Organization (WHO). Registration must be achieved through the inclusion of a research protocol that provides a clear rationale for the study and a transparent account of the study design and the intended conduct of the study. In support of this procedure, the SPIRIT (Standard Protocol Items: Recommendations for Interventional Trials) 2013 statement [16], which includes a 33-item checklist, has been endorsed internationally as the accepted standard for providing guidelines on the minimum content of protocols for clinical trials.

The importance of prospective trial registration ("pre-registration") in managing selective reporting of study results has been defended elsewhere [17, 18]. Nevertheless, a recent review has demonstrated that unregistered and retrospectively registered clinical trials continue to be accepted for publication [19, 20]. Thus, before reviewing their own research or clinical practice in the light of new trial-based findings, including RCTs, it is good practice for physicians to confirm the registration status of the published clinical trial.

It has also been noted more recently for the clinical trials register, ClinicalTrials.gov, that the capacity to state intended outcomes imprecisely at the stage of registration can allow scope for post hoc modifications at the stage of publication [21]. It is the latter finding in particular that lends greatest weight to the need for public access to the same study protocol, at least at the stage of publication of the study findings, but ideally, also at the peer review stage for a prospective publication.

Well-documented trials should also involve a statistical analysis plan (SAP). A SAP is recognised as containing "a more technical and detailed elaboration of the principal features of the analysis described in the protocol, and detailed procedures for executing the statistical analyses of the primary and secondary variables and other data" [22]. It has been noted that, historically, guidelines for the content of SAPs have been lacking [23]. Therefore, it is likely that the balance in level of detail across corresponding protocols and SAPs may differ substantially according to study. This leaves the possibility open that details concerning the metrics to be used in the assessment of an outcome, such as which scales are to be used to estimate pain severity, may be insufficiently clear in the SAP. This being the case, the content of the SAP may need to be reviewed to ensure consistency in choice of metrics prior to and after blinding to study outcomes. The guidance for implementation of the recently accepted risk-of-bias assessment tool for randomised trials (RoB2) within the context of a systematic review in turn recommends that "Review authors should ideally ask the study authors to supply the study protocol and full statistical analysis plan if these are not publicly available" [24]. On consulting the SAP, review authors ought to have concerns about bias where the SAP proposes that an outcome domain is measured in multiple eligible ways (such as via a variety of health measurement scales), while, in the absence of a valid explanation, only some of them are reported in the study findings [24]. Noting that not all published RCTs are subject to scrutiny through systematic reviews, public access to the SAP in addition to the study protocol serves as best practice in avoiding selective reporting in the above sense.

Key Learning Point 1.1

In the evaluation of an RCT, take note of the importance of trial registration and the publication of a protocol and SAP in managing selective reporting of study outcomes.

As to the question of whether the absence of at least a published protocol might have facilitated selective reporting of study outcomes in the Leibovici trial, the evidence is presented, not merely in the above research literature, but also in Leibovici's own words. In relation to choice of statistical tests, he claims that since "most of the continuous variables did not have a normal distribution, the Wilcoxon rank sum test was used for comparisons". This test is also known as the Wilcoxon-Mann–Whitney U test and the Mann–Whitney U test. As the latter name currently appears to be the more commonly used in clinical journals, we shall use it in this chapter, but it is useful to be aware, including with reference to Leibovici's paper, that all three names pertain to the same test.

Now, what, exactly, is Leibovici inadvertently telling us in his above defence of the use of this test? Notice, first of all that in his entire publication, he reports group comparisons for at most two continuous variables—*length of hospital stay* and *duration of fever*—and he applies the above hypothesis test in both cases. Supposedly, it would not have been burdensome to test these variables for Normality and proceed accordingly with appropriate hypothesis testing or indeed, if the variables were genuinely continuous, to have appealed to the Central Limit Theorem [25] to justify use of the independent samples t-test. One is left wondering, therefore, how many continuous variables were originally among the "most," given that Leibovici doesn't simply say that *length of hospital stay* and *duration of fever* were not Normally distributed. Also, why does he allude to testing more than two continuous variables for Normality, but neglect to report results of hypothesis testing from comparison of the study groups for all such variables? The logic in Leibovici's statement is also troubling, noting that we should not take a 'majority rules' approach to deciding which hypothesis test to apply for an individual variable. This does not, of course, preclude the possibility that the correct hypothesis test was used for each of *length of hospital stay* and *duration of fever*. However, it makes sense to pause to consider that for modest sample sizes, a consequence of over-generalising in this way can be a loss of statistical power. As we shall see in more detail in Chap. 3 statistical power is the probability of rejecting the null hypothesis H_0 if H_0 is false. Loss of statistical power corresponds to a percentage increase in false negatives when performing statistical hypothesis tests.

The above influence on statistical power can be explained in terms of the notion of the asymptotic relative efficiency (ARE) of one test relative to another. Here, relative efficiency is understood in terms of the relative difference (as a ratio) in sample sizes required to achieve the same level of statistical power for a given value of α in testing a given statistical hypothesis. In turn, the ARE is represented by the limit to which this ratio converges with increasing sample size, assuming that this limit exists.

It is estimated that for Normal data, the asymptotic relative efficiency of a non-parametric (ranks) test compared to the corresponding t-test is 95.5%. This means that a non-parametric test is estimated to require $1/0.955 = 1.047$ times the sample size, or, 4.7% more data, than the corresponding two sample t-test to attain the same level of statistical power [26]. Where the non-parametric test is the Mann–Whitney U test, we may assume that the corresponding t-test is the independent samples t-test.

The above estimate suggests that for small samples, at least, the wrong choice of test in comparing two groups can lead to false negatives and that where a hypothesis test is chosen for general use based on distributions for a 'majority' of cases, this can introduce bias into the assessment of the potential influence of an intervention across a range of continuous variables. Similar principles apply in relation to use of the ANOVA F test with more than two groups [27]. It is unclear to what extent these principles would have influenced Leibovici's own findings. It is helpful, however, to be aware that where excessive testing of outcomes leads to shortcuts in Normality testing or neglect of the Central Limit Theorem, the culprit may be shooting themselves in the foot relative to their own agenda through boosting the false negative rate (equivalently, the percentage of Type II errors) arising from hypothesis testing. As we shall see shortly, use of non-parametric tests where they are not needed can also carry the cost of loss of simplicity and transparency in reporting statistical findings.

Key Learning Point 1.2

In an RCT, where there is a range of continuous outcome variables of interest, laxity in choice between parametric and non-parametric hypothesis tests can lead to biased comparisons of the potential influence of an intervention across these variables. Where a non-parametric test is used in lieu of the correct choice of parametric test, this can lead to Type II errors, particularly for small sample sizes. Within-study selective reporting bias, and between-study publication bias, relating to statistical significance are likely to be influenced by the presence of these Type II errors.

1.3.2 The Importance of Effect Size Estimates

Reporting bias relating to statistical significance arises from lending too much weight to the importance of low p-values in the interpretation of scientific success. In scientific reporting, not least within medicine, the need for a more balanced perspective on the place of p-values in presenting the evidence for or against a given study hypothesis has gained increasing recognition over the past four decades [28].

The definition of a p-value from Greenland et al. is included below for ease of reference.

> The P value is … the probability that the chosen test statistic would have been at least as large as its observed value if every model assumption were correct, including the test hypothesis [29].

In the medical research literature, the test hypothesis is typically the null hypothesis and a sufficiently low p-value (not greater than α) is taken as sufficient evidence to reject this null hypothesis in favour of the alternative or study hypothesis. This approach to the evaluation of hypothesis-based research findings has become known as null hypothesis significance testing (NHST). NHST originates with the frequentist interpretation of probability in terms of relative frequencies derived from sampling data. This contrasts with the Bayesian approach, which treats probabilities as degrees of belief or certainty, or likelihoods (Chap. 3). However, it should be noted that the two approaches are often used in partnership and that it is naive to suppose that they are essentially rivals.

NHST has its critics, with many of the issues raised relating to the dichotomisation of the outcome of statistical testing into success and failure according to whether or not the p-value is at most α ($p \leq \alpha$). Indeed, more recently, in some circles, stronger views have been expressed, welcoming the apparent demise of the p-value and statistical significance more generally [30] in favour of what is considered to be a more sophisticated approach to evidence evaluation, typically Bayesian analysis and in particular, approaches involving likelihood ratios [31].

While the ground is fallowing in this area, it would seem prudent to consider a few of the basic concerns associated with the reporting of p-values that appear to be influenced by the competitive nature of academic medicine. Hypotheses are developed and often, the p-value is, at least inadvertently, treated as a logical truth function determining the fate of the hypothesis ('true' or 'false') according to which value it assumes relative to the threshold assigned to α (typically 0.05). This over-simplification of truth has played an important role in raising concerns regarding the use of p-values in scientific reporting. It is also likely to have been influenced by rather cavalier comments from Ronald A Fisher, who is commonly regarded as the founding father of NHST. For example, with reference to the cut-off of 0.05, Fisher made the following remarks: "It is convenient to take this point as a limit in judging whether a deviation is to be considered significant or not." [32] and "Personally, the writer prefers to set a low standard of significance at the 5 per cent point, and *ignore entirely all results which fail to reach this level.*" [33] [my emphasis].

Whatever basis Fisher had for adopting the terms "convenient" and "prefers", it is clear that this style of language reflects a degree of arbitrariness [34] at a foundational level in proposing a threshold for statistical significance that is complemented with a rather unscientific perspective on the fate of non-significant results. Thus, Fisher's remarks are likely to have had a major role to play in the onset of reporting and publication bias through the processes of supervision, reviewing and publication of scientific research.

Given the instances of NHST that prevail in contemporary medical literature, including high impact journals, it is necessary both to be sensitive to the limitations of p-values in decision making and to be better equipped to interpret them appropriately, both as producers and consumers of statistics, in clinical research. Indeed, a balanced foundational perspective on the importance of p-values and common misconceptions around their use and interpretation [35] ought to steer medical students, among others, towards taking a more critical approach when deciding on the practical utility of the results of statistical hypothesis testing that they encounter in clinical research, including within the context of a high-quality RCT.

This can be supported by recognition of effect size estimates. In interventional studies, the *effect size* is a measure of difference between the intervention group and one or more control groups, as expressed, for example, by the difference in means or absolute risk difference (see also Chap. 6). As these groups are typically samples from larger parent populations, we can think of the *true effect size* as pertaining to a comparison of the parent populations and an *effect size estimate* of this true effect size as arising from a comparison of the corresponding study samples from these populations. While we must be careful not to suggest that statistical findings provide standalone evidence for or against the clinical effectiveness of an intervention, the term *effect size*, does point to the idea that this measure has a contribution to make to a broader evidence base concerning such a relationship.

Effects sizes are of particular interest in a study design context where clinicians can recommend what they would consider to be a clinically important (clinically significant) effect size. Depending on the clinical context, this can be relatively straightforward where the effect size is simple to calculate and can be expressed in the same units as the individual measurements. Such is the case with the arithmetic difference between two means, which is reported on applying the independent samples t-test as a test of its significance. However, for those cases involving more complex formulae for calculating effect sizes and where what constitutes a clinically important effect is unclear, there may be a preference for relying on statistically predefined cut-offs for effect sizes that are generalisable to a range of experimental settings.

In the medical research literature, effect sizes tend to be neglected for non-parametric tests, including the Mann–Whitney U test. In the latter case, it is widely assumed that this test involves a comparison of the medians of the raw data across two groups. This is reflected by the reporting of these medians as a supplement to the p-value, as is evident in Table 2 of Leibovici's 2001 paper with reference to each of *length of hospital stay* and *duration of fever*. Thereby, the reader is implicitly invited to identify their own sense of difference between these medians and assume that the p-value represents an indicator of the probability that at least this observed difference in medians would have occurred if there was no true difference in medians between the parent populations.

The above assumption is in fact a highly prevalent fallacy, compounded by the explosion in open access learning materials in statistics. In the next KLP, it is fitting, therefore, that we consider the necessary conditions for ensuring that the Mann–Whitney U test is a comparison of medians.

Key Learning Point 1.3

The Mann–Whitney U test can be thought of accurately as a comparison of medians across two groups provided the data are continuous and the frequency distributions for the two groups are so similar that one distribution can simply be translated onto the other with almost perfect overlap. This is very rare!

While it can be argued that in some, or even many, cases the above test can approximate to a comparison of medians, this does not provide the non-specialist with a basis for extrapolating to *all cases*. Indeed, as I shall illustrate shortly with reference to Leibovici's reported findings, it is possible to provide examples where the Mann–Whitney U test result favours rejecting the null hypothesis, but the medians are identical, thus begging the question as to what difference has been found to be statistically significant!

In the statistical literature, a variety of effect sizes have been proposed for use with the Mann–Whitney U test, one of which originates with Rosenthal [36, p. 31] and is defined by the formula

$$r = \frac{Z}{\sqrt{n}},$$

where Z is a test statistic, used in a normal approximation test, which, for some statistical software packages, is generated as part of the output for this test, and n is the overall sample size. The strength of the effect, in terms of there being a difference across the two groups, is then defined according to the magnitude of r in keeping with Cohen's recommendations for a correlational metric [37].

We can also gain useful information from r^2. For an RCT, r^2 signifies the percentage of variation in the continuous or ordinal outcome data (e.g., the lengths of hospital stay) that can be explained in terms of the intervention allocation (for a simple RCT, exposure or non-exposure to the intervention).

While r and r^2, as presented above, ought to prove useful to junior and senior medical students in better understanding the reporting of findings, the material to follow in Sect. 1.3.3 is likely to appeal more specifically to readers who are of a mathematical bent. Other readers may prefer to skip to Sect. 1.3.4.

This additional content is included, however, for completeness, to demonstrate explicitly that statistical significance in the Leibovici trial is unlikely to point to anything of clinical importance. The applicability of this finding to clinical research more generally cannot be overstated. In the design of medical curricula, there is a clear need to develop opportunities for undergraduate students both to acquire an improved awareness of the distinction between statistical and clinical significance and appreciate that statistical significance per se is not a sufficient condition for clinical significance. In their critical appraisal of papers, they should also be encouraged to root out cases where statistical significance and p-values receive an emphasis to the neglect of proper use of effect size estimates as formative learning for sound reporting of their own future clinical research findings. Similar comments apply to medical education research, among other research fields applicable to health professionals, where there is a need for clearer thinking on what should constitute a meaningful effect size, both within the context of RCTs and other types of study.

1.3.3 The Probability of Superiority as an Effect Size with an Application to the Leibovici Trial

A lesser known effect size that is particularly well aligned with the above null hypothesis is the probability of superiority, the true, or population, value of which we shall denote by 'PS' [38]. In a parallel groups clinical trial, involving an intervention group, *int*, and control group, *cont*, we can define PS as the probability for a given ordinal or continuous outcome variable that, if a value is taken at random from each of *int* and *cont*, the value ('score') from *int* will represent a better outcome than that of the value from *cont*. With reference to the Leibovici trial, a shorter hospital stay and a shorter duration of fever would each be desirable. Therefore, the assessment of superiority in the above definition would involve interpreting a better outcome in terms of a randomly selected score from *int* being lower than that from *cont*. As with other effect sizes, PS is estimated using the study sample. To avoid confusion, we shall make a notational distinction by referring to the estimate of the effect size PS as $\widehat{\mathrm{PS}}$.[1]

---

[1] Statisticians often use a circumflex ("hat operator") for a given parameter to denote that this parameter is being used at the sample level to *estimate* the true value for the corresponding parent population.

Fortunately, once we have the test statistic, U, for the Mann–Whitney U test, $\widehat{PS}$ is not difficult to calculate. However, in ensuring that in any one study, $\widehat{PS}$ is calculated in terms of the superiority of the intervention rather than the control, it is important to understand how U is calculated.

As is more widely known, for two independent groups, G_1 and G_2, two statistics U_1 and U_2, are used in obtaining U. U_1 is calculated through taking one value at a time from G_1 and pairing it with all values from G_2. U_1 is then calculated as the total number of times across all such pairings that a value from G_1 exceeds the value of its partner from G_2. Through taking one value at a time from G_1, U_2 is calculated in a similar way to U_1 as the total number of times across all such pairings that a value from G_2 exceeds the value of its partner from G_1. The test statistic U used in the calculation of statistical significance and frequently reported in computer output on running the Mann–Whitney U test is the smaller of U_1 and U_2. In the above procedures for calculating U_1 or U_2, n_1n_2 pairings are possible, where n_1 and n_2 are the number of individuals in each of G_1 and G_2, respectively. This means that a sample probability, $\widehat{p}$, can be obtained as follows:

$$\widehat{p} = \frac{U}{n_1 n_2}.$$

It is possible, in a general sense, to equate $\widehat{PS}$ and $\widehat{p}$ as defined above, without consideration of the expectation and historical underpinning behind reference to a probability of *superiority*. However, we shall avoid using this simplification here, recognising that the intent is indeed to assess the probability that the intervention is superior.

To ensure that $\widehat{PS}$ remains true to its name and to provide a clear basis for the future use and interpretation of this effect size estimate within the context of an RCT, let's briefly consider the above algorithm for calculating U within the context of the Leibovici trial. This will also prove helpful in deriving approximate effect sizes for Leibovici's findings. In doing so, we shall initially focus on *length of hospital stay*. However, similar remarks apply to *duration of fever*.

Suppose, therefore, that G_1 and G_2 are *int* and *cont*, respectively. Shorter hospital stays correspond to superior events.

Suppose that $U_1 \neq U_2$. According to the definition of U, above, $\widehat{p}$ corresponds to the probability of a randomly chosen value of *int* exceeding that of *cont* (if $U = U_1$) or to the probability of a randomly chosen value of *cont* exceeding that of *int* (if $U = U_2$). This depends on whether U_1 is less than or greater than U_2, respectively. Let's consider these two cases separately.

The case $\mathbf{U = U_1}$ The case $U = U_1$ reflects that of *int* having less randomly chosen instances of strictly longer stays and the above formula for $\widehat{p}$ reflects the probability of a strictly longer hospital stay (an inferior outcome) in *int*. But we are interested in the probability of superiority, not inferiority. Therefore, in this case, we must define $\widehat{PS}$ as the complement[2] of $\widehat{p}$: $\widehat{PS} = 1 - \widehat{p}$.

The case $U = U_2$ The case $U = U_2$, reflects that of *cont* having less randomly chosen instances of strictly longer stays and the above formula for $\widehat{p}$ reflects the probability of a strictly longer hospital stay in *cont*. Equivalently, $\widehat{p}$ reflects the probability of a strictly shorter stay (a superior outcome) in *int*. Therefore, in this case we *do* have a measure of superiority and $\widehat{PS} = \widehat{p}$.

Of course, if $U_1 = U_2$, $\widehat{PS} = \widehat{p} = 1 - \widehat{p} = 0.5$. In interpreting the strength of a given value of $\widehat{PS}$, it is helpful to know that this parameter can in principle assume any value between 0 and 1 and that the null hypothesis value is 0.5, where, for our current example, the latter value pertains to the outcome that if a value is taken at random from each of *int* and *cont*, the value ('score') from *int* is just as likely to represent a better outcome as that of the value from *cont*.

For the Leibovici example, if the results were genuinely in favour of (superior for) the intervention arm, we would expect there to be a tendency for *length of hospital stay* and *duration of fever* to be lower in *int* than in *cont*. In this case, again, assuming that G_1 and G_2 are *int* and *cont*, respectively, according to the above algorithm for determining, U, U would be equated with U_1, which, as we have seen, would require $\widehat{PS}$ to be equated with $1 - \widehat{p}$. Statistical packages typically include functionality to manage the order in which two groups are assigned to the above algorithm for obtaining U. Importantly, however, whatever the order, the above principles for determining a probability of superiority as $\widehat{p}$ or $1 - \widehat{p}$ are the same—a result that will prove useful when obtaining plausible estimates for $\widehat{PS}$ from the Leibovici trial.

Optional Exercise for Mathematically Curious Students

Instructors may wish to invite their students to investigate how $\widehat{PS}$ should be defined in the above example, based on the algorithm we have already considered for calculating U, if for an RCT, superiority is represented by *higher* measurements, as would normally be the case with a quality-of-life scale, for example, and we have reason to believe that the sample data are in favour of the intervention.

Use of an effect size, such as PS, in the interpretation of output from the Mann–Whitney U test is notably absent from Leibovici's paper and is typically absent from the clinical research literature more generally. The interpretation of PS as an effect size has been defended as more meaningful in concrete terms to clinicians than one expressed in terms of a difference in standard deviations, such as Cohen's *d* [39]. For this reason, PS may serve as a useful complement to Rosenthal's *r*. The caveat should be included, however, that the term 'superiority' must be loosely interpreted

[2] A fuller treatment of the concept of complement within the context of probability theory is provided in Sect. 3.5.1.1.

insofar as further analyses would be required to assess the validity of the findings (see Chaps. 2 and 5 for consideration of the concepts of internal and external validity). Also, a variety of alternative correlation-based effect sizes exist for non-parametric tests [40], including the "simple difference formula", representing the differences in probabilities between obtaining a superior and non-superior outcome. Each available effect size can be examined in its own right in terms of the trade-off between desirable statistical properties and conceptual transparency.

Undoubtedly, a commitment to the reporting of valid effect sizes and the interpretation thereof would have proved helpful in assessing the practical relevance of Leibovici's trial findings. Instead, he advises us that "[L]ength of stay in hospital and duration of fever were significantly shorter in the intervention group ($P = 0.01$ and $P = 0.04$, respectively)" [1]. For the reader's interest, the tabulated intervention and control group medians for *length of hospital stay* are very similar, seven and eight days, respectively, and for *duration of fever* are identical: two days in each case! Clearly, to assess evidence for difference across the groups more effectively, that is, in terms of the extent of that difference, the reader needs to be equipped with a better measure of effect size. Indeed, it would be reasonable to conclude that in the above instances, *no effect size* is reported. Correspondingly, there is no indication of how to recognize clinical significance, which, in RCT design, relies on one or more pre-determined effect sizes as threshold markers for the therapeutic effectiveness of an intervention.

The reporting of p-values in isolation, with no supporting effect size, can serve to perpetuate the common misconception that "[a] statistically significant finding is clinically important" [35]. It is important to appreciate that statistical significance can occur in conjunction with low effect sizes. A very good example of this is in the use of the Pearson correlation coefficient, where low to moderate correlation coefficients can readily achieve statistical significance for modest sample sizes. Within the context of clinical trials, neglect of effect sizes is most likely to occur for secondary outcomes. Typically, protocol requirements for funded projects include, for at least one primary outcome, the calculation of a minimum required sample size to achieve statistical significance for a clinically important effect, as defined by an agreed threshold, conditional on the existence of a clinically important effect in the parent population. Nevertheless, systematic reviews may still rely on the interpretation of secondary outcomes from primary studies and indeed adopt these outcomes as the primary outcomes for the review. Where possible, the need for effect size estimates to support the evidence base for therapeutic effectiveness in these cases ought to be anticipated in primary studies through extension of the practice of reporting effect size estimates to incorporate secondary outcomes from the study protocol.

Key Learning Point 1.4

As a general rule, it is desirable to identify at least one effect size for use with the Mann–Whitney U-test and be aware of the potential to mislead through a simple presentation of medians. Suggested effect sizes include Rosenthal's r and the probability of superiority. More generally, the reporting of effect size estimates should not be limited to the primary outcomes of a study.

For each of *length of hospital stay* and *duration of fever*, the intervention and control groups sizes in the Leibovici trial are large—1691 and 1702, respectively or, if patients who died in hospital were excluded for these variables (Leibovici does not specify), 1216 and 1188, respectively. This tells us that if there is a true difference of clinical importance, even in the broadest of senses, the p-values forthcoming from comparing these two groups ought to be very small ($p < 0.0001$). This is due to the fact that, assuming that the null hypothesis is false, there is a monotonic decreasing relationship between the sample size and the p-value; yet the p-values that Leibovici reports are unremarkable, albeit indicative of statistical significance.

Let's assume for the sake of argument that the Mann–Whitney U test was the *correct* choice of hypothesis test for each of the above two variables. This allows us to assume the effect size PS, as defined above. Also, assuming consistency of approach to NHST across the above two variables, we must entertain either one of the two possibilities that Leibovici's reported p-values were:

(a) forthcoming from the application of two-tailed tests;
  or
(b) forthcoming from the application of one-tailed tests, in which case, the corresponding two-tailed equivalents are 0.02 and 0.08 for *length of hospital stay* and *duration of fever*, respectively.

Assuming also, as is typically the case for large samples, that the standardised Z-value for U has been used to obtain the p-values Leibovici has reported for each of *length of hospital stay* and *duration of fever*, it is in fact possible to work backwards from these p-values to the corresponding Z-values and in turn, an approximation to the corresponding values of U. The need to take this approach arises from lack of access to the original dataset or the relevant findings.[2] For each of possibilities (a) and (b), the different possible pairs of group sizes listed above and the corresponding derived estimate of U, Table 1.1 reports the corresponding approximations to $\widehat{p}$ and $\widehat{\text{PS}}$. Here, it is assumed that Leibovici is correct in his directional claim that "[L]ength of stay in hospital and duration of fever were significantly shorter in the intervention group". Therefore, based on the reasoning provided earlier in this section, the values of $\widehat{\text{PS}}$ in Table 1.1 have been calculated according to the equation $\widehat{\text{PS}} = 1 - \widehat{p}$.

For *length of hospital stay* and *duration of fever*, the estimated values for $\widehat{\text{PS}}$ are, to two decimal places, consistently 0.47 or 0.48 and 0.48, respectively. The approximation to U referred to above arises from our lack of information regarding ties in the data, and therefore the requirement to ignore ties in estimating the variance for U. However, on comparing the formulae for the variances, σ^2_{ties} and $\sigma^2_{no\ ties}$, assuming ties and no ties, respectively [41], it can be shown mathematically that for large sample sizes, including those of the order represented in Table 1.1,

Table 1.1 Estimation of the sample probability of superiority ($\widehat{\text{PS}}$) for reported outcomes in the Leibovici trial, based on an estimated value for the test statistic U for the Mann–Whitney U test

Variable	Sample size	Status of test	2-tailed p-value	Z	$\hat{p}$	$\widehat{\text{PS}}$
Length of hospital stay	3393	Two-tailed	0.01	2.576	0.526	0.474
	3393	One-tailed	0.02	2.326	0.523	0.477
	2404	Two-tailed	0.01	2.576	0.530	0.470
	2404	One-tailed	0.02	2.326	0.527	0.473
Duration of fever	3393	Two-tailed	0.04	2.054	0.520	0.480
	3393	One-tailed	0.08	1.751	0.517	0.483
	2404	Two-tailed	0.04	2.054	0.524	0.476
	2404	One-tailed	0.08	1.751	0.521	0.479

the resultant error of estimation will not have an appreciable effect on the actual value of U or in turn, $\widehat{\text{PS}}$. Indeed, for sample sizes of this order, the estimated value of $\widehat{\text{PS}}$ has extremely low sensitivity to percentage changes in the value of $\sigma^2_{no\ ties}$, even for percentage reductions up to 99.99%, which is well beyond the order of the corrections that would be required for *hospital stay* and *duration of fever*.

Therefore, the actual values of $\widehat{\text{PS}}$ for *length of hospital stay* and *duration of fever* in the Leibovici trial should still be very close to 0.5. The proximity of the results in the last column of Table 1.1 to 0.5 is very telling in two main senses. Firstly, noting the interpretation of PS as a measure of superiority, for each variable, the above investigation suggests that the value of $\widehat{\text{PS}}$ is not strongly supportive of an effect in favour of the intervention, if at all. In everyday language, there is approximately a 50–50 chance that retrospective prayer is beneficial, reducing the probability of benefit for the individual patient to a matter of tossing a coin. This finding ought to be weighed against Leibovici's conclusion for his study that "Remote, retroactive intercessory prayer said for a group is associated with a shorter stay in hospital and shorter duration of fever in patients with a bloodstream infection and should be considered for use in clinical practice." I shall leave the reader to draw their own conclusion. Secondly, the finding that for each of *length of hospital stay* and *duration of fever* the sample value for $\widehat{\text{PS}}$ lies so close to 0.5 (the null hypothesis value) demonstrates the capacity of NHST to identify effect size estimates suggestive of very weak statistical effects as statistically significant for large sample sizes. Thus, the Leibovici trial also serves medical education well in demonstrating the common problem of misuse of statistical significance to make a bold claim (whether in the 'festive spirit' or otherwise) about evidence for the effectiveness of a clinical intervention in the absence of essential statistical acumen.

The above observations regarding appropriateness of statistical claims also lend perspective to the earlier recommendations on trial pre-registration and even use of SAPs. While on the one hand such procedures serve as effective tools for managing selective reporting, they cannot serve as guarantors of sound statistical practice in terms of choice of statistical procedures or interpretation of the findings thereof.

1.3.4 *The Potential for False Positives*

In an RCT, randomisation is understood as involving treatment allocations that allow each person to have an equal chance of receiving the trial intervention. This is often misunderstood as meaning that there is no bias whatsoever in the distribution of patient characteristics across treatment groups. The absence of bias in this sense is desirable, as we can be reassured that a difference detected between two treatment arms was purely related to the intervention, there being no noise from confounding factors. However, it is instructive to note that by the very nature of the simple randomisation process, even for a well-designed RCT, a statistically significant difference can occur across two treatment groups, including a control group, at baseline and further to treatment allocation, each with probability equal to the chosen level of statistical significance. This is a consequence of sampling variation, which, at the sample level, produces unpredictable imbalances in factors across these two treatment groups, including measurement errors, which increase the likelihood of a false positive outcome (Type I error) through NHST. These observations give rise to the following KLP.

Key Learning Point 1.5
For a well-designed RCT, if the null hypothesis is true, simple randomisation does not preclude statistical significance on comparing an intervention and control group. Instead, it limits the proportion of instances in which we should expect this conditional outcome (a Type I error) to occur, to the significance level, α, regardless of the sample size. For this reason, α is sometimes referred to as the false positive rate (FPR).

Examples where KLP 1.5 applies include[3]:

(a) for a specific variable at baseline or specific outcome variable across multiple independent treatment allocations based on a specific treatment
and
(b) at baseline, for a specific treatment allocation, across multiple baseline variables.

As we shall consider more specifically in Sect. 1.4, should the observed FPR exceed α, we must question the credibility of the randomisation process.

---

[3] The use of the term "at baseline" is consistent with the idea that, for a variety of reasons, for some baseline variables, e.g., variables pertaining to anxiety level (see Chap. 5), the reported patient values can change at later stages of the trial, upsetting the original balance forthcoming from random treatment allocation. For completeness here, note that within the context of hypothesis testing, the FPR can be interpreted more broadly than is assumed in NHST, including from a Bayesian perspective, and as a function of the likelihood ratio favouring the null against the competing alternative hypothesis. As discussed elsewhere, such approaches can serve as contenders to NHST [42].

Sampling variation is reflected by the existence of a variety of possible samples, including instances where there is departure of the distribution of sample characteristics across patient groups from that of the parent populations that they are understood to represent. In the latter case, statistical significance can arise purely due to sampling variation, thus leading to Type I errors. This is, of course, good news for Leibovici's opponents, as it can be used to explain away 'post-treatment effects' represented by $p \leq 0.05$ without any need to entertain the possibility that retroactive intercessory prayer can make a difference to patient outcomes in an RCT. This is a plausible position to take on considering possible sources of covariate imbalance across the intervention and control groups at the sample level, purely based on sampling variation. It is feasible that, for example, the baseline variable for age and, across a range of variables, the medical history of the patients, could have served as covariates influencing duration of fever and length of stay in hospital, with the identity of the health professional who approved hospital discharge, as an additional covariate, also influencing length of hospital stay.

For example, let's consider the possibility that one physician had a tendency to keep their patients in hospital an extra day as a pre-cautionary measure. While, at the population level, the characteristics of health professionals might be balanced across patient groups, patients for whom this particular physician had been allocated responsibility could feature considerably more often in the control group than the treatment group for a given sample. As this example, illustrates, not all covariate influences on the outcome of an RCT are predictable or 'systematic'; they can also be the product of sampling variation. In the Leibovici trial, random fluctuations in the values of covariates across the sample treatment and control groups could have led to covariate imbalance purely at the sample level and contributed to the Type I error rate, as could random errors in the recording of duration of fever and length of stay. In such cases, the imbalance in the distributions of values across the two sample groups does not reflect that across the corresponding two parent populations.

Covariate imbalance at the population level is a topic for consideration in Chaps. 2 and 5. It suffices to note here that multivariable approaches do not rule out the existence of false positives in the sense discussed in this chapter and in indeed, in some cases they can lead to an inflated FPR.

The above discussion of Type I errors specifically within the context of the Leibovici trial, together with the more general observations of KLP 1.5, highlight the importance of an investigation of consistency of findings across similar studies in strengthening the evidence base for the intervention as the source of statistical significance before carrying forward study findings with clinically significant effect sizes to clinical research and practice. As we shall see in later chapters, the existence of different types of bias in RCTs adds further weight to the case against reliance on a standalone RCT for evidence-based practice.

1.3.5 Demonstrating the Influence of Sampling Variation on the Results of Hypothesis Testing Using a Fictitious Retrospective Intervention

In the teaching of statistics to non-specialists, the role of data visualisation has an important role to play. This is perhaps particularly so within the context of illustrating phenomena that conflict with traditional ways of thinking, an example of which is the inevitability of false positives within the context of RCTs.

Leibovici seems unwilling to share his study data (personal correspondence with Leibovici and rapid responder to the publication). However, this has opened the door of opportunity to explore a subset of a large dataset forthcoming from the International Stroke Trial (IST) [43], one of the largest randomised trials in acute stroke [44]. This subset was obtained from the original sample of '19,435 patients with suspected acute ischaemic stroke, based on the view of the "responsible physician", entering 467 hospitals in 36 countries' who had been recruited for the IST during the period January 1991 to May 1996. In the IST, patients were administered "up to 14 days of antithrombotic therapy". Using a factorial design, patients were allocated in roughly equal proportions to either of the treatments 'aspirin' and 'avoid aspirin' in combination with either of the treatments 'heparin' and 'avoid heparin,' resulting in a control group of 4860 control patients who received no intervention, as they were allocated the treatments 'avoid aspirin' and 'avoid heparin'. These patients comprised approximately 25% of the entire cohort.

We shall now consider the above control group as a concrete illustration of the possibility, even for a fairly large RCT, of achieving statistical significance independently of an intervention. For this purpose, we shall consider time to discharge (in days) as the measured outcome from the IST data, this being comparable to Leibovici's chosen outcome, length of hospital stay (in days). However, for educational purposes, the intervention is allocated in a purely nominal sense by simple randomisation to members of the above control group. The nominal random allocation involves random allocation of the names 'intervention' and 'control' by simple randomisation to the patient identifiers in our sample. While no intervention or control treatment has been allocated to the patients themselves, note that that this process splits the patient data into separate groups and permits analysis of these data as though the treatments were genuine. As the intervention is purely nominal in the above sense, there is no scope to appeal to retroactivity or even a post-intervention effect as a possible explanation for statistical significance, as, from the patient perspective, the intervention is fictitious.

As a special case illustrating KLP 1.5, let's consider the idealistic scenario in which for the same sample, and a fixed outcome variable—time to discharge (in days)—the results from hypothesis testing are considered separately for multiple 'treatment allocations' based on simple randomisation. The scenario is, of course, idealistic insofar as in natural settings, attempts to replicate RCTs typically require new, albeit similar, patients to test the intervention, whereas for our nominal intervention, this is not the case. Nevertheless, there is a clear carryover effect to experimental

RCT settings in that in these settings, any one treatment allocation can be thought of as having been randomly selected from infinitely many possible treatment allocations.

Tables 1.2 and 1.3, provide a frequency distribution for statistically significant results and corresponding summary statistics for an increasing number of fictitious independent random treatment allocations across our study sample. These results have been generated using the statistical package Stata (IC v.16). For each random allocation (repetition), the treatment and control groups were each of size 1300.

Table 1.2 Frequency distribution of significant differences in mean time to discharge due to Type I errors further to simple randomisation using fictitious intervention (n = 4860)

No. of random treatment allocations	Choice of significance level		
	0.10	0.05	0.01
	Frequency (%) of significant differences		
10	0 (0)	0 (0)	0 (0)
100	8 (8)	4 (4)	1 (1)
1000	96 (9.6)	48 (4.8)	16 (1.6)
10,000	1032 (10.32)	467 (4.67)	87 (0.87)
100,000	10,123 (10.123)	4996 (4.996)	928 (0.928)

The distribution of p-values in Table 1.2 illustrates that probability is an asymptotic notion, in that the percentage of instances in which statistical significance is achieved when the null hypothesis is true converges to the expected probability (the significance level, α) as the number of random allocations (column 1) increases. This is the case whatever the chosen value for α. The latter behaviour reflects the Law of Large Numbers, which, in its simplest form, tells us that for a random process, the average of the observed values will become progressively closer to the expected outcome as the number of observations increases. Of course, we know from KLP 1.5 that the expected proportion of false positives for each value of α is in fact α. Note, however, that α itself does not vary with sample size; equivalently, α *is invariant to* changes in sample size.

From Table 1.3, we can also see that, irrespective of the number of iterations, the p-value can range across either side of the threshold represented by α due to variation in random treatment allocations. This is noteworthy given that no intervention was genuinely received by patients in this experiment and that therefore there is no real intervention effect. For each p-value, there is also a corresponding confidence interval (CI) for the difference in means, which is understood to contain the true mean difference with 90, 95 or 99% certainty, depending on whether the chosen value of α was 10, 5 or 1%, respectively, relative to the parent population of stroke patients, who had been allocated neither aspirin nor heparin. As a general principle, when testing a null hypothesis using NHST, the null hypothesis value of no effect coincides with a limit of the corresponding CI if and only if $p = \alpha$ or is excluded altogether from the CI if and only if $p < \alpha$. In our current example, the null hypothesis value is 0, corresponding to no difference in means. For those instances in Table 1.2 where statistical significance has been achieved, $p < \alpha$ and correspondingly, 0 is excluded from the CI, as is illustrated in Fig. 1.1b). However, as is evident from this example, even where $p < \alpha$, the lower or upper limit of the CI can still be very close to 0. This observation in itself illustrates the added meaning from including CIs in statistical reporting.

Table 1.3 Summary statistics and corresponding p-value based on independent 'treatment allocations' on comparing mean time to discharge (in days) across treatment and control groups using a significance level of 0.05 (n = 4860)

No. of random allocations	Summary statistics for *p-value*				
	Minimum *p-value*	Maximum *p-value*	Difference in means (with intervention group mean subtracted from control group mean) for:		Standard deviation of all mean differences
			Minimum *p-value*	Maximum *p-value*	
10	0.12280	0.95888	0.94385	0.03154	0.25783
100	0.00012	0.98896	1.73462	0.00846	0.32041
1000	0.00076	0.99900	−2.05769	0.00077	0.29480
10,000	0.00006	0.99900	−2.46385	0.00077	0.29004
100,000	0.00006	0.99900	−2.46385	−0.00077	0.28863

Note i. Values are all provided to 5 decimal places for consistency and to ensure that all numbers are rounded to at least 1 significant figure. ii. Shaded cells highlight p-values less than 0.05 and corresponding differences in means

Also, taking the difference in means as a measure of effect size, Table 1.3 illustrates the general principle that in comparing pairs of groups from a parent population using NHST, the p-value is inversely proportional to the observed effect size, provided the group sizes are held constant. Notice also from Fig. 1.1 (a) that as the p-value converges towards 1, the sample difference in means converges towards 0. Based on the definition of p-value (Sect. 1.3.2), this makes sense, as for the independent samples t-test, 0 is the null hypothesis value.

More generally, Fig. 1.1 illustrates the extent of variation in mean difference that can occur for the same sample across multiple independent random treatment allocations purely on account of sampling variation and therefore independently of any treatment effect. Correspondingly, in Table 1.3, for each listed number of replications, the range in sample differences in means (effect size estimates) for the minimum and maximum p-value and the considerable magnitude of the standard deviations in sample differences in mean illustrate that effect size estimates can fluctuate markedly on account of sampling variation. Of particular interest is the variability in the sample differences in means for $p < 0.05$ illustrated in Fig. 1.1 (b). Indeed, as the cases highlighted by shaded cells in Table 1.3 illustrate, the signs, or directions, of these differences are not consistent, even where these differences have been attributed a particularly high level of statistical significance (here, $p \leq 0.001$).

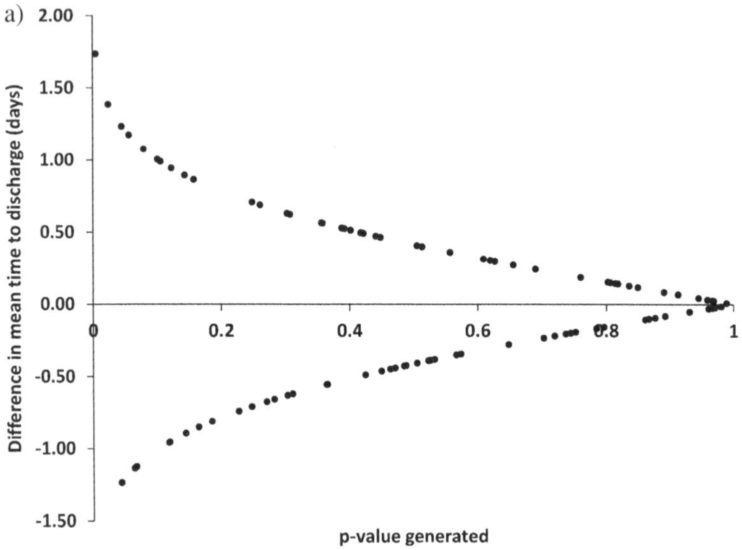

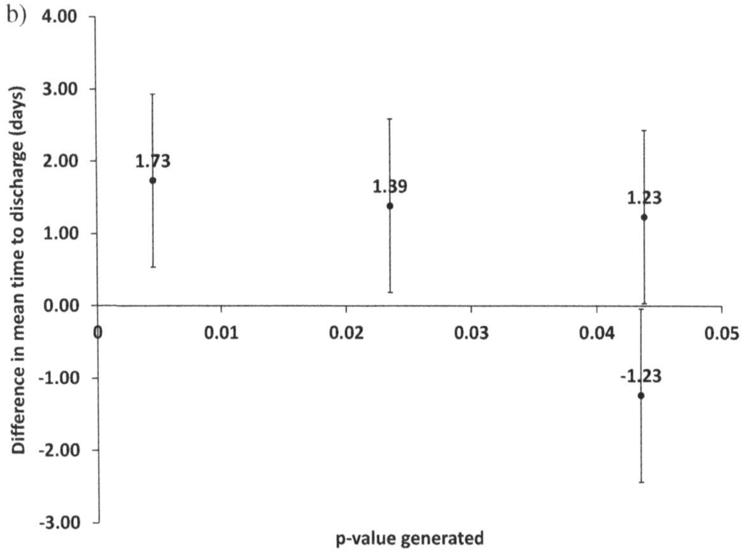

Fig. 1.1 Difference in mean time to discharge against p-value generated for stroke patients, based on 100 nominal treatment allocations by simple randomisation. *Note* i. Means have been compared, and p-values generated, using the independent samples t-test. Differences in means are based on subtracting the intervention group mean from the corresponding control group mean. ii. Part b) of figure displays increased resolution to zoom in on the region in part a) for which the p-value generated is less than 0.05. Error bars display 95% CIs

Based on the findings in Table 1.3 and Fig. 1.1, it is evident that speculating on the effectiveness of an intervention based on a one-off RCT is not a reliable foundation for clinical decision making. Thus, the status of a trial as an RCT does not provide a license for treating the outcomes of the trial as the gold standard. In the current experiment, we can see that in an RCT, Type I errors can arise alongside a considerable range of values of a given effect size estimate. Allowing for a small margin of error, the magnitude of which will depend on the nature of the measured outcome, consistency in effect size estimates across studies (for example, in terms of the clinical significance and direction of the effect) is important. As we have seen with the IST data, even for a large study, by the very nature of randomisation, neither this type of consistency nor consistency in terms of statistical significance, is guaranteed on attempting to replicate an original study. This observation reinforces the need for trainee and graduate health professionals to be cautious in their interpretations of their own and published research, even for the case of large RCTs that are of a very high quality in terms of design, conduct and reporting.

Key Learning Point 1.6

In an RCT involving treatment allocation through simple randomisation, effect size estimates for comparing the treatment and control groups for a given outcome variable can fluctuate considerably in the absence of a true treatment effect. This can extend to inconsistency as to whether the effect size estimate represents a favourable or unfavourable intervention effect. Thus, based on a single, albeit large, study, researchers should be wary of:

(a) appealing to the possible need for a larger study in order to achieve statistical significance *exclusively* for cases where a favourable effect size estimate has been obtained
and
(b) interpreting statistical significance as confirmation of a true treatment effect of magnitude similar to the given effect size estimate.

The findings in this section should help to reinforce the need for investigation of consistency of study results across similar patient samples. They should not, however, be viewed as evidence in support of the school of thought that seeks to do away with p-values. As Murtaugh conveys, the concern is arguably not so much about the utility of the p-value as its misinterpretation, including through failing to appreciate that proposed rival approaches to scientific testing share common weaknesses with those attributed to NHST [45]. Furthermore, the ongoing prolific use of NHST in contemporary medical literature lends credence to Paulo's view that retiring statistical significance would "be analogous to retiring a still very productive person" [46].

1.4 Chance and its Connection with Randomisation

In so far as the occurrence of false positives in NHST can conflict with what is true about a population or the quality of the study design process, they are frequently referred to loosely as *chance events*, with the understanding that in some sense they are accidental, spurious or 'non-deterministic,' events. In randomised experiments this is to some extent supported by the interpretation of randomisation as arranging items in an unpredictable, or unsystematic, order, as, then, any errors that arise from this process can be viewed as mere accidents arising from chaos. However, this notion of randomisation is itself contentious and historically, the concept of chance has been fluid, leaving scope to conclude that 'chance' is a purely nominal term—a mere convenience of language with no concrete phenomenon as a reference. While the idea of chance as purely nominal provides a strong basis for philosophical debate, the focus of this section is to present and challenge more traditional notions of chance. This will help us to appreciate that the occurrence of false positives is not merely an inconvenience and that, as touched on earlier in this chapter, the *regular* pattern of behaviour we should expect in the FPR for baseline variables in a well-conducted RCT can serve as a useful property in detecting problems with the authenticity of the randomisation process and potentially, the quality of the study outcomes.

Traditional notions of chance can be observed in the reflections of Alberoni within the field of psychology. Alberoni presents events with a causal explanation as those for which the observations are too far removed from the system of expectations to be merely a result of chance. To this he adds, "Chance and cause are antithetical terms and the one excludes the other. Chance cannot be said to result from the interference of several causes" [47]. From his experiments involving children, he concludes, "this idea [chance] comes into the mind of the subject when he cannot see the reason for a certain phenomenon, by referring it to some other phenomenon (cause), or by seeing in it a regular pattern (order)." With reference to Loosli-Usteri's research in developmental psychology, also involving children, he adds, "Loosli-Usteri also has showed that the idea of chance develops when the subjects have realised the impossibility of explaining an occurrence by following certain rules. The idea of chance is built upon this failure."

Here, in a few sentences, Alberoni identifies three common phenomena—a cause, the presence of a regular pattern (order) and the provision of effective rules—that cannot be associated with chance events. Within the context of Loosli-Usteri's experiments involving squirting paint on paper, it is in her view the absence of a cause for the paint producing a particular image on paper when the tube is squirted, and the absence of an evident mechanical rule for ensuring that an animal head (a successful outcome) is guaranteed, that demonstrates that the image created is the product of chance. As she sees it, the arrival of doubt in a child's mind regarding their potential for success is integral to the emergence of their appreciation of chance as an agent in the experiment. She also connects this apparent loss of childhood naivety, typically at age 13, with renouncing 'blind' belief in a cause for events, where no cause is evident. This she does without acknowledging the possible influence of social construction of knowledge within school education. Instead, she perceives the subject as entering a stage of intellectual enlightenment, whereby they view chance as a necessary explanation for the observed event. Similarly, it is consistent failure to observe the recurrence of a pattern or order when squirting paint on paper that sews the seed of chance in the mind of the developing child in explaining their inability to determine the outcome of their own efforts or that of another.

In the above representations of the discovery of chance in terms of absence of cause, effective rule and pattern or order, there is a consistent theme of human limitation, as it is at the point of conceding that they are unable to find either of these phenomena to explain an event that the observer is understood to be exposed to the need to appeal to chance. In each case, the observer must admit uncertainty and lack of control over an event. However, it needs to be stressed that this need not rule out the existence of one or more causes for an event. This can be illustrated by an analogy with our use of the vernacular.

A colleague who needs a letter to be posted may enquire, "By any chance, do you pass a post box on the way home?" and I might in turn reply, "Not as a general rule, but it just so happens that I need to do so today". Here, we are both appealing to an event that, by admission, does not arise from a rule of which either of us are aware. Nevertheless, my need for the detour is also by admission prompted by a cause, such as my decision to post a celebration card for a rare event. Therefore, arguing that the chance-cause dichotomy is foundational to human reasoning rather than a product of social construction is itself highly contentious. The above perspective is in agreement with that of Aquinas, Aristotle and the nineteenth century French philosopher and mathematician, AA Cournot, including in relation to their adherence to the "long-standing idea of intersecting casual lines" [48]. Hacking illustrates the latter idea through reference to the unplanned meeting of two persons at the market as a chance meeting, yet not through absence of causal stories for each person's arrival at the market at a specific time [48]. These ideas resonate with the nineteenth century interpretation of determinism originating with the German physician and physiologist, Emil Du Bois-Reymond, that "everything that happens constitutes a chain of causation" [49]. As such, they support the idea of chance events as deterministic events without *requiring* belief in the universal law of necessity. This law is defined by one of its opponents, the American philosopher Charles S. Peirce, as the proposition "that every single fact in the universe is determined by law" [50].

Belief in the involvement of cause in chance events is compatible with Alberoni's portrayal of such events as those that we cannot explain through an effective rule. Such a rule may involve mechanical (such as physiological) or logical operations that guarantee replicability. In the above example involving the letter, neither participant may have an effective rule to predict in a general sense when conditions will be propitious to ensure that such a meeting of needs will converge in time. Even so, this does not rule out the activity of a divine orchestrator who orders causes in time to ensure that the foreordained end is accomplished. Thus, as illustrated by the above examples, it can be argued that chance events are not cause-free events; rather, they are events for which we cannot determine an effective rule. They can at the same time, involve a mechanistic process, that, as is illustrated with reference to "the trajectory of a molecule in the bloodstream from injection to metabolism", "consists in a spatio-temporal pathway along which certain features are propagated from the start point to the end point" [51]. The appeal to divine providence transcends the accidental, and the arbitrariness, spontaneity and chaos, that are so central to everyday notions of chance. It also replaces the above version of the law of universal necessity with one requiring that all events are necessary, since foreordained, and as such, are based on causes, without the need for human capacity to discern the causes or the connections between them.

This philosophy of chance resonates with the allegory of chance provided by Puskar [52], through appealing to a 19th detective story written by Anna Katherine Green, including the introduction provided by Catherine Ross Nickerson. Of particular relevance is Puskar's account below of one of the characters catching a glimpse of the murderer in a mirror without realising it:

> It turns out that this "vision" is a glimpse of the murderer reflected in the pane of the alcove's open window, which is in turn reflected in a mirror across another room, toward which Van Arsdale inadvertently and passively "found" herself looking. Afterward, when she and the police inspector realise what she had seen, the inspector says, "It would seem that by one of those chances which happens but once or twice in a lifetime, every condition was propitious at the moment to make this reflection a possible occurrence." Had certain shades not been drawn and others left open, had the alcove's window not been opened to precisely the right angle, and had the alcove not been elevated due to the "freak of the architect," this particular glimpse would not have occurred. It follows, of course, that in most cases—and in many other instances in this novel—no such insight does occur, not because there are sights that are overlooked but because the world fails to come together in precisely the right way to make those sights available.

Under the novelist's direction, the above 'accidental' sighting, which initially seemed irrelevant, is essential to the plot, as this sighting serves later as a vital clue to identification of the murderer. Outside of the plot, the sighting is, as such, transformed from the accidental to the intentional and necessary. There is no insistence on an effective rule for making every condition propitious at the desired moment. However, the conditions of the sighting and, under the novelist's pen, the relationship between them that gives rise to the sighting, constitute an orchestrated mechanism and thus the cause of the sighting.

As we have seen, for RCTs based on simple randomisation, Type I errors occur, irrespective of the quality of the trial at any stage. The coexistence of a probabilistic law predicting that this possibility will be realised 5% of the time (where $\alpha = 0.05$) and the involvement of sampling variation as the explanation need not be challenged. Instead, unpredictable fluctuations at the sample level in the values of particular types of variables, such as those representing patient age, or indeed, measurement error, can be assumed to act in harmony with the above prediction independently of any machination of the experimenters. This harmony exists, irrespective of the apparent chaotic nature of the randomisation processes. Such behaviour is not, however, restricted to procedures involving hypothesis testing. Notice, for example, the bewilderment expressed by Blastland and Speigelhalter in the following account of dangerous events:

> Up close, life can appear chaotic. Every murder is unique and unpredictable, every fall and crash laced with infinite chance. Seen from above, people move in patterns with spooky regularity.
>
> This is the great puzzle about danger: that a million stories describe it, feelings inform it, and a million occasions conspire for or against every incident—and yet there are a relatively consistent number [53].

The undeniable appearance of order in apparent chaos can also be recognised in (a) the phenomenon of regression to the mean, according to which, for a variable ranging over values that cannot be predicted exactly, the greater the deviation of a value from its mean, the greater the probability that the next measurement for that variable will deviate less far, and (b) the Central Limit Theorem [25].

In all of these cases, given sufficiently many observations, there is a sense of prescribed order arising out of apparent disorder. It is indeed contentious to universally portray chance events as those that probabilistic laws must occasionally step aside for, as there is arguably a probabilistic law in place regarding the regularity of the occurrence of such events in randomised experiments, with the accuracy of this law increasing with the number of iterations (Table 1.2).

It is noteworthy that Leibovici does not reflect on the potential influence of chance on his study findings. As is implicit from observations made earlier in this chapter, including KLP 1.6, what may seem appealing in isolation (by virtue of a single study) may eventually turn out to be anomalous based on future studies. This does not relegate the original outcome to the realm of the accidental or render it cause-free.

Based on the above reflections, within the context of RCTs, the everyday idea of chance events as accidental and unplanned is a rather inaccurate one. The intention of the randomisation process is to design a fair comparison of one or more study outcomes across the intervention and control groups. Nevertheless, as we have noted, Type I errors can occur in this context, although not without a cause. As such, arguably, they constitute errors relative to expressing the truth about a population, not by virtue of an experimental accident. For well-designed RCTs, simple randomisation controls the probability of such errors asymptotically, as is illustrated in Table 1.2. In comparing case and control groups in an RCT, the usefulness of this property is illustrated as follows with respect to baseline comparisons:

> [F]or example based on the application of the independent samples t test, it is appropriate to ask, do the means differ significantly for individual variables more often than would be expected simply by chance—that is, more than 5 % of the time—and do these differences tend to involve p values which are markedly below the 5 % significance level? Clearly, such phenomena are at odds with what is essential to randomisation and as such, ought to arouse suspicion in terms of (a) the authenticity of the data reported to have been used in the study or (b) the credibility of the study as a bona fide RCT [54].

Key Learning Point 1.7

In simple randomisation processes, including random allocation of treatments, there is a regularity in the pattern of unexpected events that can be observed after sufficiently many iterations. This regularity, which governs the false positive rate, ensures that whatever the nature of the irregular causal pathways involved in these events, they are governed by a probabilistic law. In referring to such unexpected events as chance events, we should therefore challenge everyday notions of chance that appeal to the accidental and unintentional, and resist the temptation to justify use of these notions in terms of the seemingly chaotic nature of random processes. Within the context of NHST, it should not be assumed that a statistically significant difference across the treatment groups at baseline points to poor study design; rather, the prevalence, in combination with the strength, of statistical significance at baseline may serve as an indicator of malpractice in research design or conduct.

While more sophisticated approaches to random sampling, including stratified random sampling (see Chap. 2), can be helpful in generating more accurate effect size estimates, these are already well-covered from a methodological perspective in educational and research literature and are not, as such, the focus of this volume.

1.5 Conclusions

Neither the scientific reputation of the researchers nor that of their chosen journal should stand in the way of the need for accountability in approaches to reporting study outcomes and presenting statistical findings for an RCT. In this chapter, we have noted the importance of public access, not only to the protocol, but also an SAP, as best practice in guarding against selective reporting of outcomes, although this also applies to enhancing the integrity of trial conduct. It is also important to note, however, that adherence to these practices cannot guarantee rigorous *implementation* of RCT procedures. In relation to procedures assumed for random allocation of treatment, for example, it has been noted that, "Of the two essential components of unbiased allocation—genesis of an unbiased sequence, and unbiased implementation of the sequence—the former remains a trivially easy task, while the latter will continue to pose challenges" [55].

We have also considered the importance of choosing statistical hypothesis tests based on assumptions testing at the level of the individual variable to avoid bias, particularly with small sample sizes. Furthermore, we have reflected on the importance of better use of effect sizes, particularly in relation to non-parametric tests, where there is a clear need for better choice and interpretation of effect sizes for reporting of appropriate findings alongside p-values.

These statistical principles are not sufficient, however, as an adjunct to the above more general study principles in the pursuit of scientific truth. With the help of the stroke trial data, we have seen that simple randomisation is unable to fully eliminate false positives arising from (unpredictable) sampling variation. It should be understood, however that RCTs are not in this sense inferior to other study types. For well-designed RCTs, within the context of NHST, a false positive rate equal to that of α is expected purely on account of the nature of simple randomisation. Where this is seen as a design feature rather than a shortcoming, the researcher is better equipped to appreciate the importance of confirming the consistency of findings across multiple RCTs prior to deciding on the potential effectiveness of a health intervention.

It is my intention that these considerations, and in particular, the KLPs in this chapter, will prove supportive to medical educators in persuading trainee doctors, and other health professionals, that adoption of an RCT design, albeit as a gold standard, does not guarantee the reliability of the findings from a single study. Adherence to guidelines for study design together with trial registration and the publication of a protocol and an SAP are fundamental steps in ensuring the credibility of RCT findings. However, even here, the importance of consistency of findings across multiple trials needs to be recognised in the interpretation of RCTs and in the translation of these findings into clinical practice. Furthermore, the use of a rigorous approach to statistics in the reporting and in the critical appraisal of published RCTs remains an essential ingredient in progressing towards what can more justifiably be described as a gold standard in medical research. As is evident in the clinical literature there is a danger that non-specialists with an aspiration for using medical statistics in their own research may overlook the nuances in choice of effect sizes together with the limitations of a stand-a-lone 'achievement' of statistical significance in contributing to an evidence base for the effectiveness of an intervention.

A shortcoming of statistical software packages in current use by medical and other health professionals is the absence of such effect sizes in standard output. This lack of accessibility in identifying appropriate statistics for reporting purposes can and should be addressed through suitable prior training aimed at addressing the questions, What statistics are needed, other than the standard output from hypothesis testing, and why? The obvious need for such training or for collaboration with a statistical expert should not justify the omission of appropriate effect sizes and the interpretation thereof from the reporting of clinical research findings. As educators, we have a duty to drive this message home in the supervision and training of junior researchers. The importance for patient safety must not be overlooked, including in medicine, where the pressure to impress in a highly competitive work environment is particularly intense.

References

1. Leibovici L. Effects of remote, retroactive intercessory prayer on outcomes in patients with bloodstream infection: randomised controlled trial. BMJ. 2001;323:1450–1.
2. Eveleth R. The ethics of sarcastic science. The Atlantic [Internet]. 2014. https://www.theatlantic.com/technology/archive/2014/12/the-ethics-of-sarcastic-science/383988/.
3. Harris W, Gowda M, Kolb J, Strychacz C, Vacek J, Jones P. A randomized, controlled trial of the effects of remote, intercessory prayer on outcomes in patients admitted to the coronary care unit. Arch Intern Med. 1990;159:2273–8.
4. Byrd R. Positive therapeutic effects of intercessory prayer in a coronary care unit population. South Med J. 1988;81:826–9.
5. Astin J, Ernst E. The efficacy of "distant healing": a systematic review of randomized trials. Ann Intern Med. 2000;132:903–10.
6. Olshansky B, Dossey L. Retroactive prayer: a preposterous hypothesis? BMJ (Hist Mystery). 2003;327:1465–8.
7. Dummett MA. Symposium: can an effect precede its cause? (I). Proc Aristot Soc. 1954;Supplementary volume 28:27–62.
8. Dummett M. Bringing about the past. Philos Rev. 1964;73(3):338–59.
9. Flew A. Symposium: can an effect precede its cause? (II). Proc Arist Soc. 1954;Supplementary volume 28:27–62.
10. Ben-Yami H. The impossibility of backwards causation. Philos Q. 1950;57(228):439–55.
11. Black M. Why cannot an effect precede its cause? Analysis. 1954;16(3):49–58.
12. Agassi J. Causality and medicine. J Med Philos. 1976;1(4):301–17.
13. Howick J. The philosophy of evidence-based medicine. Wiley-Blackwell, BMJ Books; 2011.
14. Murad M, Asi N, Alsawas M, Alahdab F. New evidence pyramid. BMJ Evid-Based Med. 2016;21(4).
15. Association WM. World Medical Association declaration of Helsinki ethical principles for medical research involving human subjects. JAMA. 2013;310(20):2191–4.
16. Chan A-W, Tetzlaff JM, Altman DG, Laupacis A, Gøtzsche PC, Krleža-Jerić K, Hróbjartsson A, Mann H, Dickersin K, Berlin J, Doré C, Parulekar W, Summerskill W, Groves T, Schulz K, Sox H, Rockhold FW, Rennie D, Moher D. SPIRIT. Statement: defining standard protocol items for clinical trials. Ann Intern Med. 2013;2013(158):200–7.
17. Chan A-W, Tetzlaff J, Gøtzsche P, Altman D, Mann H, Berlin J, et al. SPIRIT 2013 explanation and elaboration: guidance for protocols of clinical trials. BMJ (Res Methods Rep). 2013;346:e7586.
18. Baker M. Is there a reproducibility crisis? Nature. 2016;533:452–4.
19. Loder E, Loder S, Cook S. Characteristics and publication fate of unregistered and retrospectively registered clinical trials submitted to The BMJ over 4 years. BMJ Open. 2018;8(e020037).
20. Gopal AD, Wallach JD, Aminawaung JA, Gonsalves G, Dal-Re R, Miller JE, et al. Adherence to the International Committee of Medical Journal Editors' (ICMJE) prospective registration policy and implications for outcome integrity: a cross-sectional analysis of trials published in high-impact specialty society journals. Trials. 2018;19(448).
21. Perlmutter A, Tran V-T, Dechartres A, Ravaud P. Statistical controversies in clinical research: comparison of primary outcomes in protocols, public clinical-trial registries and publications: the example of oncology trials. Ann Oncol. 2017;28(28):688–95.
22. ICH harmonised tripartite guideline: statistical principles for clinical trials E9. In: International conference on harmonisation of technical requirements for registration of pharmaceuticals for human use. London: European Medicines; 1998.
23. Gamble C, Krishan A, Stocken D, Lewis S, Juszczak E, Doré C, et al. Guidelines for the content of statistical analysis plans in clinical trials. JAMA. 2017;318(23):2337–43.
24. Sterne J, Savović J, Page M, Elbers R, Blencowe N, Boutron I, et al. A revised tool for assessing risk of bias in randomized trials (RoB 2). BMJ. (in press).
25. Petrie A, Sabin C. Medical statistics at a glance. 4th ed. West Sussex: Wiley; 2020.

26. Riffenburgh RH. Statistics in medicine. San Diego: Elsevier Science & Technology; 2012.
27. Healy M. Non-normal data. Arch Dis Child. 1994;70:158–63.
28. Gardner MJ, Altman DG. Confidence intervals rather than P values: estimation rather than hypothesis testing. BMJ. 1986;292:746–50.
29. Greenland S, Senn SJ, Rothman K, Carlin J, Poole C, Goodman SN, et al. Statistical tests, P values, confidence intervals, and power: a guide to misinterpretations. Eur J Epidemiol. 2016;31:337–50.
30. Pike H. Statistical significance should be abandoned, say scientists. BMJ. 2019;364(1374).
31. Wasserstein R, Schirm A, Lazar N. Moving to a world beyond "p < 0.05". Am Stat. 2019;73(SUP 1):1–19.
32. Fisher R. Statistical methods for research workers. Edinburgh: Oliver & Boyd; 1925.
33. Fisher R. The arrangement of field experiments. J Minist Agric. 1926;33:503–13.
34. Cowles M, Davis C. On the origin of the 0.05 level of statistical significance. Am Psychol. 1992;37(5):553–8.
35. Goodman S. A dirty dozen: twelve p-value misconceptions. Sem Haematol. 2008;45:135–40.
36. Rosenthal R. Meta-analytic procedures for social research. London: Sage Publications; 1984.
37. Cohen J. Statistical power for the behavioural sciences. 2nd ed. New York: Lauren Erlbaum Associates; 1988.
38. Grissom RJ, Kim JJ. Effect sizes for research: univariate and multivariate applications. 2nd ed. New York: Routledge; 2012.
39. Acion L, Peterson JJ, Temple S, Arndt A. Probabilistic index: an intuitive non-parametric approach to measuring the size of treatment effects. Stat Med. 2006;25:591–602.
40. Kerby DS. The simple difference formula: an approach to teaching nonparametric correlation. Compr Pyschol. 2014;3(1).
41. Gibbons J. Nonparametric statistical inference. 6th ed. New York and Basel: Routledge and CRC Press; 2020.
42. Colquhoun D. The false positive risk: a proposal concerning what to do about p-values. Am Stat. 2018;73:192–201.
43. Sandercock PA, Niewada M, Czlonkowska A. The International Stroke Trial database. Trials. 2001;12(101).
44. International Stroke Trial Collaborative Group. The International Stroke Trial (IST): a randomised trial of aspirin, subcutaneous heparin, both, or neither among 19435 patients with acute ischaemic stroke. Lancet. 1997;349(9065):1569–81.
45. Murtaugh PA. In defense of P values. Ecology. 2014;95(3):611–7.
46. Paulo S. Should we retire statistical significance? Braz J Pain. 2019;2(3):211–2.
47. Alberoni F. Contribution to the study of subjective probability. I. J Gen Psychol. 1962;66(2):241–64.
48. Hacking I. The doctrine of necessity. The taming of chance. Cambridge: Cambridge University Press; 1990. p. 11–4.
49. Hacking I. The taming of chance. Cambridge: Cambridge University Press; 1990.
50. Pierce C. The doctrine of necessity examined. Monist. 1892;2:321–37.
51. Parkinnen V-P, Wallmann C, Wilde M, Clarke B, Illari P, Kelly MP, et al. An introduction to mechanisms. In: Valleriani M, editor., et al., Evaluating evidence of mechanisms in medicine: principles and procedures. Switzerland: Springer Nature; 2018. p. 11–21.
52. Puskar J. Detecting "absolute chance": Charles Peirce and Anna Katharine Green. In: Puskar J, editor. Accident society: fiction, collectivity, and the production of chance. Stanford, CA: Stanford Scholarship Online; 2013. p. 108–47.
53. Blastland M, Speigelhalter D. The norm chronicles: stories and numbers about danger. Suffolk: Profile Books Ltd.; 2013.
54. MacDougall M. Assessing the integrity of clinical data: when is statistical evidence too good to be true? Topoi. 2014;33(2):323–37.
55. Chalmers I. Why the 1948 MRC trial of streptomycin used treatment allocation based on random numbers. J R Soc Med. 2011;104(9):383–6.

Margaret MacDougall is a senior lecturer in medical statistics and researcher in education at the Usher Institute, Edinburgh Medical School. She developed an interest in philosophy early on in her career through her Ph.D. research in mathematical logic, constructive set theory and the philosophy of mathematics. She has conceived and led eleven funded projects in medical education and statistics and supervised research staff and students at undergraduate and postgraduate levels through various spin-off projects. Margaret seeks to use her research to promote curriculum innovation and enhance student learning in statistics at undergraduate and postgraduate levels and has published on an international scale in the fields of education, medicine, philosophy and statistics. She has served as an invited speaker, chair and conference organiser at various local, national and international events relating to statistics, and learning and teaching in higher education. Currently, she serves as Vice Chair of the Teaching Statistics section of the Royal Statistical Society.

Chapter 2
Answering Criticisms of RCTs

Gillian Raab

Abstract In recent years, there have been several papers published, mainly by philosophers, criticising randomised controlled trials (RCTs). These include criticisms of their logic and structure as well as of individual published studies. In order to address these criticisms, this chapter reviews the history of RCTs and how their methodology has developed in recent years and illustrates aspects of their design and analysis with real examples. Against this background, it examines the three main perspectives from which philosophers criticise RCTs: objections to randomisation, promotion of observational studies and objections to lack of generalisability. It is argued that some of the objections to randomisation do not withstand scrutiny. The claim that RCTs can be replaced by observational studies is also disputed, while it is acknowledged that observational studies can contribute to treatment decisions. The criticism that the results of RCTs do not always transfer to other settings is one that no one could disagree with. Recommendations on how the results of trials should be registered and reported need to be followed to help decisions as to whether the result could work elsewhere.

2.1 Introduction

Section 2.2 of this chapter reviews the history of randomised controlled trials (RCTs) and their logic and structure. This section also reviews methods of avoiding bias in assessing the outcomes for patients in RCTs. The term *bias* here refers to the situation when the outcomes assessed in a trial differ from those that would be obtained if the treatment were used in normal clinical practice. Section 2.3 uses examples from real trials to illustrate how biases of this sort can arise and how the design of trials can attempt to minimise them. To provide evidence for how to treat future patients, findings must be made available to inform others, usually via medical journals. This

G. Raab (✉)
Administrative Data Research Centre – Scotland, Edinburgh, UK
e-mail: gillian.Raab@ed.ac.uk

M. MacDougall (ed.), *A Medical Educator's Guide to Thinking Critically about Randomised Controlled Trials: Deconstructing the "Gold Standard"*,
https://doi.org/10.1007/978-3-031-25859-6_2

can be another source of bias if results are reported selectively. Continuing from Chap. 1, reporting bias is discussed briefly in Sect. 2.4 with an outline of national and international initiatives to ensure fair reporting of the results of RCTs. Section 2.5 then examines the criticisms that have been levelled at RCTs, most commonly by philosophers. This provides a useful basis in Sect. 2.6 for reviewing evidence in recent years for established RCT results having been used for the benefit of the health of populations in developed countries.

2.2 History and Structure of RCTs

2.2.1 History of RCTs

A clinical trial is an experimental study using human subjects to compare the outcomes of treating the subjects in different ways; the purpose is thus to inform treatment decisions for future patients. James Lind's experiment to investigate treatments for scurvy is often considered to be the first clinical trial, although the James Lind Library[1] traces the history of clinical trials to a much earlier time. In 1753 [1], Lind reported treating twelve seamen with scurvy by allocating to them uniquely in pairs one of the six treatments:

- 2 L of cider;
- 25 drops of elixir vitriol (dilute sulphuric acid) three times per day;
- 18 ml of vinegar three times throughout the day before meals;
- half a pint of sea water;
- two oranges and one lemon continued for six days only (when the supply was exhausted);
  and
- a medicinal paste made up of garlic, mustard seed, dried radish root and gum myrrh.

He in turn reported the following results from this experiment:

> The most sudden and visible good effects were perceived from the use of oranges and lemons; one of those who had taken them being at the end of six days fit for duty... The other was the best recovered of any in his condition; and being now deemed pretty well, was appointed nurse to the rest of the sick.

As far as we can judge, Lind's trial was controlled in that all twelve seamen were treated in a similar setting, but it seems unlikely that it was randomised, since we have no evidence about how the treatments were allocated. The first RCT is generally agreed to be the Medical Research Council (MRC) 1948 trial of streptomycin for the treatment of tuberculosis. We read in the report of the study [2]:

[1] This library comprises a website illustrating the development of fair tests of treatments in health care https://www.jameslindlibrary.org/, accessed 2 December 2023.

> Determination of whether a patient would be treated by streptomycin and bed-rest ... or by bed-rest alone ... was made by reference to a statistical series based on random sampling numbers drawn up for each sex at each centre by Professor Bradford Hill; the details of the series were unknown to any of the investigators or to the co-ordinator and were contained in a set of sealed envelopes, each bearing on the outside only the name of the hospital and a number.

The method of sealed envelopes has been used to allocate treatments in thousands of RCTs up to the present. Before the first patient is entered, a list of the treatment allocations for all future patients is prepared. Historically, this would have been done from a table of random numbers, while now a computer-generated series of random numbers is used. If there are to be just two treatments, then the odd random numbers could be assigned to one treatment and even ones to another. Cards with the patient number and the assigned treatment are prepared and inserted into envelopes. The outside of each envelope has just the patient number and, usually some details of the trial as well as space for details of the patient to be entered once the envelope is opened. The envelopes, which must be fully opaque, are then securely sealed and delivered to the person entering the patients into the trial. The envelope is not opened until the patient has been assessed for eligibility for the trial and has agreed to take part with the understanding that they may receive either treatment.

The MRC study was the first RCT, but by no means the first controlled trial. Before the MRC study, most trials used alternate allocation of treatments in the order in which the patients presented [3]. With this method, those caring for the patient know what treatment the next patient will receive and this can influence the decision as to whether to enter the patient into the study. For example, a doctor entering patients where one of the treatments involves a complex regime might decide not to enter a patient into a trial who might not be able to follow it, whereas they would have been entered into a trial with a simpler regime. Such selection bias—henceforth, entry bias—introduces differences between patient groups that are not a consequence of their original treatment allocations. If each sealed envelope remains unopened until the patient's details have been written on the outside, then the patient's entry into the trial is recorded without prior knowledge of the treatment allocation and this defines comparable randomised groups.

With larger, multicentre, trials, patients may be entered online or by telephoning the trial centre. The treatment allocation is not given until patient details have been entered. In this case, the list of treatment assignments is held securely at the trial coordinating centre, with no access made available to staff entering the patients. An approach to addressing a more subtle variation of entry bias is discussed in Sect. 2.2.2 for the design of the BASIL trial [4]. Entry bias is represented under the more general header of "Selection bias" in the Cochrane Collaboration's tool for assessing trials [5]. We shall consider three of the other specified forms of bias from this tool—detection bias, performance bias and attrition bias—at Key Learning Points (KLPs) 2.5, 2.6, 2.7 and 2.9. The remaining specified form of bias—reporting bias—involving the selective reporting of outcomes, was considered in Chap. 1.

Key Learning Point 2.1 A special type of selection bias, known as entry bias, can occur when the decision to enter a patient into a trial is made based on knowledge of the treatment they will receive.

RCTs are now an important part of medical research. In 1962, the Food and Drug Administration in the USA made evidence from clinical trials mandatory for drug regulation, and other countries soon followed. The second half of the twentieth century saw an exponential increase in the number of RCT reports published. The number of trials has continued to increase through the twenty-first century[2] with over 60 thousand trials registering on the World Health Organization Trials Registry in 2018. The 1990s saw the birth of the movement for Evidence-Based Medicine (EBM) that aims to educate clinicians in using research results to inform their practice. EBM prioritises conclusions from RCTs as a source of evidence, but not to the exclusion of other types of research. Methods and practice for the design and conduct of trials are now well developed. Shortly, we shall consider the key concepts. The first step, however, is to define what we hope to learn from a clinical trial.

2.2.2 What Do We Want to Learn from a Clinical Trial?

In the broadest sense, the aim of a clinical trial must be to discover the best treatments for future patients, but there is more than one way to achieve this. Schwartz, Flamant and Lellouch [6, 7] distinguish between explanatory aims and pragmatic aims for clinical trials. An explanatory aim seeks to understand whether the treatments could work and how. It is akin to a laboratory experiment in which it is extremely important to control other factors that might affect the outcome of the trial. In contrast, a pragmatic aim seeks to answer the question, which of these treatments will provide the best outcomes for future patients? An explanatory trial measures *efficacy*, the benefit of one treatment compared to another in ideal, controlled conditions. In contrast, a pragmatic trial measures *effectiveness*, the benefit of one treatment compared to another for all eligible patients.

This can be illustrated for drug treatments for hypertension. An explanatory aim poses the questions, what is the difference between the fall in blood-pressure produced by drugs A and B, and what is the mechanism by which it is achieved? To answer these questions, a trial would select subjects who would be expected to follow the dosage regime carefully, often excluding those with diseases, other than hypertension, that might cause changes in blood pressure unrelated to the treatments. The fall in blood pressure produced by each of the treatments would be the outcome to be compared across the two corresponding treatment groups. The subjects would usually

[2] International Clinical Trials Registry Platform https://www.who.int/clinical-trials-registry-platform/ accessed 2 December 2023.

undergo extensive pharmacokinetic tests to understand the mechanisms involved. In contrast, patients for a pragmatic trial would be those presenting with hypertension and recommended for drug treatment. Appropriate outcomes for comparison across groups would be their subsequent health and survival. Drug A might produce a greater fall in blood pressure than drug B in the explanatory trial. If a proportion of patients could not tolerate drug A and then failed to take the medication, those on drug B might have better health and survival. Here efficacy has not led to effectiveness. Effectiveness is what is needed to inform future treatment policies.

Key Learning Point 2.2 We can distinguish between exploratory and pragmatic aspects of a clinical trial. The former look at efficacy through comparing the benefits of treatments in ideal conditions and the mechanisms by which they are achieved. The latter measure effectiveness through comparing treatment outcomes for patients in real clinical situations.

Trials are seldom completely distinguishable as explanatory or pragmatic in the sense represented in KLP 2.2. In practice, pragmatic trials will usually carry out more detailed monitoring of patients than is usual in routine care and the test results may be used to answer explanatory questions. Longer follow-up of explanatory trials may assess effectiveness. The Cochrane Collaboration have developed a tool [8] to assess trials on the continuum between exploratory and pragmatic attitudes. They argue that all trials lie somewhere along this continuum. Exceptions are first-in-human studies. These studies are typically carried out with healthy volunteers rather than patients and as the name suggests, involve the administration of an intervention to humans for the first time. These are purely explanatory and can have practical and ethical problems, as highlighted by a recent case when volunteers became severely ill. This is discussed in a Royal Statistical Society report [9]. This chapter will exclude this type of study. It will only discuss trials where participants could benefit from the treatment(s) and it will focus largely on how their pragmatic aims can be met.

2.2.3 *The Structure of a Clinical Trial*

Figure 2.1 represents the two-armed, double-blinded, parallel group, individually randomised controlled trial introduced in Chap. 1. Here, A and B are the available treatments. Therefore, by virtue of the design, patients are randomly assigned to treatment A or B and as such assigned to one of two separate groups corresponding to each of these two treatments. However, the principles apply to other, more complex, designs, including those involving more than two treatments. The left-hand panel background represents subjects who could be offered the treatments in the trial. The backgrounds of the other two panels represent future patients whose treatment may be influenced by the results of the trial. The background for future subjects is deliberately

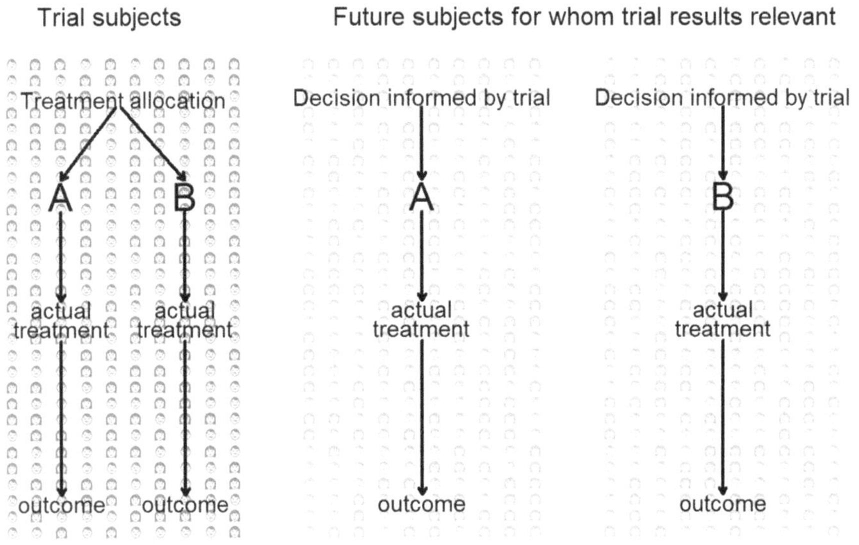

Fig. 2.1 Illustration of the structure of a two-treatment, parallel-group trial with the aim of informing a policy of offering treatment A or B to future patients

vague as their circumstances will often be uncertain when a trial is being planned and even after it has been completed. External validity, the generalisability of the trial results to other settings outside the trial, typically involving alternative patients, is illustrated by these two panels, as they represent how the treatment decisions would influence outcomes in such settings. They are shown separately to emphasise that the treatments are being used outside the clinical trial setting.

Note that it is the allocation of treatments rather than the actual treatment received that defines the two treatment groups. The management of the patients in the two groups should be similar, except for differences that are a consequence of the treatment allocation. The assessment of the outcome should be carried out in a similar way in both groups.

The details of how these procedures are carried out will determine whether the trial will give an unbiased answer to the question, *How do these treatments compare for the group of patients from whom the participants were selected?* This question asks about the internal validity of the study in contrast to its external validity, for which the corresponding question is, *Will the findings from this study apply to patients offered similar treatments in different settings?* [10].

Key Learning Point 2.3 The internal validity of a study is the extent to which it is an unbiased comparison of the treatments for the subjects recruited to the study. In contrast, external validity refers to whether the conclusions reached by the study will apply to future patients offered the same treatments in different settings.

Usually, trials allocate individuals to only one of two or more treatments.[3] How, then, can we determine internal validity? We need to introduce the What if? question concerning the counterfactual outcome (henceforth, counterfactual)—the outcome for a patient if they had received a different treatment (see also Chap. 5). We can never observe the counterfactual in parallel group trials, so we must rely on drawing conclusions by comparing treatment groups. The results of a trial are the product of the comparison of specified outcomes across the treatment groups. Statistical inference methods define the conclusions that can be drawn from such data about the distribution of outcomes that would be expected for similar patients experiencing each treatment. There are two main approaches to statistical inference: frequentist and Bayesian. Frequentist approaches were developed at the start of the twentieth century by RA Fisher and his colleagues within the context of agricultural experiments. Their use in RCTs is discussed in some detail by Pocock [11]. Bayesian approaches are more commonly used for predictions, such as diagnosing disease types, as well as in meta-analyses that attempt to combine results from many different trials. While Bayesian approaches are progressively becoming more popular in the design and conduct of clinical trials, frequentist approaches remain the dominant paradigm in inferential statistics, including within the context of clinical research. Although a single RCT can be analysed by Bayesian methods, this is more difficult, as we shall see below.

From a frequentist approach, at the end of the trial, a test for evidence of a treatment difference is carried out, typically by testing a null hypothesis as defined in Chap. 1. The null hypothesis, H_0, can be expressed in terms of one or more parameters being identical across treatment groups, where these parameters pertain to previously chosen outcome measures.

Bayesian methods provide an alternative approach to analysing data from clinical trials. They are less commonly used than frequentist methods in analysing a single trial, although they have become more common in recent years [12]. They answer the question we would like to answer, namely, what is the probability that treatment A is better than treatment B? But that question cannot be answered by the data from the trial alone. A Bayesian analysis must start with the initial belief as to the relative effectiveness of the two treatments, the prior probability. The results from the trial can then be used to update this belief, using a quantity called the likelihood, calculated from the trial data. A Bayesian analysis requires a more detailed specification than a frequentist one, including probabilistic models to describe the outcome in the treatment groups and a specification of prior belief for their parameters.

As noted in Chap. 1, for a well-designed RCT, the false positive rate is defined by the value of α that we assume, while the false negative rate is controlled by the level

[3] The exceptions are treatments that can be applied to more than one part of a person (e.g., skin creams) or treatments that can be given sequentially in cross-over trials.

of statistical power. At the design stage of an RCT, the level of statistical power can be decided on, conventionally, as 0.80 or 0.90, and typically, the minimum required sample size to achieve this level is estimated by means of a sample size calculation. This then determines the numbers to be recruited into the different arms of a trial to provide statistical evidence for a specified difference between the treatments[4] if that difference exists at the population level. Details of the numbers of subjects to be accrued before the results are analysed must be specified in the trial protocol.

Trial results may be used to inform treatment decisions in many different settings. In Sect. 2.3.1, we will consider an example of where similar trials in different settings gave different results and how such differences may be reconciled.

2.3 Examples from Real Trials

In this section, we will consider examples illustrating how aspects of the design of trials can influence internal and external validity. External validity is discussed first, because informing policies for future patients is the ultimate goal of all trials and should determine how a pragmatic trial is conducted.

2.3.1 External Validity—Generalisability

Generalisability refers to whether the results from a clinical trial will apply to future patients receiving the same treatments as the trial subjects. In terms of Fig. 2.1, this is whether the outcomes from the two arms of the left-hand panel could be used to inform the best choice of the two treatments for future patients. All the aspects of a trial, including the selection of subjects, how treatments are allocated and how the outcome is measured, as discussed below, will contribute to its level of generalisability. Judgements about a trial's generalisability are made by those treating future patients and considering whether the results will apply to them. The outcome of a treatment always depends on the patients who received the treatment and the setting in which they were treated.

Two trials set in sub-Saharan Africa illustrate this. Both aimed to answer the question, Should an HIV-positive mother breast-feed? The treatments, breast-feeding versus formula-feeding, and the choice of measurement outcomes, HIV infection and survival of the infants, were the same in both trials but the results were very different. The authors of the first trial, set in Nairobi, Kenya [13] conclude, "The use of breast milk substitutes … was associated with significantly improved HIV-1-free survival."

[4] More complex designs, which are beyond the scope of this book, can allow the data to be checked sequentially during the trial for a difference in treatment outcomes while maintaining the overall levels of statistical significance and power.

However, the authors of the second trial, set in towns and villages in Botswana [14], report that "Both strategies had comparable HIV-free survival at 18 months."

The reports of the trials each give extensive details of the subjects, and of the settings and the follow-up of the infants. Two factors that could explain the differences emerge from a close reading of the trial reports and an understanding of factors that influence HIV infection and survival in such infants. In the Kenyan study, patients did not have access to anti-retroviral treatment (ART), which has been shown to reduce the transmission of HIV infection. In the second trial, mothers and infants had access to ART,[5] so the benefits of avoiding transmission by using formula feed would be less. It is likely that a second difference between the trials was even more important to the survival of infants. The first study was set in an urban area, where the mothers had access to clean water. This was not the case in the second study. Therefore, the formula-fed infants may have been at greater risk from other infections. The survival advantage of avoiding HIV transmission from breast milk would be negated by other infections from the water used to prepare the formula. This example illustrates the importance of context in generalising the results of clinical trials, and thus of detailed reporting of the design of trials. To inform decisions on how to advise mothers as to whether they should breast-feed their infants, the reports from many trials need to be reviewed. The output of one such review is the guidelines produced by the World Health Organization [15] for infant feeding policies in those poor countries where HIV is endemic.

Key Learning Point 2.4 Reports of trials must contain details of the treatments being compared, any other aspects of how the patients are managed and the setting where the treatments are compared. This allows readers to assess the external validity of the results for future patients.

The setting is not the only factor that can threaten the external validity of trial results. For example, if only a small proportion of possible subjects take part, the participating subjects may be particularly unrepresentative of patients who are offered the treatments in future. The CONSORT guidelines on the reporting of clinical trials [16] require details of the recruitment and retention of patients. In the Kenyan infant-feeding trial, 41% of mothers who tested HIV positive and were eligible for the trial were allocated to feeding regimes [13]. In Botswana [14], the percentage was even lower, with the main reason for not entering the trial being an unwillingness to consider using formula feeding. Low recruitment rates are also a feature of trials in the UK [17]. Those excluded because they did not meet eligibility criteria will not influence external validity when generalising to settings where future patients would be judged to be ineligible in a similar way. But trial results could be used to inform choices for future patients who would not have met the eligibility criteria. Clinical trials, especially those run by the pharmaceutical industry, are often criticised for

[5] This is simplified somewhat here since the trial design was changed part-way through recruitment when information on the benefits of ART in preventing transmission became available.

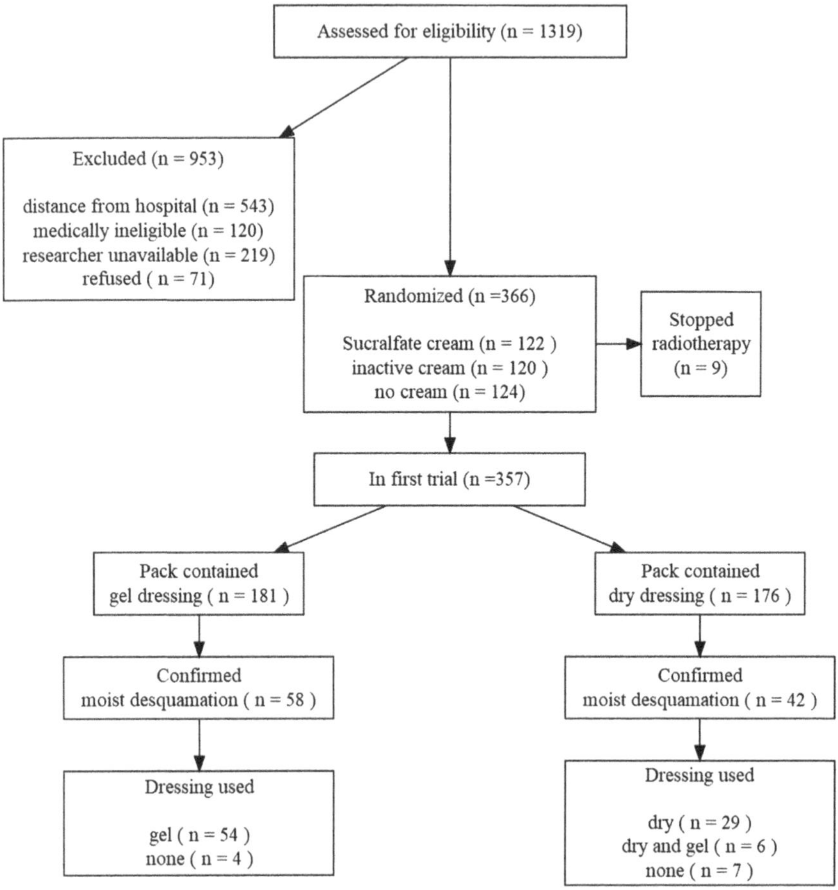

Fig. 2.2 CONSORT diagram illustrating patient numbers in the trial of skin reactions [18, 19]

using patients selected in a way that would make them different from future patients, for example in their willingness to comply with their treatments allocations.

To allow assessment of external validity, details of the numbers of patients considered for eligibility and the reasons they were excluded should be reported, as is illustrated in Fig. 2.2 through a flow diagram for a trial of treatments for skin reactions.

2.3.2 *Internal Validity: Subjects and Treatment Allocation*

The aim of a pragmatic trial is to inform a decision about how to treat a subject. The treatments should be as close as possible to what a subject would receive outside a trial and the subjects should comprise individuals for whom the clinical decision would be relevant. The BASIL trial [4] investigated two treatments for severe ischaemia of the leg, a condition where obstruction of the arteries causes pain and ulcers and may lead to amputation. The two treatments were a surgical bypass of the artery and an angioplasty that opens the artery by inserting and inflating a balloon. The clinical decision investigated in the BASIL trial was which of these two treatments to attempt first on these patients. Patients were considered for the trial if their physicians thought they could be eligible for either of these treatments, so that patients thought to be too frail to sustain surgery or unsuitable for angioplasty were excluded from the trial.

Subjects were allocated to treatments using sealed envelopes, with the assigned treatment from random number sequencing opened only after the patient consented to take part and was registered as a subject in the trial. Registration ensured that a record was kept of all randomised subjects. Separate sets of sealed envelopes were used for each of the 27 UK hospitals contributing patients to the trial.

A sequence of truly random numbers can include subsequences of consecutive numbers pertaining to the same treatment allocation. This represents local imbalance in the probabilities of individual treatment allocations within the general sequence. A possible solution is to create a sequence of treatment blocks of size 4 where every block contains two of each type of treatment but with the order within the block generated at random, typically through computerised randomisation. But if this is done, staff entering patients, knowing the size of the block, are able to predict the treatment at least for every fourth patient, which could lead to entry bias. To ensure balance in treatment groups over time, while avoiding entry bias, the size of treatment blocks can be varied. In the BASIL trial, a mixture of treatment blocks of size 4 and of size 6 were selected in a random order. This method known as 'randomly permuted blocks' ensures a balance in treatment allocations, while preventing entry bias because the person entering the patients does not know whether they are observing a block of size 4 or one of size 6 and is therefore unable to predict patient treatment allocations.[6]

Important predictors of the outcome are often well understood and known when treatments are being allocated. This information can be recorded at the point of entering patients. The trial design may stratify the subjects into groups (strata) based on, for example, disease severity, and use stratified randomisation. To achieve this, a separate list of treatment allocations is randomly generated for each stratum of patients and, at entry, a given patient's treatment allocation is selected from the appropriate list. The above study design ensures a balance in treatment allocations across strata. The stratification factors are included as covariates in a statistical analysis (see Table 5.3, Chap. 5). The overall results from comparing both treatment groups are then equivalent to those obtained from averaging the different differences found in each stratum.

---

[6] Even with this algorithm, rare cases can occur where the next treatment would be determined by someone observing the sequence with a knowledge of how it had been created. For example, if the first three treatment allocations pertained to the same treatment, then it would reveal that the first block was of size six and the next three treatment allocations would pertain to the other treatment. It is normal practice to check a sequence of allocations before using it and, if necessary, select a different one.

When data have been collected at baseline, that is, prior to randomisation, but not used to balance the randomisation by stratification, the variables ranging over these data—the baseline variables—can still be used for covariate adjustment. The above lack of balance will nevertheless tend to reduce the benefits of covariate adjustment, especially for small trials with many covariates, where large imbalances can arise. Baseline variables can also be included in a statistical model to investigate whether the effect of treatment differs in different subgroups of patients. This type of relationship, also discussed later in Chap. 5, is sometimes referred to as a treatment-covariate interaction. It is important that the intention to test any such interactions are specified before the analysis (see Sect. 2.4) to avoid finding a spurious misleading result based on testing interactions with many possible factors.

In the BASIL trial, patients were recorded at entry as being in one of four groups, based on the severity of their disease and the treatment allocations were stratified by severity of disease within each centre. The analyses also took into consideration a range of factors measured at entry, including age, sex, smoking status, disease severity and other clinical measures, through covariate adjustment.

The actual treatments received by the patients in the BASIL trial were often different from their randomised allocation. A few patients received no treatment or the opposite treatment to that assigned and substantial numbers had two or more procedures carried out in their first year after entry into the trial. The policy evaluated in the trial pertained to which treatment should be attempted first. For some patients in the trial, as well as for some future patients, the attempted procedure would prove impossible. For example, a bypass is not possible if a suitable vein cannot be found. The analysis correctly compared the groups according to their intended treatment as defined by their random allocation (intention-to-treat (ITT) analysis), rather than the treatment received (treatment received analysis), since excluding patients who did not receive treatments as originally allocated could have compromised the existing level of balance in disease severity between the groups.

In a pragmatic trial, the outcomes must be those that would be relevant to the subjects, and the way they are measured should not differ by treatment group. When outcome variables are recorded by someone who knows the treatment allocations or treatments received, the same outcome can be reported differently for different treatments [20]. The Cochrane collaboration term this ‘detection bias’. Detection bias can be avoided if the patient is assessed by someone unaware of the patient’s treatment allocation or treatment received. During the trial, patients must not know which treatment they are receiving (a single-blind study) and no member of the trial team capable of influencing the results should know which treatment the patients are receiving or have received (a double-blind study). If one arm of a trial is the absence of treatment, a placebo (a treatment resembling the treatment of interest, but containing no active substance), often a pill with no active ingredient, may be used to make the study blind.

Key Learning Point 2.5 Within the context of an ITT analysis or treatment received analysis, detection bias occurs when, based on their awareness of treatment allocations or treatments received, the person assessing treatment outcomes reports the same outcome differently. It can be avoided if the assessor is ignorant of the treatment allocation or treatment received (blinded assessment).

Another reason for making a trial single-blind or double-blind is that a knowledge of the allocated treatment, rather than the treatment itself, may influence how staff or patients behave—an influence known as performance bias. For example, patients who know they have been assigned to an experimental drug may be more likely to report side effects, and staff may be more likely to identify them. In contrast, the expectation that a new treatment will be successful may influence staff and patients to report favourable outcomes. Performance bias is especially likely in trials where the outcome depends on patient perceptions, such as in the case of pain or psychological mood. There is ample evidence for changes in outcome produced by placebo treatments [21]. A variety of mechanisms, both psychological and physiological, can be involved, depending on the setting. Detection and performance bias are more likely when the outcomes are subjective, compared to what are often described as 'hard endpoints' such as reoperation or death, In the infant feeding study and the BASIL trial, it would have been impossible for subjects, and difficult for the medical team, to be blinded to the treatments being offered, but both trials had hard (non-subjective) outcomes, so these biases were unlikely.

Key Learning Point 2.6 Performance bias occurs when the staff caring for patients, or the patients themselves, act differently or feel different because of knowledge of the treatment that has been allocated or received.

A reported trial of treatments for skin reactions [18, 19] caused by radiation therapy for cancer illustrates the practical difficulties of blinding. Two research questions were addressed. The first was how to treat the skin from the start of radiotherapy to avoid skin damage from the radiotherapy, including broken and weeping skin (technically known as 'moist desquamation'). The second question was how best to treat this moist desquamation when it occurred. The trial comprised two sub-trials, one nested within the other. The first sub-trial involved three treatment groups, separated according to the following treatments: application of a cream with an active ingredient that was thought to reduce skin problems, application of a cream made up to appear identical to the first but with no active ingredient and, finally, non-use of any cream.

The outcomes for the first sub-trial required patient and researcher assessments of skin damage—clearly subjective, 'soft' outcomes—including the appearance of moist desquamation. In addressing the exploratory aim of learning whether sucralfate cream or aqueous cream benefitted healing, only the comparisons between the two cream groups were blinded. Staff training in the assessment of skin reactions due to radiation and a standardised diary format for patients were used to attempt to overcome bias in the unblinded comparisons. The appearance of moist desquamation was judged initially by the patients themselves because they needed to use their allocated dressings to treat the condition as soon as it appeared.

The second sub-trial compared the healing of moist desquamation with a gel dressing versus a dry dressing. Treatment assignments for both trials were made at the start of the patient's radiotherapy and after the initial skin assessment was carried out. All patients were issued with a pack including the dressings assigned to them for the second trial. Advice on when to start using the dressings was provided for all patients, although, the decision on this was their own and they recorded the actual time as part of the study. Moist desquamation was confirmed by medical staff at the next clinic visit. All patients were issued with a pack to use if they experienced moist desquamation along with instructions on how to use the enclosed dressings. The allocation for this sub-trial was made for all patients who had been balanced within the arms of the first sub-trial. But only 100 of the 366 patients entered the second trial. The second trial was vulnerable to *entry bias* (KLP 2.1), since the patients decided when their reaction was severe enough to need a dressing and it proved impossible to make up the dressing packs without revealing their contents. The main outcome for the second sub-trial was time to healing assessed at clinic visits by a staff member unaware, as far as possible, of the treatment assignment.

The lack of blind assessment in the first sub-trial might have biased the comparison between the two cream groups and the no-cream group, possibly by giving higher scores to the cream groups if patients believed they were effective. However, the results showed that all three groups had very similar outcomes with no evidence of differences in results of the magnitude assumed for the sample size calculations at the stage of study design.

The lack of blinding in the second sub-trial might have influenced some of the outcome assessments and caused entry bias. This sub-trial compared the 58 with moist desquamation assigned to hydrogel dressings with the 42 with moist desquamation assigned to dry dressings. The imbalance in these numbers might be an indication of entry bias due to patients being more likely to decide they had moist desquamation if they knew that had a gel dressing, perhaps because they believed that it would be effective. Moist desquamation was confirmed by medical staff at the next clinic visit.

Figure 2.2 shows the numbers of patients in the second sub-trial [19], presented as recommended in the CONSORT guidelines [16]. A higher proportion—58/181 (32%)—of those assigned to the hydrogel dressings group were confirmed to have moist desquamation, compared to 42/176 (24%) of those assigned to the dry dressings group. This is inconsistent with the idea of there being entry bias in the above sense. On the other hand, noting that the sub-trial was unblinded, it could be argued that the above discrepancy across groups suggests that enthusiasm for the "novel" hydrogel treatment may have led to a bias towards identifying moist desquamation, due to believing the gel dressings to be effective [19], including through the influence of *performance bias* (see KLP 2.6). If present, the latter form of bias would have led to bias in the study findings in favour of the efficacy of the hydrogel treatment. Patient enthusiasm for the hydrogel treatment may also have contributed to this source of performance bias. Indeed, the idea of patient preference for the hydrogel treatment is consistent with the observed higher rate of patient adherence to this treatment allocation (see bottom two boxes of Fig. 2.2). On the other hand, the absence of blinding may also have led to entry bias (see KLP 2.1), with patients with more severe wounds being assigned to the dry dressing group on clinical grounds. Arguably, these forms of bias could have led to a lesser likelihood of healing in the dry dressing group. However, In an ITT analysis, the reported hazard ratio (HR) for healing versus non-healing for the 58 randomised to gel dressings versus the 42 randomised to dry dressings favoured the latter treatment (HR: 0.64; 95% CI: (0.42, 0.99)). It seems evident, therefore, that neither *performance bias* nor *entry bias* had influenced the conclusion that dry dressings should be used for future patients.

Key Learning Point 2.7 When outcome scoring without the knowledge of the treatment is impossible, bespoke training in assessment of outcome measures and, if possible, use of objective outcome measures, can help to mitigate levels of detection bias.

Ethical issues are often cited as problematic in blinded or placebo-controlled trials. For treatment assignment to be ethical, there must be uncertainty as to which treatment would be best for the subject. There are examples of placebo-controlled trials where drugs, thought to be beneficial, have been shown to be useless or harmful, in some cases even causing serious harm. An example of the latter is the CAST trial of a drug that it was hoped would improve survival of patients with arrythmias [22]. This trial found an increased death rate for the treated patients. Other forms of blinded treatments, such as sham procedures, may be more problematic and will rarely get ethical approval now. A sham procedure is one where the patient, and sometimes other personnel, have been led to believe, from what they have observed, that a particular procedure has been given when it has not.

Let's consider an interesting historical example of a trial involving a sham procedure. This trial is a small explanatory trial of mammary artery ligation [23], a surgical procedure that was thought to reduce angina—a subjective outcome—by reducing the blood flow through the cardiac artery. The trial, involving 18 patients, compared those with surgery to a group that underwent sham surgery—an anesthetic and a skin incision—with the real procedure. It was able to provide evidence to support the case that the litigation procedure did not reduce blood flow, improve cardiac function or reduce the pain of angina, thus sparing future patients useless surgery.

Key Learning Point 2.8 Within the context of RCTs, new treatments previously thought to hold promise of better outcomes have sometimes been found to be useless or harmful.

The role of placebos and blinding in clinical trials is still the subject of much debate (see, for example, Chaps. 6–8 of Howick's book [24]). It could even be argued that placebos are never appropriate for trials with pragmatic aims, since they do not represent a treatment that would ever be recommended for future patients. The *WMA Declaration of Helsinki—Ethical Principles for Medical Research Involving Human Subjects* initially stated that placebos could only be used "where no proven prophylactic, diagnostic or therapeutic method exists". However, a revision in 2000[7] included additional circumstances where the decision to use a placebo might be reached, but with the caution that this should be with extreme care. In a drug trial with a placebo control, all subjects might be subjected to medical checks to ensure that those in the treatment group were not suffering from possible side-effects of the drug. These checks in themselves might prove beneficial specifically for patients in the treatment group. However, excluding these checks from the placebo group, without breaking the blinding, could prove difficult and be considered unethical. In planning such a trial, or any trial, it is important to define exactly which aspects of treatments are being evaluated.

2.4 Reporting of Clinical Trials

To inform future treatments, clinical trials need to be reported in a manner that will allow critical evaluation of their conclusions. The CONSORT Statement [16] spells out guidelines for this. Detailed accounting of all patients and their outcomes is part of their recommendations. Figure 2.2 is an example of what is often called a *CONSORT diagram* and shows that there were few withdrawals in the first sub-trial of this skin treatment trial, and that treatment allocations were evenly balanced across groups. When patients withdraw from a trial or are lost to follow-up, we may not know their outcomes and the internal and external validity of the trial may be compromised. This is known as attrition bias. Trials that look at long-term outcomes can, with the subjects' consent at entry, obtain follow-up data for all patients by linking to medical records. The BASIL trial obtained complete outcome data by linking to mortality records and to hospital episode data to identify amputations.

[7] WMA Declaration of Helsinki Ethical Principles for medical research involving human subjects https://www.wma.net/policies-post/wma-declaration-of-helsinki-ethical-principles-for-medical-research-involving-human-subjects/ accessed 2 December 2023.

Key Learning Point 2.9 Attrition bias can occur when some randomised patients are omitted from the analysis. The report of a trial should always include a CONSORT diagram, like Fig. 2.2, that accounts for all randomised patients.

Another area that is emphasised in the CONSORT guidelines is the need for trials to define details of their recruitment targets, the outcomes, and how these will be compared at the time a trial is being planned. This is best achieved by having the trial reported to a public register (see Chap. 1). The Helsinki Declaration referenced above (see footnote 7) states, "Every clinical trial must be registered in a publicly accessible database before recruitment of the first subject." This is important because of the need to define the outcomes that will be used to compare treatments. There are often many different outcomes that could be evaluated by a clinical trial. If the choice of outcome is made during analysis and reporting, authors will tend to pick out those that show the greatest differences or perhaps point to subgroups of patients where differences were found. Selective reporting of trials (publication bias) or selective choice of outcomes for reporting have been shown to provide misleading results for future patients [25–27]. Reports of trials need to identify those defined as primary outcomes when the trial was planned and make any conclusions about unplanned comparisons with caveats about their possible reporting bias [28]. Then need for such caution is illustrated through consideration of the first report of the BASIL trial [4], where there was no evidence of an overall difference in survival between randomised groups, although there was evidence that medical costs were much higher for patients assigned to surgery first. Examination of the survival curves suggested a benefit for surgery in the final years of follow-up compared to very similar survival patterns in the initial years. This was noted, with caution, in the initial report, but extending the follow-up time showed that the apparent difference was not confirmed [29].

2.5 Criticisms of Randomised Trials

A critique of RCTs was published by philosopher Peter Urbach [30] in 1985. This was followed by a series of related papers in the philosophical, medical, economic and sociological literature that criticise the logical or practical ability of RCTs to guide treatment decisions. Most of the authors of the books and papers have been philosophers. In this section, their arguments are reviewed by topic and related to the rationale for use of RCTs in Sect. 2.2 and to the examples in Sect. 2.3.

2.5.1 Objections to Randomisation

Urbach argues [30] that when preliminary evidence has suggested that a treatment may be beneficial, it is unethical to deny it to "randomly selected sufferers [who] are exposed to the established and probably inferior treatment." That this argument can be false is exemplified by the trial of mammary artery ligation [23], the CAST trial [22] and many other examples of proposed treatments. When a trial is being planned there must necessarily be uncertainty about the outcome, as is indeed evident from Urbach's own use of the term "probably". The possible benefits of using a placebo in getting a better evaluation of the outcome must be balanced against the ethics of ever using a placebo treatment especially when a patient is being misled. This process must be followed for each trial according to the biases that might arise in a particular study. In the case of the CAST trial, patients who fulfilled the entry criteria gave written informed consent to be randomised to treatment options that included a placebo arm [22]. As soon as the excess mortality in the treated group was identified, recruitment to the active arm was terminated. Here it could be argued that it would be unethical *not* to use a placebo treatment. The case of mammary artery ligation is more difficult. The trial showed it had no benefit, so those in the placebo arm were not disadvantaged by being exposed to the probably inferior treatment, here absence of mammary artery ligation. But the use of a sham procedure may have disadvantaged them. Such a trial would be unlikely to receive ethical approval now, but it may have spared future patients from undergoing the useless and invasive procedure.

Alternative proposals for experimental studies are suggested such as "the selection of patients might be entrusted to an independent, non-medical person, or it might be performed according to the order in which they present themselves, or by whether their hair is parted on the left or the right" [30]. Such methods would make it impossible for trials to be single- or double-blind, including where lack of blinding would be likely to lead to entry bias (KLP 2.1) or to bias the outcome. For these and other reasons relating to control of treatment allocation, they are no more or less ethical than randomised assignment.

Urbach and Worrall note that random assignment of patients can, by chance, lead to erroneous conclusions [30, 31]. They also note that a trial result may be biased because randomisation does not guarantee that the treatment groups will be balanced on every possible factor that predicts the outcome [30, 31]. Neither of these shortcomings regarding simple random-isation are denied by the authors of the current edited book (see, for example, Chaps. 2, 5 and 7).

More detailed criticisms of RCTs in relation to imbalance are presented in a review paper by Deaton and Cartwright [32] (henceforth D&C). They define the individual treatment effect as the difference between the two treatments for a subject where one of the outcomes is the unobserved counterfactual, the outcome for the treatment the subject did not receive. In a parallel group study, this is the difference for an individual patient in what their outcome would be on one treatment compared to another. Although we can use this idea to think about a trial, it is not something that can ever be measured.

In D&C, the term 'random confounding' is used for the contribution to the treatment difference purely due to imbalance in covariates across treatment groups in a particular trial. The authors go on to argue that claims that RCTs eliminate bias due to all observed and unobserved differences between the treatment groups are false. The authors state that "the RCT strategy is only successful if we are happy with estimates that are arbitrarily far from the truth, just so long as the errors cancel out over a series of imaginary experiments." They also claim that the only difference between treatments that can be estimated from an RCT is the arithmetic average of the individual treatment effects. Furthermore, they claim that randomisation tests cannot even test whether the average treatment effect is non-zero, but only evaluate the null hypothesis that all individual treatment effects are non-zero. One of the commentaries [33] on D&C demonstrates formally that these last two claims are false, as explained below, and similar arguments can be found in [24].

Within the context of RCTs, a statistical significance test tests the null hypothesis, typically, by seeking evidence that the distribution of outcomes differs by treatment group. The result is then a yes/no answer as to whether there is evidence that the distribution of outcomes as represented by an aggregate effect size, such as a mean, differs by treatment. The null hypothesis for this test is that there is no difference between the distributions of the outcome by treatment group, not that individual patient treatment effects are zero. From this model-theoretic perspective, we think of the subjects as a sample of an ideal population of subjects who might have been treated in the trial. Very few, if any,[8] trials select their subjects as random samples from a population (Chap. 5), but a population perspective allows likelihood inference that can estimate differences in outcome between treatment groups, when each group is considered as a sample from its population distribution.

In an RCT, the term 'confounder' is commonly used for a variable associated with an outcome that is imbalanced between the groups being compared and 'confounding' for the difference in the outcome that this imbalance could produce. More detailed definitions are provided in Chap. 5 (see Table 5.1).

If there has been no attempt to balance possible confounders in the design, then differences between subjects arising from these confounders will contribute to the variability of the treatment outcomes. The range of values of the treatment effect estimate that we could obtain through indefinite repetitions of the study is likely to be larger for an unbalanced comparison; but, for a well-designed RCT, it will always be a finite range that cannot be *arbitrarily* far from the truth. In likelihood inference, the estimated treatment effect comes with a confidence interval that can be interpreted as the range within which the treatment effect may lie with a given probability.[9] This limits the probability that the treatment effect is outside this range to one minus the size of the confidence level, or equivalently, the statistical significance level, α.

---

[8] The exception may be group-randomized trials that can identify a sample of all villages, schools or communities in an area and select a sample to be offered participation in the trial.

[9] Strictly, this interpretation is only correct for a Bayesian analysis with a non-informative prior over the possible range of the outcome. The distinguished statistician and ardent anti-Bayesian, DJ Finney, once agreed with me that he would interpret a confidence interval in this way.

What should be done to minimise the effect of random confounding on the results of a clinical trial? An obvious choice is to balance the treatment groups with respect to important confounders, usually by allocating treatments within strata. Both significance testing and likelihood inference can be used for a trial with stratified randomisation. Here, in keeping with what we have noted earlier concerning stratified randomisation, treatment allocations are balanced within individual strata. Correspondingly, within each stratum, the outcome for patients on the same treatment is less variable. The comparison of outcomes is in turn more precise because it is a combination of the within-stratum comparisons. We can think of this as reducing "random confounding" because the results do not depend on comparisons between patients in different strata and random confounding due to stratification factors does not contribute to the variability of the outcome.

If data for possible confounders is available for all units before treatments are allocated, an alternative to stratification is to use adaptive allocation that retains an element of randomisation but minimises the difference between groups on a range of factors [34]. In cluster randomised controlled trials, where treatment allocation is to units such as schools, villages or hospitals, possible confounders for all units may be known before treatment allocations are assigned. Methods of design and analysis that provide a balance on several factors can be used for the design and analysis of group randomised controlled trials [35]. Data on potential confounders is not always available when treatments are allocated and subdivision into many strata may not achieve balance if the numbers of treatment units is small. If covariate data measured at recruitment are available when the trial is analysed, statistical models allow the treatment estimates to be adjusted for possible confounders or for interaction effects to be estimated, as we noted earlier for the BASIL trial. This raises the question as to how to select possible confounders from the available covariates. Covariates balanced in the design, for example, through stratification, should be included. To select covariates to be used to adjust the treatment comparison by the extent of their imbalance between treatment groups in the trial is not only illogical [36], but it has also been shown to reduce the power of the trial to find a real treatment difference [37]. The relationship between the covariate and the outcome is the crucial factor and evidence from other sources, usually observational studies, can be used to inform this process [38] so that information on covariates to use for adjustment can be included in the analysis plan, thus limiting potential bias arising from post-hoc selection of covariates through trial and error [37].

2.5.2 *Observational Studies Can Be Used Instead*

Several critiques of randomised trials have suggested that a Bayesian analysis of a clinical trial using a causal model does not require randomisation [30, 31, 39]. The philosophers' claim is that a Bayesian analysis allows all of what is already known or believed about how the relevant treatments work to be combined with evidence from a non-randomised controlled trial or an observational study to calculate the probabilities of outcomes for future patients. One problem with this argument is that it implies that the ways in which treatments will work for different patients is fully understood. This cannot be assumed to be the case, especially for new treatments.

Furthermore, randomisation *is a requirement* for the likelihood approach to statistical inference, involving degrees of belief, that forms part of the Bayesian analysis of a clinical trial. The philosophers' claim is contrary to recommendations from eminent Bayesian statisticians writing about clinical trials; for example, Spiegelhalter et al. write, "We fully accept the need for randomization." [40] and Rubin, "…randomization plays a central role in Bayesian inference for causal effects" [41]. The literature on the use of causal models sees the goal of observational studies as attempting to reproduce what might be found in a randomised experiment; yet, in their recent book on causal inference [42], Herman and Robins state in Chap. 1 that, "The question is … under which conditions real world data can be used for causal inference." and go on in Chap. 2 to provide one answer: "conduct a randomized experiment".

Key Learning Point 2.10 Eminent and respected Bayesian statisticians agree on the importance of random treatment allocation for Bayesian inference from clinical studies.

Randomised trials may, however, be unnecessary or inappropriate for evaluating *some* treatments. For example, for a few treatments, the effectiveness may be so obvious that results from only a few patients may suffice. Edward Jenner prepared a report on his smallpox vaccine based on treating one subject, the son of his gardener.[10] His conclusion could be justified by a Bayesian argument. There was prior knowledge that smallpox was highly infectious and that milkmaids who had suffered from cowpox did not contract smallpox. This could then be combined with evidence that the gardener's son, who had been inoculated with pus from a cowpox blister, did not contract smallpox when exposed to it. But this is hardly evidence that could be generalised to recommend it as a safe treatment for additional patients. Wisely, the Royal Society did not accept Jenner's report until he had tested a further 23 cases.

Few treatments are as effective as this, but those that improve patients' chances of good health are still valuable. Ethical concerns may preclude some evaluations of treatments. For example, to evaluate whether lead exposure in childhood affected children's abilities it would be unethical to randomly allocate children to exposure to lead. Nevertheless, there was need for evidence to inform decisions on whether to treat water supplies to lower lead levels and to decide whether to remove lead additives from petrol. Evidence in support of these changes was based on observational studies, such as the Edinburgh Lead Study [43], that adjusted for as many confounding factors as possible.

---

[10] See https://en.wikipedia.org/wiki/Edward_Jenner#cite_note-34 accessed 2 December 2023.

Some treatments, such as blood transfusion for blood loss or anaemia, may be so well established and evidently effective that clinical trials could be considered unethical. We could not envisage conducting a trial to assess the overall benefit of blood transfusion compared to not giving blood transfusion or intraoperative blood salvage to any patients. A review of observational studies, adjusted for all known confounders [44], concluded that blood transfusion *increased* mortality, but commented, "Whether this is a true effect remains uncertain as it is possible that even the best conducted adjustments cannot completely eliminate the impact of confounding." But RCTs have been used to investigate aspects of blood transfusion practice. Following concern that blood transfusions were overused, a series of RCTs, reviewed in [45], compared the use of different haemoglobin thresholds for deciding on blood transfusion. These studies revealed that patients fare just as well, and sometimes better, when doctors use lower thresholds, a finding confirmed by a recent review [46].

There is ample evidence that where a treatment can be evaluated by an RCT, the results from corresponding observational studies can be misleading [47]. A relevant example is evidence of whether high β-carotene levels might be protective against lung cancer. Several observational studies, including a longitudinal study of workers at the Western Electric Company [48], had found an association between dietary intake of β-carotene and lung cancer. When I discussed this study as part of a course on critical appraisal for medical students, shortly after its publication, they could not fault its design. There was a proposed mechanism, good follow-up, and extensive control for other factors. It was only when prompted to consider other constituents that might be present in foods high in β-carotene that they had any doubts about the causal relationship between β-carotene and *protection from* lung cancer. But in three large, well-conducted clinical trials, ß-carotene supplementation either showed no effect on the incidence of lung cancer or showed, that supplementation was associated with an increased incidence of lung cancer (see [49] for a recent review of the evidence). A single example like this does not mean that observational studies always get the wrong answer. Every medical study needs to be evaluated critically. For RCTs this means evaluating the possibility of the biases discussed in KLPs 2.1, 2.5, 2.6, 2.7 and 2.9. In addition, observational studies need to be evaluated for potential confounding factors which could have influenced the supposed effect of the intended treatment.

Studies involving assessment of relationships between conventional dietary intake and health outcomes are especially vulnerable to confounding effects from nutrient exposure arising from intake of dietary supplements or other aspects of an individual's lifestyle which have not been accounted for. There have been several comparisons of treatment evaluations by RCTs compared to well-designed observational studies, e.g., [49–53]. In many cases, the two types of study agreed, on average, although there were sometimes large differences in treatment effects both between individual RCTs and individual observational studies. Such comparisons are considered in more depth in Chap. 6.

D&C cite a trial of hormone replacement therapy (HRT) [54], to treat menopausal symptoms, as a case where the results from an RCT were wrongly privileged over those from an observational study. Observational studies showed a protective effect of HRT for cardiovascular disease (CVD), whereas a placebo-controlled RCT, intended to confirm this, was stopped early because of excess cardiovascular deaths in the HRT group. D&C acknowledge that it was still not clear if the conclusion that HRT increased CVD deaths was "correct" and comment that "widespread abandonment of the therapy ... might (or might not) have been a mistake", but fail to explain the details. Re-analyses of both the observational studies [55] and extended data from the RCT [56] show that the effect of HRT on CVD risk depends on several factors including length of use, time since menopause, the type of HRT used and the different outcomes within CVD. After allowing for these differences, the RCT and the observational studies no longer disagree. The topic is still controversial and further data is becoming available [57].

Contrary to D&C's accusation of RCT's being wrongly privileged, the above research need not be viewed from the perspective of one type of study design giving the "correct" answer. That observational studies cannot replace all RCTs does not deny the contribution from observational studies to EBM. Bradford-Hill is well known for his list of criteria to evaluate the strength of causal conclusions from observational studies [58]. He was also the co-author of the major observational study [59] that provided evidence that smoking causes lung cancer. Dr Spock, in his popular book *Baby and Child Care* [60], advised that infants should not be put to sleep on their backs so they would not choke if they were to vomit. This advice, based on no evidence, was common in manuals of childcare published from the 1960s until 1988. The observational studies that showed this advice to be wrong were case–control studies that compared the sleeping position of children who died of sudden infant death syndrome (SIDS) with those of matched controls. A review of the evidence [61] identified 40 well-conducted observational studies that, taken together, showed that infants put to sleep on their front were five times more likely to die of SIDS. These examples represent successful high-profile studies, the success of which is unlikely to be questioned based on lack of opportunity to perform an RCT.

2.5.3 Criticisms of Trials Due to Lack of Generalisability

All of the critiques of RCTs discussed above criticise them for their lack of generalisability. Several other articles focus exclusively on this criticism [62–65], as does the book *Evidence-Based Policy: A Practical Guide to Doing it Better* by Cartwright and Hardie [66]. The approach taken in the book is to cite studies where "it worked somewhere" was shown not to predict "it will work here". The studies quoted are largely group-randomised controlled trials of social interventions. These involve random allocation of interventions to *groups* of individuals (e.g., clinics or school classrooms) rather than to individuals, but the unit of observation is individuals. In this type of study, the setting in which the intervention is tested can be crucial in determining the relative effectiveness of different interventions.

The conclusion of the book, stated in the preface, is the following:

> You are told to use policies that work. And you are told that RCTs will tell you what these are. That's not so, RCTs are great, but they do not do that for you. They cannot alone support the expectation that a policy will work for you.

There is nothing to disagree with in what is being concluded here. The problems of generalisation and the external validity of trials are well understood (Chap. 5). *Realistic Evaluation* by Pawson and Tilley [67] also emphasises the need to understand how interventions work before implementing them in another setting.

The articles criticising RCTs cite many examples where the results of RCTs have failed to be upheld when the interventions are used in other populations. Why have these examples been selected from the literature? The articles do not cite any examples of RCTs of social interventions where the conclusions have proven or potential value for policymaking. But these do exist, as exemplified by trials in development economics discussed in the next section.

2.6 Long-Term Effects of RCTs on Policies

One way to assess whether RCTs appear to have been effective is to examine trends in populations where the RCTs have affected policy. This is not a simple deduction, since trends may be produced by other factors that change over time and conclusions must be modified by these. Like observational studies, RCTs must be very well informed about the treatments, the health outcomes that are being compared and the factors that might influence their reporting in population data. Two recent examples that appear to meet these requirements are treatments for breast cancer and for CVD.

A study in the USA asked why breast cancer mortality rates have declined by 36% from 1975 to 2011 [68]. They ruled out prevention by screening as a major cause and concluded that the increased use of chemotherapy was the most likely cause. Periods with a sharp fall in mortality mapped onto the time points when interventions, shown to be effective in RCTs, started to be used. This was most evident at the times during the 1970s when Cyclophosphamide Methotrexate Fluorouracil (CMF) and Tamoxifen, each shown to reduce breast-cancer mortality in RCTs, began to be prescribed.

Another disease area where mortality has declined is CVD, where the age-adjusted mortality rates "throughout the industrialized world" have been reported to have "declined to about one third of their 1960s baseline by 2000" [69]. The reasons for this decline are complex and include changes in risk factors such as the decline in smoking. A detailed analysis of data from the US [70] concludes that approximately half of the decline could be attributed to evidence-based policies. Interventions that have become established from the results of RCTs include drug treatments such as statins as well as medical procedures, including thrombolysis, and the use of coronary stents to make angioplasties more durable.

Outside the health area, it is not so easy to point to how RCTs have affected policy over such a long period. Trials of educational, social or economic interventions have only started to become common since the mid-1990s [71]. Much of the increasing use of RCTs in development economics has been due to initiatives like the Abdul Latif Jameel Poverty Action Lab (J-PAL),[11] which sponsors randomised studies to fight poverty across the world. Two of J-PAL's founders, Abhijit Banerjee and Esther Duflo, and their long-term associate Michael Kremer received the Nobel Prize for Economics 1919 "for their experimental approach to alleviating global poverty."[12] The J-PaL website discusses many initiatives that J-PaL are evaluating by RCTs. Among them is a set of initiatives to help the very poorest to become self-employed. This has been evaluated in a series of RCTs in six different countries [72] with very positive results. The results from these RCTs and many more of the above initiatives have the potential to improve the lives of some of the poorest people in the world. We must wait to see to what extent these results will be implemented by policy makers.

2.7 Conclusions

In clinical research, a key purpose of RCTs is to provide evidence relevant to treatment choices for future patients. This is never going to be easy, as it depends crucially on future patients being similar to those in the RCT. Furthermore, although the principles of RCT design may be clear, it may not be easy to apply them in practice and the recruitment of patients into a trial can prove difficult. Relatedly, when an RCT is being planned and conducted, the benefits of the different treatments for the patients in the trial are not fully known; otherwise, it would be unethical to run such a trial. To add to these difficulties, as will be considered at more length in Chaps. 5 and 7, there are many sources of bias that can influence the results from trials and the way that they are interpreted.

Some of the criticisms that have been levelled at RCTs are clearly wrong. For example, it has been claimed that the failure of an RCT to achieve exact balance on all predictors of outcome can produce results that are "arbitrarily far from the truth". As this chapter has explained, the way that trials are analysed and reported, with their results presented with confidence intervals, explicitly prevents this. In other cases, statements are made that can be refuted from the experience of running trials. It is claimed that it is unethical to run trials to evaluate treatments where there is preliminary evidence of benefit. Examples are cited where such preliminary evidence has proved wrong. A further claim is that the analysis of observational studies by Bayesian methods makes RCTs redundant. We have seen that this claim is denied by eminent Bayesian statisticians and considered examples where observational studies can mislead, while emphasising that this is not always the case.

---

[11] See https://www.povertyactionlab.org/ accessed 2 December 2023.

[12] See https://www.nobelprize.org/prizes/economic-sciences/2019/summary/ accessed 2 December 2023.

One of the most difficult aspects of using evidence from trials is knowing if their results are relevant to the current decisions. Are those for whom the current decision is being made, and their circumstances, sufficiently similar to those in the RCT for the RCT results to be applicable? This is a difficult question and there are examples throughout this chapter of how treatments decisions can depend on differences in the types of patient and the settings where they are treated. Recent books and articles have used this as a means to criticise the use of evidence from RCTs. Nevertheless, issues relating to generalisability do not undermine the position that RCT evidence should be prioritised over other forms of evidence when both the RCT results and the other form of evidence are equally timely and relevant to the current decision.

References

1. Lind J. A treatise of the scurvy. In three parts. Containing an inquiry into the nature, causes, and cure, of that disease. Together with a critical and chronological view of what has been published on the subject. Edinburgh: Printed by Sands, Murray and Cochran for A. Kincaid & A. Donaldson; 1753.
2. Medical Research Council. Streptomycin treatment of pulmonary tuberculosis. BMJ. 1948;2:769–82.
3. Yoshioka A. Use of randomisation in the Medical Research Council's clinical trial of streptomycin in pulmonary tuberculosis in the 1940s. BMJ. 1998;317:1220.
4. Adam DJ, Beard JD, Cleveland T, et al. Bypass versus angioplasty in severe ischaemia of the leg (BASIL): multicentre, randomised controlled trial. Lancet. 2006;366:1925–34.
5. Cochrane Statistical Methods Group. The Cochrane Collaboration's tool for assessing risk of bias in randomised trials. BMJ. 2011;343.
6. Schwartz D, Lellouch J. Explanatory and pragmatic attitudes in therapeutic trials. J Chron Dis. 1967;20:637–48.
7. Schwartz D, Flamant R, Lellouch J. Clinical trials. London: Academic Press; 1980.
8. Thorpe KE, Zwarenstein M, Oxman AD, et al. A pragmatic explanatory continuum indicator summary (PRECIS): a tool to help trial designers. J Clin Epi. 2009;2:464–75.
9. Royal Statistical Society Working Group chaired by Senn, S. Statistical issues in first-in-man studies. J R Statist Soc A. 2007;170:517–79.
10. Patino CM, Ferreira JC. Internal and external validity: can you apply research study results to your patients? J Bras Pneumol. 2018;44:183.
11. Pocock SJ. Clinical trials: a practical approach. Chichester: Wiley; 2013.
12. Lee JJ, Chu CT. Bayesian clinical trials in action. Statist Med. 2012;31(25):2955–72.
13. Nduati R, Jotn G, et al. Effect of formula feeding on transmission of HIV-1. JAMA. 2000;296(7):794–805.
14. Thior I, Lockman S, Shapiro RL, et al. Breastfeeding plus infant Zidovudine Prophylaxis for 6 months vs formula feeding plus Infant Zidovudine for 1 month to reduce mother-to-child HIV transmission in Botswana JAMA. 2006;283(9):1167–74.
15. World Health Organization, United Nations Children's Fund. Guideline: updates on HIV and infant feeding: the duration of breastfeeding, and support from health services to improve feeding practices among mothers living with HIV. Geneva: World Health Organization; 2016.
16. Schulz KF, Altman DG, Mohe D, for the CONSORT Group. CONSORT 2010 statement: updated guidelines for reporting parallel group randomised trials. BMC Med. 2010;8–18.
17. Walters SJ, Henriques-Cadby IBDA, Bortolami O, et al. Recruitment and retention of participants in randomised controlled trials: a review of trials funded and published by the United Kingdom Health Technology Assessment Programme. BMJ Open. 2017;7:3.

18. Wells M, Macmillan M, Raab G, et al. Does aqueous or sucralfate cream affect the severity of erythematous radiation skin reactions? A randomised controlled trial. Radiother Oncol. 2004;73:153–62.
19. Macmillan M, Wells M, Raab G, et al. Randomized comparison of dry dressings versus hydrogel in management of radiation-induced moist desquamation. Int J Radiat Oncol Biol Phys. 2007;68(3):864–72.
20. Noseworthy JH, Ebers GC, Vandervoort MK, et al. The impact of blinding on the results of a randomized, placebo-controlled multiple sclerosis clinical trial. Neurology. 1994;44:16–20.
21. Price D, Finniss D, Benedetti F. A comprehensive review of the placebo effect: recent advances and current thought. Annu Rev Dev Psychol. 2008;59:565–90.
22. The Cardiac Arrhythmia Suppression Trial (CAST) Investigators. Preliminary report: effect of encainide and flecainide on mortality in a randomized trial of arrhythmia suppression after myocardial infarction. N Engl J Med. 1989;321:406–12.
23. Grey D, Kittle F, Crockett JE. Comparison of internal mammary artery ligation and sham operation for angina pectoris. Am J Cardiol. 1960;5:483–6.
24. Howick J. The philosophy of evidence-based medicine. Hoboken, NJ: Wiley-Blackwell, BMJ Books; 2011.
25. Sun X, Briel M, et al. Credibility of claims of subgroup effects in randomized controlled trials: systematic review. BMJ. 2012;344:e1553.
26. Chan AW, Krleza-Jeric K, Schmid I, Altman DG. Outcome reporting bias in randomized trials funded by the Canadian Institutes of Health Research. CMAJ. 2004;171:735–40.
27. Easterbrook PJ, Berlin JA, Gopalan R, Matthews DR. Publication bias in clinical research. Lancet. 1991;337:867–72.
28. Freeman SC, Fisher D, Tierney JF, Carpenter JR. A framework for identifying treatment-covariate interactions in individual participant data network meta-analysis. Res Syn Meth. 2018;9:393–407.
29. Bradbury AW, Adam DJ, Bell J, et al. Bypass versus angioplasty in severe ischaemia of the leg (BASIL) trial: an intention-to-treat analysis of amputation-free and overall survival in patients randomized to a bypass surgery-first or a balloon angioplasty-first revascularization strategy. J Vasc Surg. 2010;51(5):5S–17S.
30. Urbach P. Randomization and the design of experiments. Philos Sci. 1985;52(2):256–73.
31. Worrall J. Evidence: philosophy of science meets medicine. J Eval Clin Pract. 2010;16(2):356–62.
32. Deaton A, Cartwright N. Understanding and misunderstanding randomized controlled trials. Soc Sci Med. 2010;210:2–21.
33. Dahabreh IJ. Response to understanding and misunderstanding randomized controlled trials. Soc Sci Med. 2010;210:41–4.
34. Pocock SJ, Simon R. Sequential treatment assignment with balancing for prognostic factors in the controlled clinical trial. Biometrics. 1975;31(1):103–15.
35. Raab GM, Butcher I. Balance in cluster-randomized trials. Statist Med. 2001;20:351–65.
36. Senn S. Testing for baseline balance in clinical trials. Statist Med. 1994;13:1715–26.
37. Schluchter MD, Forsythe AB. Post-hoc selection of covariates in randomised experiments. Commun Stat Theory Met. 1985;14:679–99.
38. Raab GM, Day S, Sales J. How to select covariates to include in the analysis of a clinical trial. Control Clin Trials. 2000;21(4):330–42.
39. Cartwright N, Munro E. The limitations of randomized controlled trials in predicting effectiveness. J Eval Clin Pract. 2010;16(2):260–6.
40. Spiegelhalter DJ, Freedman LS, Parmar MKB. Bayesian approaches to randomized trials. J R Statist Soc Ser A. 1994;157(3):357–416.
41. Rubin DR. Bayesian inference for causal effects: the role of randomization. Ann Statist. 1978;6(1):34–58.
42. Hernán MA, Robins JM. Causal inference: what if. Boca Raton: Chapman & Hall/CRC; 2020.
43. Fulton M, Raab G, Thomson G, et al. Influence of blood lead on the ability and attainment of children in Edinburgh. Lancet. 1987;1(8544):1221–6.

44. Hopewell S, et al. A systematic review of the effect of red blood cell transfusion on mortality: evidence from large-scale observational studies published between 2006 and 2010. BMJ Open. 2013;3(5):e002154.
45. Anthes E. Save blood, save lives. Nature. 2015;520(7545):24.
46. Holst LB, et al. Restrictive versus liberal transfusion strategy for red blood cell transfusion: systematic review of randomised trials with meta-analysis and trial sequential analysis. BMJ. 2015;350:1354.
47. Pocock SJ, Elbourne DR. Randomized trials or observational tribulations? N Engl J Med. 2000;342:1907–9.
48. Pandey DK, Shekelle R, Selwyn B, et al. Dietary vitamin C and β-carotene and risk of death in middle-aged men. Am J Epidemiol. 1995;142(12):1269–77.
49. Cortés-Jofré M, Rueda JR, Asenjo-Lobos C, et al. Drugs for preventing lung cancer in healthy people. Cochrane Database Syst Rev. 2020;3:1465–858.
50. Benson K, Hartz AJ. A comparison of observational studies and randomized, controlled trials. N Engl J Med. 2000;342(25):1878–86.
51. Anglemyer A, Horvath HT, Bero L. Healthcare outcomes assessed with observational study designs compared with those assessed in randomized trials. Cochrane Libr. 2014;(4):MR000034.
52. Soni PD, Hartman H, Dees RT, et al. Comparison of population-based observational studies with randomized trials in oncology. Clin Oncol. 2019;14:1209–16.
53. Smeeing DPJ, van der Ven DJC, Hietbrink F, et al. Surgical versus nonsurgical treatment for midshaft clavicle fractures in patients aged 16 years and older: a systematic review, meta-analysis, and comparison of randomized controlled trials and observational studies. Am J Sports Med. 2017;45(8):1937–45.
54. Rossouw JE, Anderson GI, Prentice RL, et al. Risks and benefits of estrogen plus progestin in healthy postmenopausal women: principal results from the women's health initiative randomized controlled trial. JAMA. 2002;288(3):321–33.
55. Hernan MA, Alonso A, Logan R, et al. Observational studies analyzed like randomized experiments: an application to postmenopausal hormone therapy and coronary heart disease. Epidemiology. 2008;19:766–79.
56. Prentice RL, Langer R, et al. and International Women's Health Initiative. Combined post-menopausal hormone therapy and cardiovascular disease: toward resolving the discrepancy between observational studies and the Women's Health Initiative clinical trial. Am J Epidemiol. 2005;162:404–14.
57. Vandenbroucke JP. The HRT controversy: observational studies and RCTs fall in line. Lancet. 2009;373(9671):1233–5.
58. Hill AB. The environment and disease association or causation. Proc R Soc Med. 1965;58:295–300.
59. Doll R, Hill AB. The mortality of doctors in relation to their smoking habit. BMJ. 1954;328:1529–33.
60. Spock B. Baby and child care. 165th printing. New York: Pocket Books; 1966. p. 163–4.
61. Gilbert R, Salanti G, Harden M, See S. Infant sleeping position and the sudden infant death syndrome: systematic review of observational studies and historical review of recommendations from 1940 to 2002. Int J Epi. 2005;34:8748–87.
62. Cartwright N. Are RCTs the gold standard? BioSocieties. 2007;2:11–20.
63. Cartwright N. Presidential address: will this policy work for you? Predicting effectiveness better: how philosophy helps. Philos Sci. 2012;79:973–89.
64. Cartwright N. What is this thing called 'efficacy'? In: Mantzavinos C, editor. Philosophy of the social sciences: philosophical theory and scientific practice. Cambridge: Cambridge University Press; 2009.
65. Cartwright N. A philosopher's view of the long road from RCTs to effectiveness. Lancet. 2011;377(9775):1400–1.
66. Cartwright N, Hardie J. Evidence based policy: a practical guide to doing it better. Oxford: Oxford University Press; 2012.

67. Pawson R, Tilley N. Realistic evaluation. London: Sage Publications; 1997.
68. Narod SA, Iqbal J, Miller AB. Why have breast cancer mortality rates declined? J Cancer Policy. 2015;5:8–17.
69. Mensah GA, Wei S, Sorlie PD, et al. Decline in cardiovascular mortality: possible causes and implications. Circ Res. 2017;120(2):366–80.
70. Earl S, Ford MD, Umed A, et al. Explaining the decrease in U.S. deaths from coronary disease, 1980–2000. N Engl J Med. 2007; 356:2388–98.
71. Cameron DB, Mishra A, Brown AN. The growth of impact evaluation for international development: how much have we learned? J Dev Eff. 2015;8(1):1–21.
72. Banerjee A, Duflo E, Goldberg N, et al. A multi-faceted program causes lasting progress for the very poor: evidence from six countries. Science. 2015;348(6236):1260799.

Gillian Raab has been involved in the design and analysis of a wide range of randomised clinical trials. These have included trials to evaluate treatments for tibial fractures, and for the prevention of skin damage after radiotherapy, and group-randomised trials for educational interventions, among others. She has published papers on trial methodology in the journals *Controlled Clinical Trials* and *Statistics in Medicine*. For 17 years, she taught statistical methods, including the design of clinical trials, to medical students at the University of Edinburgh. She then took up the position of Chair of Applied Statistics at Edinburgh Napier University, where she was involved in teaching courses on practical statistical methods. Since her retirement, she has remained active in statistical research projects. She is currently working with the Administrative Data Research Centre—Scotland and with the Scottish Longitudinal Study, mainly on ways of making administrative data more accessible to the research community.

Chapter 3
Logical Fallacies and the Misuse of RCT Findings: Lessons from Diagnostic Statistics

Margaret MacDougall

Abstract This chapter offers an excursion through many of the important analogies encountered on comparing the conceptual frameworks for diagnostic statistics and null hypothesis significance testing (NHST). We shall meet three common logical fallacies that can prove influential in overestimating the probability of an intervention being effective, given a statistically significant outcome in favour of the intervention. In keeping with previous work, this conditional probability will be referred to as the positive predictive value (PPV) of the hypothesis test for the given study hypothesis. Relatedly, we shall see how researchers can be duped into a false reliance on statistical significance from a single randomised controlled trial (RCT) as an indicator of the reputability of the study hypothesis, while overestimating, and neglecting the importance of, the PPV. Recognition of the need for a higher PPV points to the fallibility of the RCT, in keeping with other study designs, as a source of scientific truth. In addition to training students to recognise and circumvent common misapplications of conditional probabilities in diagnostic statistics and hypothesis testing, the content of this chapter warns against some misconceptions in traditional representations of the relationships between *prevalence* and diagnostic statistics in medical textbooks and related expository papers. While this ought to support modern teaching on the theme of diagnostic testing, the aim in view is that of preparing the way for consideration in Chap. 4 of the parameterisation and management of the PPV using Bayes' rule specifically within the context of NHST.

3.1 Overview

In Sect. 3.2, the concepts of statistical significance level and p-value as conditional probabilities and the role of these probabilities in null hypothesis significance testing (NHST) will be introduced. This will provide a foundation stone in Sect. 3.3 for us to

---

M. MacDougall (✉)
Usher Institute, Edinburgh Medical School, Edinburgh, UK
e-mail: margaret.macdougall@ed.ac.uk

M. MacDougall (ed.), *A Medical Educator's Guide to Thinking Critically about Randomised Controlled Trials: Deconstructing the "Gold Standard"*,
https://doi.org/10.1007/978-3-031-25859-6_3

consider two forms of the p-value fallacy, one of which will allow us to question the logical soundness of NHST. As in Chap. 1, we shall be able to see a sense in which the randomised controlled trial (RCT) study design must succumb to the limitations faced by other study designs within particular contexts—here, by virtue of applying NHST, which remains the backbone of statistical findings across a wide range of clinical research literature. In Sect. 3.4, we shall then consider an adapted form of a diagnostic problem originally set for medical students at Harvard University as an important tool in identifying fallacious diagnostic reasoning. In so doing, we shall recognise some common ground with the inferential reasoning explored in the previous section. In Sect. 3.5, we shall consider the base rate fallacy and confusion of the inverse and the influences these fallacies can have on the understanding of individual types of diagnostic statistic and on the tendency to overrate statistical significance within the context of NHST. In Sect. 3.6, we shall meet Bayes' rule and investigate the traditional model that employs this rule in diagnostic testing specifically to explain the respective relationships between the positive and negative predictive values, as conditional probabilities, and prevalence. This will include exploring the reasons for challenging the above model. These findings will prove important for Chap. 4 in acknowledging the need not to stretch the analogy too far in comparing the usages of Bayes' rule to represent the positive predictive value (PPV) in diagnostic statistics and NHST.

3.2 Null Hypothesis Significance Testing as an Inferential Method for RCTs

In this chapter, we shall take the two-armed, double-blinded, parallel group, individually randomised controlled trial to be the paradigm RCT design, as was the case in Chap. 1. In Chap. 1, we saw that, based on the simple randomisation procedure associated with this design, when the null hypothesis of no treatment effect is true a treatment effect will be detected through statistical significance testing purely due to unpredictable imbalance arising from sampling variation (or, "chance") with probability α, where α is the statistical significance level.

α can be expressed as a conditional probability, which, in the above context, takes the following form, involving the null hypothesis, H_0:

α: *The probability in a hypothesis test of rejecting H_0 if H_0 is true.* *(Pr1)*

As such, α is commonly referred to as the probability of a Type I error, the Type I error rate, or, occasionally, as the false positive rate (FPR).

In employing α to assess the evidence for a treatment effect in an RCT, NHST assumes an inferential method whereby the p-value is compared with α as the threshold value.

Greenland's definition of *p-value* was provided in Chap. 1. Where, for a given treatment outcome, the study hypothesis involves the claim in terms of averages that there is an arithmetic difference between two groups (an intervention and control group), it is commonplace for H_0 to involve the assumption that this value is zero and standard practice to interpret the p-value in this particular context as follows:

p-value: *The probability of obtaining an observed value for the difference at least as extreme as that found for our given sample if H_0 is true.*[1] *(Pr2)*

This leads to the key statement *(KS)* below.

A very small p-value (less than or equal to the significance level, α) is telling us that it is very unlikely that one would get this observed difference (or one more extreme) if H_0 were true and that it is reasonable, therefore, to reject the null hypothesis. (KS)

Using NHST, where a p-value forthcoming from the application of a hypothesis test is sufficiently small, in the above sense, we conclude on the basis of *KS* that it is probably the case that there is a true difference between the intervention and control groups.

As we shall see shortly, the contributions that statistical significance and relatedly, the concept of *p-value*, make to NHST have a bearing on the logical validity of NHST.

3.3 The P-Value Fallacy

In this section we shall look at two types of p-value fallacy. For convenience, throughout this chapter, we shall adopt the practice commonly used in clinical research of assuming the value 0.05 for α.

A logical fallacy is a common form of reasoning that, although it looks reasonable, is in fact logically unsound and liable to lead to error. In its simplest form, the p-value fallacy can be understood as reasoning as though the value obtained on subtracting the p-value from 1 is the probability that the study (or, alternative) hypothesis, H_a, is true, when, in fact, the p-value is calculated conditional on the hypothetical assumption that H_0 is true. Such reasoning is based on the false premise that the p-value is the probability that the null hypothesis, H_0, is true. A second type of p-value fallacy can be identified, however, by noting that use of the p-value in NHST circumvents the fact that, since the p-value is a probability, it ought to be calculated on the basis of many repetitions of a study, not a single study. *KS*, above, is a product of this particular misuse of the p-value within the context of NHST. To see this, it is helpful to compare the form of inference in NHST with that of a logically valid rule of inference, namely Modus Tollens.

In classical logic, Modus Tollens is a rule of inference which, for propositions *p* and *q*, can be expressed informally as follows:

1. If *p* is true, *q* is true.

---

[1] Note that while α and the p-value can both be expressed as conditional probabilities, the definition of *p-value* refers to a range of possible values for the observed effect size including and beyond the observed value. It should not, therefore, be confused with α and as such, misrepresented as the FPR.

2. q is false.
3. Therefore, p is false.

Line 1 represents a conditional statement with antecedent p and consequent q, while lines 2 and 3 collectively tell us that if we deny the consequent, we must also deny the antecedent.

The following example illustrates an instance of Modus Tollens:

1. If the patient was properly vaccinated, he does not have measles.
2. The patient has measles.
3. Therefore, the patient was not properly vaccinated.

Let 'ΔO' and 'd' denote the variable for the observed difference and the value of the observed difference, respectively, between an intervention and control group for outcome O in an RCT. We can construct a similar but flawed (and thus, invalid) form of reasoning to represent NHST for the particular case where, further to hypothesis testing, the p-value is found to be at most 0.05, and thus, the requirement for statistical significance and rejecting H_0 has been met, as follows:

Rule of inference 1 (IR1)

1. True statement 1: *If H_0 is true, $P(\Delta O \geq d) \leq 0.05$. (Simplified form: p-value ≤ 0.05)*

2. True statement 2: $\Delta O = d$

3. Therefore, H_0 is probably false (equivalently, H_a is probably true).

Based on IR1, the decision is then made to reject H_0 in favour of H_a.

True statement 2 is derived easily based on the sample data and in practice, this may occur prior to hypothesis testing. On the other hand, *True statement 1* rests on the derivation of a p-value, based on the sample data, through hypothesis testing, that confirms statistical significance. From a logical point of view, in arriving at the conclusion in step 3, this order of events does not matter. *True statement 1*, in its non-simplified form, can be seen to be a conditional statement of the form 'if p then q', which is expressed in the equivalent form 'if p, q'. In the above example, the consequent, q, *is* the proposition '$P(\Delta O \geq d) \leq 0.05$'. This proposition informs us that it is very unlikely that the observed difference (ΔO) is of value at least d. But according to true statement 1, the observed value for this single study actually is d. Therefore, it is concluded that the consequent is probably false, and in turn that the antecedent is probably false, to the end that the antecedent is rejected.

Concluding that the consequent '$P(\Delta O \geq d) \leq 0.05$' is probably false because an instance ($\Delta O = d$) of the (highly unlikely) outcome $\Delta O \geq d$ has been observed is contentious. Correspondingly, in IR1, the conclusion is not logically guaranteed. Therefore, through comparing IR1 with the logically valid rule, Modus Tollens, the logical invalidity of IR1 can be exposed. This is useful, as we can see that, taking d to represent a lower limit for the observed difference across the trials, what is actually needed to correct this reasoning is a sufficiently large number of repeated trials, such that $P(\Delta O \geq d)$ can be seen to converge to a fixed number. If this number were greater than 0.05, we would then have more compelling evidence than that presented in Truth statement 2, above to refute the above consequent through an application of Modus Tollens, and in turn, reject the null hypothesis.

This, of course, would not entail a precise application of Modus Tollens, as we would be relying on strong probabilistic evidence rather than fact to treat the consequent, $P(\Delta O \geq d) \leq 0.05$, as false. However, noting that probability is an asymptotic notion, the logical soundness of arriving at the above conclusion concerning the consequent, which would in turn lead to refuting H_0, should increase with increasing number, N, of RCTs, provided the RCTs are well-designed. Correspondingly, given these conditions, we should expect $P(\Delta O \geq d)$ to be close to 1 for sufficiently large N. Of course, we know that in practice, there are constraints on the number of repetitions that can be carried out for a given RCT. The point to be emphasised here, however, is that examining NHST from a logical point of view offers an important contribution to the ongoing debates on the use of this method for scientific hypothesis testing, including in the design and reporting of RCTs. That is not to say that a suitable alternative to NHST has been devised to justify its removal altogether from clinical, and other, research reporting.

Key Learning Point 3.1

The p-value fallacy can be thought of as comprising two forms. It is frequently understood in terms of reasoning that H_a is the probability forthcoming from subtracting the p-value from 1 based on the misunderstanding that the p-value is the probability that H_0 is true while, *in fact*, the definition of *p-value* requires us to *assume* that H_0 is true. It can also be understood as consisting in the misapplication of a statement about a hypothetical series of repetitions of a study to a unique member of that series, thus confusing the short-run and long-run views [1]. An attempt to capture the inferential structure behind the application of NHST when statistical significance has been achieved reveals that the latter aspect of the p-value fallacy renders NHST logically unsound even if we allow 'probably false' to replace 'false' (cf. Modus Tollens and IR1, above).

An awareness of the second form of the p-value fallacy highlighted in Key Learning Point (KLP) 3.1 is important in appreciating the limitations of NHST as a basis for scientific truth, including in the design and reporting of RCTs.

3.4 The Harvard University Problem

As we shall now see, the inferential reasoning identified above for NHST can resurface in a different guise elsewhere. A problem similar to the following one was set for medical students at Harvard University [2]:

How likely is it that an individual (without other indications or risk factors) who tested positive for a rare disease in fact has the disease. The disease has a prevalence of 1 case per 1000, which is clearly low. The test is 95% reliable.[2]

For ease of reference, we shall henceforth refer to the adapted version of the original problem presented above as the Harvard University problem. This particular version of the problem will allow us to consider an illustration of the fallacy of confusion of the inverse within the context of diagnostic statistics that ought to be of interest in the training of future health professionals. The above problem and associated misleading forms of reasoning are critical to medical student training, as analogous scenarios are likely to present themselves in clinical practice with the expectation that clinicians are able to respond accurately when communicating findings to their patients. To understand the problem more fully, let's consider some relevant terminology.

The reliability of a diagnostic test refers equally to two important statistics for assessing diagnostic accuracy, namely the sensitivity and specificity.

The sensitivity and specificity of a diagnostic test may be represented as conditional probabilities as follows:

sensitivity: *The probability of an individual obtaining a positive diagnostic test result for a disorder if they have the disorder according to the gold standard.* *(Pr3)*

specificity: *The probability of an individual obtaining a negative diagnostic test result for a disorder if they do not have the disorder according to the gold standard.* *(Pr4)*

Correspondingly, *1-sensitivity* and *1-specificity* can be defined as conditional probabilities *Pr5* and *Pr6*, respectively as follows:

1-sensitivity: *The probability of an individual obtaining a negative diagnostic test result for a disorder if they have the disorder according to the gold standard.* *(Pr5)*

1-specificity: *The probability of an individual obtaining a positive diagnostic test result for a disorder if they do not have the disorder according to the gold standard.* *(Pr6)*

For the Harvard University problem, the reliability value informs us that the sensitivity and specificity of the diagnostic test are both 0.95. Based on the above definitions of *sensitivity* and *specificity*, we can conclude, therefore, that for this case, 5% of individuals who test positive and 5% of individuals who test negative, according

[2] In the original statement of the problem, the authors specify a false positive rate of 5% rather than a percentage value for the reliability of the test. Without due care, this can lead to confusion among students, as in diagnostic testing, the false positive rate could be understood as referring to two different types of error, depending, for example, on whether we are focused on specificity or the positive predictive value of the test (see later). With a focus on specificity, the authors were specifying the value that we shall identify shortly as *1-specifcity* (see Sect. 3.5.1.1 of this chapter). From this we can see that they were indirectly declaring specifically that the specificity of the test was 0.95. Also, in the current version, the qualifier regarding absence of indications and risk factors is important. As we shall see later (see Sect. 3.6.2), although this has not traditionally been made clear in medical textbooks, the type of diagnostic statistics considered in this chapter can be treated as multifactorial statistics.

to the gold standard, are wrongly diagnosed by the diagnostic test; equivalently, both *1-sensitivity* and *1-specificity* are equal to 0.05.

On tackling the Harvard University problem, some of the students may have been tempted to reason as follows:

1. If the individual does not have the disease, it is unlikely (the probability being 0.05) that they will test positive.
2. The individual has tested positive.
3. Therefore, the individual probably has the disease.

While the particulars differ, this form of inference follows that of IR1, above, and, as such, is logically invalid.

Assuming that they took *1-specificity* to be 0.05 at step 1, above, these observations could go partly towards explaining why almost half of the Harvard University students wrongly concluded that the probability that the above individual had the disease was high (see step 3, above), offering 95% as the answer to the problem, with additional students offering answers ranging up to 99%. As we shall see shortly, alternative forms of logically invalid reasoning may also have been involved, however, at a more fundamental level, in shaping student belief that 95% was the correct answer.

3.5 The Base-Rate Fallacy and Confusion of the Inverse

3.5.1 The Case of Diagnostic Statistics

3.5.1.1 The Base Rate Fallacy

The p-value fallacy is closely related to another fallacy, the fallacy of ignoring (neglecting) the base rate (the *base rate fallacy*). The latter can occur when, if presented with specific information pertaining to an individual together with the base rate in making a judgement, the recipients fail to acknowledge the general information captured by the base rate, and arrive at a false conclusion.

This situation can arise in diagnostic test situations when the prevalence of the condition being diagnosed is low, with the effect that there are considerably more false positive than true positive diagnostic test predictions. For the Harvard University problem, the potential role of this special case of the base rate fallacy in such a high percentage of students wrongly opting for 95% as their answer can be seen on comparing this answer, as the reliability value with which they had been presented, with the additional information, pertaining to the prevalence of the disease, that only one person in a thousand had the disease. The latter information pertains to the prevalence of the disease, also known, as *the base rate* of the disease, for the study population. This rate is equivalent to 100 in 100,000 and, to keep the figures simple, we shall use this fact shortly for our calculations through assuming that the study sample consists of 100,000 individuals.

On comparing *Pr1* and *Pr6*, we can see that *1-specificity* for diagnostic testing has an analogous interpretation to α for RCTs in NHST. This analogy is, of course, not restricted to RCTs, but we are focusing on this study design for our present purposes. We can also see that *Pr6* refers to the probability of obtaining a false positive in an analogous sense to α in NHST. Correspondingly, like α, *1-specificity* is sometimes referred to as the false positive rate (FPR).

Consider Table 3.1, below. From the prevalence (or, base rate) of the disease in the statement of the problem, we already know the entries for the last row in this table. As we do not have the actual sample for different combinations of test result and disease status, we can at best estimate the observed frequencies based on what we would expect their values to be on average. This serves our purposes sufficiently well to address the problem in hand.

The required calculations for frequencies *a* and *c* of outcome column 1 involve multiplying *sensitivity* and *1-sensitivity*, respectively by the corresponding number of patients in the last row of outcome column 1. Likewise, the required calculations for frequencies *b* and *d* of outcome column 2 involve multiplying *1-specificity* and *specificity*, respectively by the corresponding number of patients in the last row of outcome column 2. These calculations are shown in Table 3.1.

Table 3.1 Table of frequencies for Harvard University problem

Diagnostic test result	Outcome based on gold standard		
	Has the disease	Does not have the disease	Total
Tests positive	*a* 95 (100 × 0.95)	*b* 4995 (99,900 × 0.05)	5090
Tests negative	*c* 5 (100 × 0.05)	*d* 94,905 (99,900 × 0.95)	94,910
No. of patients in 100,000	100	99,900	100,000

Note '*a*' to '*d*' are labels for the frequencies calculated in the corresponding cells

Based on *Pr3* and *Pr4*, *sensitivity* and *specificity* can be calculated from the frequencies in Table 3.1 using the formulae $\frac{a}{a+c}$ and $\frac{d}{b+d}$, respectively. Correspondingly, *1-sensitivity* and *1-specificity* can be calculated using the formulae $1-\frac{a}{a+c}$ and $1-\frac{d}{b+d}$, respectively, which, by simple algebra, simplify to $\frac{c}{a+c}$ and $\frac{b}{b+d}$, respectively.

For the Harvard University problem, however, a *different* diagnostic statistic is required, namely, the positive predictive value (PPV) for the diagnostic test, which can be defined as a conditional probability, *Pr7*, as follows:

> ***PPV:*** *The probability that, according to the gold standard, a patient has the disorder they are being tested for if they have received a positive diagnostic test result for that disorder.* *(Pr7)*

From row 1 of Table 3.1, a total $(a + b)$ of 5090 patients tested positive. In turn, from cell *a*, only 95 of these 5090 patients have the disease according to the gold standard. Thus, based on *Pr7*, our required probability is obtained from the formula $\frac{a}{a+b}$ by dividing 95 by 5090 to obtain 0.019 (to 3 decimal places).

Note that in a similar way we can see from Table 3.1 that the probability that if the patient tests positive, they do not have the disease, according to the gold standard, is $\frac{b}{a+b}$. This probability is sometimes referred to as the false discovery rate (FDR), and can be defined formally as a conditional probability using *Pr8*, below.

> ***FDR:*** *The probability that, according to the gold standard, a patient does not have the disorder they are being tested for if they have received a positive diagnostic test result for that disorder.* (*Pr8*)

The PPV and FDR are complements to one another. This means that they represent the probabilities of the occurrence of each of two complementary events—separate events, each of which occurs if and only if the other does not occur. When two events are complementary, we can obtain the probability of one of these events occurring by subtracting the probability of the other event occurring from 1; we can also say that, as such, the two events are complements to one another. As the FDR and PPV represent the probabilities of complementary events, we can obtain the FDR by subtracting the PPV from 1. The FDR can therefore be obtained from the formula $1 - \frac{a}{a+b}$, which, with a little algebra, can be simplified to the formula $\frac{b}{a+b}$ which we took directly from Table 3.1, earlier. Also, as we shall see more explicitly later on considering Bayes' rule, the PPV is directly proportional to the prevalence of the disorder being tested for.

Note that both through using this formula and through subtracting the above value of 0.019 for the PPV from 1, we arrive at the value 0.981 (to 3 decimal places) for the FDR. This represents a very high error rate, demonstrating that if an individual tests positive, they are much more likely not to have the disease than to have it.

For completeness here, observe that the negative predictive value (NPV) and false omission rate (FOR) are also recognised in the assessment of diagnostic test quality. The NPV can be conveniently defined by replacing *has the disorder* by *does not have the disorder* and *positive* by *negative* in *Pr7*. Likewise, the FOR can be conveniently defined by replacing *does not have the disorder* by *has the disorder* and *positive* by *negative* in *Pr8*.

It may serve as a useful exercise for instructors to invite students to derive the corresponding definitions in these ways and to confirm that they represent the probabilities of complementary events (so that $FOR = 1-NPV$ and $NPV = 1-FOR$).

On considering the above definitions, it is important to note that the term 'gold standard' can be understood as referring to the best available standard of reference in terms of accuracy and reliability. Typically, this cannot involve the assumption that the gold standard is, according to its definitive criterion, a perfect one. Correspondingly, students should be encouraged to reference the gold standard when defining the above terms.

Nevertheless, in what follows, the expression 'according to the gold standard' will be omitted for simplicity and ease of explanation, meaning that it should be implicitly assumed that the gold standard is being referenced.

3.5.1.2 The Fallacy of Confusion of the Inverse

Misunderstandings can also arise in diagnostic statistics through confusing the concepts of *sensitivity* and *PPV*. The example below, involving presence or absence of spots as the diagnostic test and measles as the disease, illustrates the distinction.

Sensitivity: The probability that someone has spots if they have measles.

PPV: The probability that someone has measles if they have spots.

Notice that for this particular example, we would expect the sensitivity to be very high and, in most contexts, the prevalence of measles, and hence the PPV, very low, which is consistent with the properties reflected in the Harvard University problem as presented above.

In the language of conditional probability that we considered earlier, the concepts of *sensitivity* and *PPV* may be written as *P(individual tests positive | they have the disorder)* and *P(individual has the disorder | they test positive)*, respectively. Thus, each concept may be transformed into the other by inverting the arrangement of the events on either side of the conditional sign '|'. As such, the sensitivity and PPV are examples of *probabilities for two conditional statements that are the inverse of one another*.

Confusion of the inverse is the fallacy of conflating the separate probabilities pertaining to the truth of each of two conditional statements that are the inverse of one another, leading to the assumption that these probabilities take at least approximately the same value. Confusing the sensitivity and PPV, as diagnostic test statistics, is an example of this fallacy. Confusion of the inverse could also contribute to our students responding wrongly to the Harvard University problem, with the sensitivity of 0.95 being misconceived by virtue of this fallacy as the PPV.

Key Learning Point 3.2

The *base rate fallacy* is the fallacy of neglecting the base rate (e.g., disease prevalence) when presented with this general information together with more specific information (e.g., the sensitivity or specificity of a diagnostic test for this disease).

Confusion of the inverse is the fallacy of conflating the separate probabilities pertaining to the truth of each of two conditional statements that are the inverse of one another, leading to the assumption that these probabilities take at least approximately the same value. An example of this is conflating the PPV and the sensitivity of a diagnostic test.

The base rate fallacy and the confusion of the inverse can be jointly responsible for errors in estimating probabilities, such as when estimating the PPV of a disorder through focusing specifically on the individuals with the disorder and in turn, the known sensitivity of the diagnostic test to the neglect of the prevalence of the disorder.

3.5.2 *The Case of NHST*

3.5.2.1 Some Conceptual Analogies

So far, we have considered three logical fallacies: the p-value fallacy, the base rate fallacy and confusion of the inverse. We have considered only the first of these within the context of NHST. In so doing, we recognised a potential connection between NHST and diagnostic testing. This was in relation to the logical form of reasoning represented by IR1 through its involvement in the p-value fallacy and possibly, the difficulties students encountered in solving the Harvard University problem. We are now going to investigate the relationships between NHST and diagnostic testing further through consideration of the base rate fallacy and confusion of the inverse within the context of NHST.

3.5.2.2 The Base Rate Fallacy

Recall that for our version of the Harvard University problem, Table 3.1 displays the relationship between diagnostic test result and disease status. By way of analogy, we can form a similar table for hypothesis testing, although, here, we shall use different percentage values for consistency with conventional practice in statistical hypothesis testing. Within the context of diagnostic testing, we noted that the base rate is the prevalence of a disorder. Within the context of hypothesis testing, we can in a similar fashion refer to the base rate as *the percentage of those published hypothesis test results favouring an intervention which, as such, reflect true study hypotheses.* This type of base rate can at best be estimated, for example, by consideration of replicability studies, that is, studies involving an investigation of whether results of the above sort can be replicated. Much could be said here about what it is, exactly, statistically speaking, that requires to be replicated and whether this should be achieved using the original data and methods. However, based on the evidence we currently have and using the term *replication* broadly speaking to refer to generation of similar results on repetition of a study, although not requiring use of the same data (see also Chap. 7), we can estimate that in basic and pre-clinical medicine, as little as 10% of the published study hypotheses supporting the efficacy of an intervention may be replicable [3, 4]. In medicine more generally, this estimate is likely to vary considerably according to a wide range of factors, including the clinical speciality and the academic credibility of the researchers and the publisher or journal. It is also contentious to treat replicability in the above sense as both necessary and sufficient for the truth of the corresponding study hypothesis. In practice, truth may be pursued through a more sophisticated approach, such as through a programme of RCTs of increasing size designed to assess correspondence, or consistency, of study findings, as is commonly used in drug development, the idea being that one is able to postulate what the infinitely large experiment should show and judge the plausibility of the previous study results based on how well they are supported by larger studies. There is also a call for extensive research aimed at obtaining reliable base rate estimates, even for clinical research broadly speaking.

Nevertheless, to gain some insight into the possible influence of a very low base rate on the PPV, for the example to follow involving RCTs, we shall adopt the estimate of 10% for the probability that our chosen study hypothesis is true with specific reference to RCTs. We can express this formally as $P(\boldsymbol{H_a}) = 0.1$ or equivalently (assuming, as in classical logic, that $P(\boldsymbol{H_a}) = P(not\ \boldsymbol{H_0})$), as $P(not\ \boldsymbol{H_0}) = 0.1$. We will also assume that a positive result involves, not only statistical significance but also, a direction in the difference across the two study groups that favours the effectiveness of the intervention. Let's now consider a series of independent placebo-controlled RCTs, each based on a study hypothesis relating to the treatment of 1000 patients infected with COVID-19. Purely for ease of explanation, we shall assume that the effect size estimate is the observed difference, ΔO, in the proportion of patients with COVID-19 at a recommended follow-up time on comparing the intervention and placebo groups. In this context, H_0 and H_a are the hypotheses that there is no difference and that there is a difference, respectively in the proportions of COVID-19 patients at the recommended follow-up time.

In NHST, β, is used to denote the probability of making a Type II error (not rejecting H_0 given that H_0 is false), commonly referred to as the Type II error rate, or occasionally, as the false negative rate (FNR). The statistical power of a hypothesis test is in turn denoted by '1−β' to signify that β and statistical power are the probabilities of complementary events. In particular, statistical power is the probability of rejecting H_0 given that H_0 is false.

Within the context of an RCT, we can express this conditional probability more explicitly as follows:

1−β: *The probability that that there is a statistically significant difference across the intervention and control groups if H_0 is false.* *(Pr9)*

On comparing *Pr3* and *Pr9*, we can see that *sensitivity* for diagnostic testing has an analogous interpretation to 1−β, that is statistical power, for RCTs in NHST. Those embarking on medical research involving statistical hypothesis testing should also note that statistical power is a pre-study probability and that it is not statistically sound to conduct retrospective power calculations (calculations after the study has been conducted) [5, 6].

We shall now move on to explore conditional probabilities for NHST with the help of a clinical example. For this purpose, we shall assume that α and 1−β take the frequently assumed values of 0.05 and 0.90, respectively. In an analogous fashion to the diagnostic test example illustrated in Table 3.1, we can present the frequencies for the relationship between the hypothesis test findings based on the study sample and the truth about the population as shown in Table 3.2.

To fully appreciate the analogy, however, an important difference should be noted in terms of the types of individuals involved. While our diagnostic statistics were calculated based on comparing the outcome of a diagnostic test with the disease outcome for individual patients, with the RCTs, we are comparing the outcome of a hypothesis test across multiple repetitions of a trial (in practice, multiple similar trials) with what is true about the population, using a particular null hypothesis. Also, while *sensitivity*, as an indicator of diagnostic testing accuracy, is estimated

Table 3.2 Relationship between hypothesis-based decisions and truth

Decision based on hypothesis test applied to sample	Truth about population		
	True difference (H_0 is false)	No true difference (H_0 is true)	Total
$\Delta O = d$ and $P(\Delta O \geq d \mid H_0$ is true$) \leq 0.05$ **Reject** H_0	*a* 90 (100 × 0.90) (True positive)	*b* 45 (900 × 0.05) (False positive)	135
$\Delta O = d$ and $P(\Delta O \geq d \mid H_0$ is true$) > 0.05$ **Do not reject** H_0	*c* 10 (100 × 0.10) (False negative)	*d* 855 (900 × 0.95) (True negative)	865
No. of RCTs in 1000	100	900	1000

Note Calculated values pertain to frequencies across 1000 placebo-controlled RCTs involving COVID-19 patients. '*a*' to '*d*' are labels for the frequencies calculated in the corresponding cells

as a relative frequency, based on the sample data, the desired statistical power of a hypothesis is decided theoretically prior to hypothesis testing. A similar distinction should be made between *1-specificity* as an indicator of diagnostic test accuracy and α in NHST.

For convenience and simplicity, we are assuming in this chapter that all hypothesis tests are two-tailed (two-sided). However, for completeness, it is appropriate to acknowledge that frequently, RCTs are superiority trials in which the hypothesis test is interpreted as a one-tailed (one-sided) test in the direction of an improvement in the intervention group, with an α value of 0.025 replacing that of 0.05 from the corresponding two-tailed test. In this context, for release of the intervention, it is common to require statistical significance for each of two trials [7, 8], that is, that the intervention satisfies the two-trials rule [9].

Let's now return to the analogies we observed earlier between *1-specificity* and α and between *sensitivity* and $1-\beta$, as conditional probabilities. These analogies are reinforced by observing that α and $1-\beta$ as defined in *Pr1* and *Pr9* correspond with the proportions $\frac{b}{b+d}$ and $\frac{a}{a+c}$, respectively in Table 3.2 just as is the case for *1-specificity* and *sensitivity*, respectively in Table 3.1.

It also important, however, to consider the probability that if, based on the p-value, H_0 is rejected in favour of H_a, there is a true difference between the proportion of patients in the intervention and placebo groups. Equivalently, for our current example, we want to know the probability that if a significant difference in percentages of patients across these two groups arises from hypothesis testing, there is a difference in these proportions between the corresponding parent populations for these groups. This conditional probability can be expressed in an analogous way to the PPV in diagnostic testing (*Pr7*). For this, we can start with the following general case for RCTs:

> ***PPV for NHST:*** *The probability that H_a is true (or H_0 is false) if there is a statistically significant difference across the two study groups.* (*Pr10*)

For the above case study, *Pr10* then gives rise to the following special case:

The probability that H_a is true if there is a statistically significant difference in percentages of individuals with COVID-19 at a pre-determined follow-up time across the intervention and placebo groups.

Now suppose we want to know for the above RCTs the probability that if a significant difference in percentages of patients across the intervention and placebo groups in an RCT arises from hypothesis testing, there is *no* difference in these proportions between the corresponding parent populations for these groups. In this case, the conditional probability can be expressed in an analogous way to the FDR in diagnostic testing (*Pr8*). For this, we can start with the following general case:

FDR for NHST: *The probability that H_a is false (or H_0 is true) if there is a statistically significant difference across the two study groups.* (*Pr11*)

Note that within the context of NHST, the FDR is sometimes referred to as the false positive report probability (FPRP) [10]. However, we shall stick to use of the term FDR so as not to lose sight of the conceptual analogies linking the FDRs in diagnostic testing and NHST.

For illustrative purposes, our special case of the above definition of the FDR can be written as follows:

The probability that H_a is false if there is a statistically significant difference in percentages of patients with a COVID-19 diagnosis at a pre-determined follow-up time across the intervention and placebo groups.

Thus far we have seen that within the context of NHST, we can define a PPV and corresponding FDR using a similar approach to that for diagnostic testing. These similarities are reinforced through noting from Table 3.2, that *Pr10* and *Pr11* are equivalent to the proportion of true positives, $\frac{a}{a+b}$, and the proportion of false positives, $\frac{b}{a+b}$, respectively, relative to a positive hypothesis test result from statistical significance testing. Also, using similar algebra to that for diagnostic testing, the above probabilities can be seen to be the complements of one another.

Using the above fractions, we can obtain values for the PPV and FDR from the frequencies in Table 3.2, as follows: $PPV = \frac{90}{135} = 0.67$ and $FDR = \frac{45}{135} = 0.33$.

Thus, if we take the above estimate of 10% as the base rate, only about two thirds of positive outcomes are true positives and hence, approximately, the remaining third are false positives, leading to the false assumption in such cases that the intervention at least has the potential to be effective. We do not know the base rate for COVID-19 trials of the above sort. However, it is not unreasonable to suppose that the estimated value of 10% and the resultant value of 0.67 for the PPV could apply in a number of RCT contexts regardless of how carefully and scrupulously the RCTs have been designed and conducted.

Despite the ongoing debates about the role of NHST in scientific research, it remains commonplace when interpreting the results of clinical studies to neglect the base rate. This is influenced, no doubt, by a lack of available evidence to accurately estimate this value. However, no less influential are the misconceptions arising from the base rate fallacy and the resultant focus on statistical significance at the expense of the PPV, and RCTs are not immune from this difficulty.

Within the context of hypothesis testing, the base rate fallacy arises through placing too much of an emphasis on the achievement of statistical significance to the neglect of the probability of the study hypothesis being true (the base rate). This misunderstanding can lead to the PPV being over-estimated through inadvertently attributing too high a status to statistical significance. As we shall see explictly in Chap. 4 using a similar argument to that for diagnostic testing, the PPV in NHST is directly proportional to the base rate. Developing an awareness of this property together with the relevance of the PPV in the pursuit of scientific truth ought to form a key component of Evidence-Based Medicine (EBM) curricula.

Noting that the base rate fallacy is likely to have a considerable impact on promotion of ineffective treatments, the need is indisputable for a greater emphasis on increasing the PPV in clinical trials, including through a better awareness of the base rate and its potential impact on the FDR, and hence the PPV, on applying NHST for a given null hypothesis. As we shall see shortly, within the context of NHST, the fallacy of confusion of the inverse can also lead to an over-estimation of the PPV.

3.5.2.3 The Fallacy of Confusion of the Inverse

For the Harvard University problem, we saw how conflating the PPV with its inverse, *sensitivity*, could contribute to the difficulties that students might experience in arriving at the correct answer. We also recognised that this was an illustration of the fallacy of confusion of the inverse.

Within the context of NHST, conflating α with the FDR is also an example of the *fallacy of confusion of the inverse*. This involves the misunderstanding that $\alpha = 0.05$ informs us that conditional on achievement of statistical significance, the probability that the corresponding null hypothesis is true is 0.05. This in turn leads to the assumption that "the PPV" (corresponding to the complementary event) is 0.95.

3.5.2.4 The Impact of the Three Fallacies on RCTs

We shall now gain a general overview of the influence of the three logical fallacies we have considered in this chapter on the interpretation of treatment outcomes in RCTs. This will include revisiting the p-value fallacy considered earlier in this chapter. Let's start with a short summary of where we have reached.

Key Learning Point 3.3

The *base rate fallacy* and *the fallacy of confusion of the inverse* can each have a role to play in overestimating the PPV based on NHST. One form of the p-value fallacy points to a mismatch between what can be concluded from a single study and the asymptotic nature of probability upon which the definition of *p-value* relies. Within the context of NHST, all three fallacies yield an over-reliance on statistical significance at the expense of accurate conclusions on the effectiveness of an intervention. As such, they may be influential in the misrepresentation of treatments as effective within the context of well-designed RCTs, that is, trials that *do not* involve questionable research practices (QRPs), such as scientific fraud and selective reporting of study outcomes. Highlighting these fallacies is therefore key to supporting the development of sound critical reasoning among trainee health professionals.

In response to KLP 3.3, there is a clear need in medical education for the PPV to be precisely defined as a conditional probability, and for its importance to be highlighted in the interpretation of positive outcomes from individual RCTs. Awareness should also be raised regarding the deceptive nature of the above three fallacies in the interpretation of clinical research findings.

Through duping researchers into inflating the PPV, the base rate fallacy and the fallacy of confusion of the inverse can contribute to explaining the apparent conundrum that so many breakthrough treatments classified as such on the basis of statistical significance fail to live up to expectations in clinical practice. As is noted elsewhere, [11] there may be occasions where the practicing clinician would wish to reconsider the administration of a treatment to a patient based on study findings if they had a better estimate of the PPV for RCTs relating to the effectiveness of the treatment they are considering. For example, in responding to a statistically significant outcome suggesting that Remdesivir is effective, this would give the clinician a better estimate of the likelihood that this outcome can be trusted. Contrast this approach with deciding between a dichotomy of choices for the patient through focusing on the result of applying α as a threshold for the p-value within an individual Remdesivir trial.

As we have already noted, the PPV is directly proportional to the proportion of true study hypotheses (the base rate for NHST). Our earlier value of 0.67 for the PPV is therefore likely to vary according to a variety of factors that influence this base rate, including the quality of the science behind the hypothesis.

We must also recognise that by virtue of the p-value fallacy, p-values are commonly misunderstood and misused.

The concern over the logical validity of NHST raised in this chapter offers a fresh contribution to the longstanding debate over the use of p-values, which escalated in 2016 further to the publication of the American Statistical Association (ASA) statement on p-values [12]. The statement itself highlights the inadequacy of p-values as self-contained measures of evidence in responding to a null hypothesis. It also lists, but does not show preference for, alternative approaches to NHST, including those involving likelihood ratios, Bayes factors and other forms of Bayesian statistics.

Within the context of well-designed RCTs, our concern need not be so much as to ban p-values. However, we certainly need to take a more tentative approach to their use purely at the level of the individual study as a measure of evidence of treatment effectiveness and give due importance to the use of statistical effect sizes in this context (see Chap. 1).

We should also remain open to the possible emergence of publishing trends in clinical journals whereby Bayesian methods for hypothesis testing are used at least as frequently as NHST. The challenge we therefore face as educators is to make the fundamentals of Bayesian statistics accessible to undergraduate medical students, while guarding against its oversimplification and misuse in no less sense than in the case of NHST.

If only 10% of study hypotheses are true, then, as we have seen, only about two thirds of positive results based on statistical significance are true positives. The immediate issue in terms of patient safety is that where clinical decisions about which treatments to recommend are influenced by positive outcomes from individual trials, up to a third of those recommendations (e.g., to use an expensive more recent treatment) could be relying on misleading evidence.

3.6 The Relationship Between Prevalence and Sensitivity and Specificity

3.6.1 The PPV as a Special Instance of Bayes' Rule

In 1959, Ledley and Lusted defined the PPV, in their own terminology, as a special instance of Bayes' rule for conditional probabilities [13].

For events A and B, Bayes' rule is defined as follows:

$$P(A|B) = \frac{P(B|A)P(A)}{P(B)}, \tag{3.1}$$

where generally speaking, for events X and Y, 'X | Y' can be interpreted as 'X given Y' or, equivalently, 'X conditional on Y'.

In the language of Bayesian statistics, in the above equation, *P(A)* is the prior probability (*the prior*), *P(B | A)* is the likelihood of B given (that) A (has occurred), *P(A | B)* the posterior probability (*the posterior*), and *P(B)* is the marginal probability. The posterior is also understood to be obtained through updating the prior based on the likelihood.

Bayes' rule can be used to derive the following relationship, where *D+* is the event that an individual obtains a positive diagnostic result and ***H*** is the event that the hypothesis, *H*, that an individual has the disease, is true:

$$P(\boldsymbol{H}|D+) = \frac{P(D+|\boldsymbol{H})P(\boldsymbol{H})}{P(D+)}. \tag{3.2}$$

$P(\boldsymbol{H})$ is understood to serve as a plausible estimate, in keeping with the above Bayesian language, a prior, for the (true) prevalence of the disorder of interest, and we can deduce from *Pr3* and *Pr7* that $P(D+|\boldsymbol{H})$ and $P(\boldsymbol{H}|D+)$ represent the sensitivity and PPV, respectively. Further, $P(\boldsymbol{H}|D+)$ is the posterior, which is updated based on the likelihood, $P(D+|\boldsymbol{H})$, while $P(D+)$ is the marginal probability. (Recall from earlier in this chapter that the true prevalence is also the baseline probability for the disorder.)

Therefore, Eq. 3.2 may be expressed as follows:

$$PPV = \frac{sensitivity \times prevalence}{P(D+)}. \tag{3.3}$$

According to the law of total probability, for an outcome, B, $P(B)$ can be expressed in terms of the distinct (or, mutually exclusive) events according to which that outcome can occur. Expressed simply, this law states that if A_1, A_2, A_3, ... is a well-defined set of the possible distinct events according to which outcome B can occur, then:

$$P(B) = P(B|A_1)P(A_1) + P(B|A_2)P(A_2) + P(B|A_3)P(A_3) + \ldots \tag{3.4}$$

Using summation notation, Eq. 3.4 can be written more concisely as follows:

$$P(B) = \sum_i P(B|A_i)P(A_i) \tag{3.5}$$

Using '*not* ***H***' to denote the event that the hypothesis H, that an individual has the disease, is false, Eq. 3.4 or, equivalently, 3.5, can be used to expand the denominator of Eq. 3.3 to obtain:

$$P(D+) = P(D+|\boldsymbol{H}) \times P(\boldsymbol{H}) + P(D+\mid not\ \boldsymbol{H}) \times P(not\ \boldsymbol{H}) \tag{3.6}$$

From *Pr6*, we know that $P(D+ \mid not\ \boldsymbol{H})$ represents *1-specificity*, so we can now express Eq. 3.3 as follows:

$$PPV = \frac{sensitivity \times prevalence}{(sensitivity \times prevalence) + (1 - specificity)(1 - prevalence)} \tag{3.7}$$

In a similar fashion, the NPV can be expressed as $P(not\ \boldsymbol{H} \mid D-)$, where $P(D-)$ is the probability of an individual obtaining a negative test result. We then obtain the following special case of Bayes' rule, with, strictly speaking, $P(not\ \boldsymbol{H})$ serving as the prior:

$$P(not\ \boldsymbol{H}|D-) = \frac{P(D-|not\ \boldsymbol{H})P(not\ \boldsymbol{H})}{P(D-)}. \tag{3.8}$$

Equivalently,

$$NPV = \frac{(specificity)(1 - prevalence)}{P(D-)} \tag{3.9}$$

On using the law of total probability to expand *P(D−)* in the denominator, we obtain:

$$NPV = \frac{(specificity)(1 - prevalence)}{(1 - sensitivity)(prevalence) + (specificity)(1 - prevalence)} \tag{3.10}$$

In each of the above equations for *PPV* and *NPV*, *sensitivity* and *specificity* are assumed constant across populations, with *prevalence* varying (for example, across a changing population throughout the course of a pandemic) [14] and in turn, *P(D+)* and *P(D-)* varying. As we shall see in the next section, experimental evidence suggests that combining the above assumptions concerning *prevalence, sensitivity* and *specificity* in any given clinical context can be misleading. Focusing on these equations for now, however, notice that as *prevalence* converges to 1, *1 − prevalence* converges to 0 and therefore the second summand, *(1-specificity) (1-prevalence),* in the denominator on the right-hand side (RHS) of Eq. 3.7 converges to 0 so that overall, the RHS of this equation converges to 1. Thus, the PPV increases with increasing *prevalence.* As the FDR is the probability of the complement of the event represented by the PPV, we can conclude that the FDR decreases with increasing *prevalence.* Using Eq. 3.10, it can also be deduced that the NPV decreases with increasing prevalence, although less rapidly so than the FDR.

It is important to be aware of the above relationships with *prevalence* in estimating patient risk of having a disorder. For example, with infectious diseases, such as malaria, in moving from one geographical region to another, a decline in prevalence can lead to a particularly low PPV which is overlooked by a visiting health professional due to a failure to appreciate the importance of the base rate (in this case, prevalence) in determining the probability of a positively diagnosed individual being a true malaria case (that is, the PPV). This example illustrates the base rate fallacy in the case of diagnostic testing. In the given clinical scenario, unnecessary alarm could be raised for a considerable proportion of healthcare providers and their patients based on a lack of awareness of the low PPV for the given diagnostic test.

3.6.2 Some Limitations of Bayes' Rule for Modelling Diagnostic Statistics

In taking the PPV to be a special instance of Bayes' rule, Ledley and Lusted set a precedent for modelling the PPV (and the NPV) that has continued across many decades. They also assumed that the prevalence of a disorder was *determined by* non-clinical characteristics, "such as age and place of residence", and, in particular, on "the calendar time of the patient's presentation", but not on any of their clinical characteristics. By contrast, they regarded *sensitivity* and *specificity* as being generalisable to other populations, while remaining *invariant to* differing non-clinical characteristics across these populations. [15] Further, relying on the law of total probability, they expressed Eq. 3.7 in terms of summations across "various conceivable [alternative] illnesses" which could explain the patient's clinical profile, and a set of symptoms rather than a single outcome from a diagnostic test, which was more indicative of, although certainly not unique to, diagnostic practices in their era. [13]. While, for this and other reasons, Eq. 3.7 is disguised by the complexity of the technical details in their paper, it is of interest from a historical perspective to be aware of the source of this application of Bayes' rule, and by inference, the corresponding application of this rule in the parameterisation of the NPV.

As we shall see shortly with a few illustrations, the above segregation of probabilities according as to whether they are determined by non-clinical or generalised clinical characteristics cannot be assumed in general.

We shall also see that the segregation between pre-test and post-test probabilities, which was introduced without a precedent from Ledley and Lusted and remains popular today, is open to debate. Based on the latter segregation, it has traditionally been assumed that the PPV is always deduced after the prevalence. It is also typically assumed from this perspective that the PPV is dependent on pre-test clinical and non-clinical characteristics of the study population. The PPV is thus taken to be temporally posterior to *prevalence*, *sensitivity* and *specificity*, with Bayes' rule being applicable for defining the PPV in terms of *sensitivity* and *specificity* (Eq. 3.7). Similar remarks apply to the NPV (Eq. 3.10). Note, however, that if Bayes' rule is foundational to the parameterisation of the PPV and NPV, this should also hold for the parameterisation of *sensitivity* and *specificity*, noting their status as conditional probabilities. *Sensitivity* and *specificity* can be expressed as follows by applying a little algebra to rearrange the terms in Eqs. 3.3 and 3.9, respectively[3]:

$$sensitivity = \frac{PPV \times P(D+)}{prevalence} \tag{3.11}$$

$$specificity = \frac{NPV \times P(D-)}{1 - prevalence}. \tag{3.12}$$

In entertaining these equations, we are allowing for the possibility that there may be contexts in which, rather than assuming *sensitivity* and *specificity* to be experimentally stable, we should instead apply this assumption to the PPV and NPV and correspondingly, treat them as constants in the parameterisation of *sensitivity* and *specificity*. In this context, one can argue that *P(D+)* and *P(D−)* are the priors, and *prevalence* and *1-prevalence*, respectively, are the marginal probabilities. From an experimental perspective, it would not be unsound to envisage scenarios where

[3] To see that (3.11) and (3.12) are special cases of Bayes' rule, first note that *sensitivity* and *specificity* can be expressed as *P(D+ |* ***H****)*, and *P(D− | not* ***H****)*, respectively and then rearrange Eqs. 3.2 and 3.8, respectively so that these terms are on the left-hand side of the corresponding equation. Also, note that *not* ***H*** and ***H*** are complementary events and that therefore if *P(****H****)* is equated with *prevalence*, *P(not* ***H****)* is equated with *1-prevalence*.

diagnostic test results (and hence *P(D+)* and *P(D−)*) are determined prior to consideration of outcomes referencing the gold standard. This may occur, for example, in cases where the diagnostic test is intended to serve as a potential surrogate for a more invasive gold standard test.

Bayes' rule has a simple derivation within the calculus of the basic laws of probability. Therefore, it is a valid rule and as such, logically guaranteed, unlike, for example, IR1 of Sect. 3.3. Therefore, it should be reasonable to ask, why should *sensitivity* and *specificity*, as conditional probabilities, be generally assumed to circumvent this logically guaranteed rule of probability, noting that the PPV and NPV are each defined by means of this rule?

While a sound debate in response to this question would be instructive in reviewing what we are currently teaching health professionals, we shall leave this particular debate for another day, noting that some fresh discussion on the topic has already emerged, leaving the ground fallow for future contributions. The interested reader may wish to consult the relevant post by Senn [14] for details of earlier work challenging the traditional textbook approach to the presentation of the PPV and NPV using Bayes' rule originally proposed by Ledley and Lusted.

Key Learning Point 3.4

Use of Bayes' rule as a foundational basis for defining the PPV or NPV as a function of *sensitivity* and *specificity*, but not vice versa, has become widely accepted in medical textbooks and "expository papers" [16], a key underlying assumption being that *sensitivity* and *specificity* can each be held constant while *prevalence* varies for a given population type. However, Bayes' rule is intended to serve as a general rule of inference for deriving conditional probabilities. Thus, it is contentious to rely on Bayes' rule exclusively in this way, given that arguably, the prevalence could itself first change for a given population type, giving rise to a change in *sensitivity* and *specificity*. Within this context, *sensitivity* and *specificity* should be defined according to this rule as functions of the PPV and NPV, respectively. From this perspective there is no justification for assuming that *sensitivity* and *specificity* are always pre-determined prior to diagnosis.

Given KLP 3.4, we have a right to question within a given clinical context whether *sensitivity* and *specificity* are any more stable than the PPV and NPV and thus, for example, whether Eqs. 3.11 and 3.12 can be understood to be less reliable than Eqs. 3.3 and 3.9. In challenging traditional teaching approaches to the application of Bayes' rule in diagnostic testing, we need to move beyond Bayes' rule, however, by noting that *sensitivity* and *specificity* can in fact covary with *prevalence*, including in a context where *prevalence* can vary across subgroups within the original population. To see what this means, we need to appreciate that, as is already established in the medical research literature, though not readily taught, *sensitivity* and *specificity* can be influenced by explanatory variables not explicitly included in the above parameterisations, some of which may also influence *prevalence*. In these contexts, rather than being regarded as functions of *prevalence*, *sensitivity* and *specificity* are seen to vary alongside *prevalence*.

This can be illustrated with reference to the diagnosis of coronary artery disease (CAD) through electrocardiography. In a reported study, patients were subdivided into groups according to diagnosis of definite angina, probable angina and nonischaemic pain, for which the corresponding prevalences of CAD were 89, 70 and 22%, respectively. In the order shown, these groupings, as defined, also reflect symptoms of severe CAD with decreasing likelihood, which gave rise to decreasing levels of *sensitivity*. However, specificities covaried inversely with *prevalence*, as presence of angina presents a challenge in deciding on absence of CAD through diagnostic testing, and more so in cases of severe angina [17]. This example reflects how *sensitivity*, *specificity* and *prevalence* for a given disorder can vary with the stage of one of the underlying illnesses that determined patient selection for a study. A closely related point is that the influence of treatment in reversing a disorder may serve to reduce *pre-test prevalence* in later testing. In a separate study [18], angina type (typical or atypical), number of diseased vessels, age, gender and maximal exercise heart rate, were found to be related to *sensitivity* in the use of exercise electrocardiography for diagnosis of coronary disease. It has also been noted that characteristics of a population that do not pertain to patient clinical features per se can also lead to prevalence changes and *sensitivity* and *specificity* covarying with these changes, an example of which is the referral filter used to identify eligible study participants. For example, one study showed that patients selected through positive mammography yielded a poorer sensitivity result with ultrasound, there being a similar effect for *specificity* [17].

These and other findings reflect the non-stability of *sensitivity* and *specificity* for a given diagnostic test and corresponding disorder. This in turn precludes the possibility of making the universal claim for any diagnostic test that the sensitivity or specificity is transferable across study populations that on face value, are similar, irrespective of the prevalence of the disorder being studied. In fact, for a given study population, they allow us to treat these statistics as variables ranging over the individual patients of a population [15] and as capable of varying within a given population over time, including through the course of a disease that is clinically related to the disorder being diagnosed. They also give rise to multivariable approaches to modelling these statistics involving categorical and continuous explanatory variables. In such cases, sensitivity or specificity outcomes can be modelled through binary logistic regression models, with the subgroup for analysis comprising gold standard positives or negatives, respectively and a case being recognised as a positive or negative diagnostic test outcome, respectively. Similar remarks apply to the PPV [15] and, by analogy, the NPV, with the subgroup for analysis comprising those patients testing positively and negatively, respectively and a case being defined as confirmed presence or absence, respectively, of the disorder according to the gold standard.

Based on the above, we should be clear that use of Eq. 3.3 as a special case of Bayes' rule, and the underlying assumptions concerning *prevalence*, *sensitivity* and *specificity*, to model the PPV may in some clinical settings be an oversimplistic approach, leading to inaccuracies in diagnostic statistics. An advantage of the multivariable regression approach is that of being able to generate case-specific probabilities for *sensitivity* and *specificity* based on different patient presentations. The same applies for the PPV and NPV. While the above findings do not contradict the age-old belief that *sensitivity* and *specificity* are mathematically independent of *prevalence*, they do illustrate that differences in *prevalence* can be a "marker" for differences in these measures of diagnostic test accuracy across patient populations with different clinical profiles [17].

Key Learning Point 3.5

The long-standing use of Bayes' rule for defining the PPV in medical textbooks has typically involved a range of assumptions that are not generally valid. These assumptions include:

- the prevailing disease prevalence for an entire study population can serve as a proxy for the prior probability of a patient having the disease irrespective of a) the clinical profile of the individual patients in a study or b) variation in the classification of a disease according to its progress or patient response to treatment;

  and

- the sensitivity and specificity of a diagnostic test are stable irrespective of variation in patients' clinical and non-clinical profiles.

Awareness of this limitation in the use of Bayes' rule and the availability of multivariable models for estimation of diagnostic statistics is foundational to accurate diagnostic decision making.

For the Harvard University problem, we were able to estimate the observed counts pertaining to Table 3.1 using the prevalence value provided. Assuming that this value was plausible, a Bayesian statistician might be satisfied to use it in Eq. 3.7, above in order to estimate the PPV. However, more generally speaking, researchers may be relying on observed frequencies from a single experiment to complete a similar table. In Chap. 1, we saw that random sampling does not preclude the selection of non-representative samples. Therefore, the prevalence that *emerges* from such a table (the sample prevalence) is subject to sampling error. From this perspective, having a good prior estimate of the population prevalence is important and ought to be just as important to the frequentist statistician (see Chap. 2), who places an emphasis on observed data, as the Bayesian statistician, who advocates for the use of plausible likelihoods. In this sense, there should be a meeting of minds with both statisticians arriving at similar results using Eq. 3.7, for example, provided the Bayesian statistician is willing to use data as evidence to inform their choice of prior.

Diagnostic statistics are frequently defined, as in this chapter, with reference to a gold standard. However, access to such a standard can in some contexts be inhibited. This can be due to resource, including financial, constraints for purchasing the test and the need to rely on multiple tests, as in the case of some infectious diseases. In such contexts, the estimation of *disease prevalence* and the above diagnostic statistics can prove particularly challenging. Methodology has been presented in the clinical literature for addressing such challenges through frequentist and Bayesian approaches. While consideration of such methodology is beyond the scope of this book, it is mentioned here, as an awareness of its existence may prove helpful to current or future medical graduates in planning collaborative work requiring the input of a statistician [19–21].

3.7 Conclusion

In the teaching of statistics to undergraduate medical students, it is commonplace to focus on basic epidemiological concepts of risk, including those of *absolute* and *relative risk* and related concepts. However, it is equally important that the topic of *conditional probability* is adequately covered in undergraduate medical curricula. This is evident from the challenges students can face in differentiating between different types of diagnostic statistic and in making sense of a statistically significant result arising from an RCT or any other type of well-designed study for that matter. As we have seen, the p-value fallacy, base rate fallacy and fallacy of confusion of the inverse can, in such contexts, strongly influence how students respond to a positive outcome from an RCT. This is important because misunderstandings arising from these fallacies are transferrable to physician–patient encounters requiring the communication of disease risk and the potential benefit of a new intervention. Therefore, it should be a valuable exercise to expose medical students to these fallacies at appropriate stages of their training. Hopefully, content from this chapter will prove inspirational in this area.

We have also examined the application of the p-value within the context of NHST and noted a sense in which NHST is logically unsound as a form of argument. As we shall see in Chap. 4, Bayesian approaches to hypothesis testing also have their limitations. Nevertheless, it is beyond the scope of this volume to aspire to defending the case that NHST should be superseded by an alternative approach in the teaching of undergraduate medical students. Furthermore, the above considerations have no bearing on the status of the RCT as a gold standard per se; rather, they highlight the different ways in which reasoning about clinical research findings can readily become flawed, regardless of its position as a research design on a traditional EBM pyramid.

In relation to the status of the RCT as a gold standard, we have also seen, however, that, as with other types of study, RCTs can give rise to false positive results and that in particular, within the context of NHST, the PPV can be considerably compromised where the base rate is low. Within this context, positive results from RCTs may be less reliable than those from non-RCT clinical studies involving hypotheses with a comparatively stronger scientific rationale and thus a higher probability of being true (a higher base rate). These observations, together with our consideration of Bayes' rule in this chapter, form the basis for focusing on the management of the PPV in Chap. 4. The importance of doing so should be evident from the fact that, on obtaining a positive RCT finding through hypothesis testing, we should want the probability of an intervention being genuinely effective to be as high as possible, thus contributing reliable statistical evidence to a developing evidence base in favour of the effectiveness of the intervention rather than appealing to statistical significance per se.

References

1. Goodman SN (1999) The p-value fallacy. Ann Intern Med 130(12):995–1021
2. Casscells W, Schoenberger A, Graboys T. Interpretation by physicians of clinical laboratory results. N Engl J Med. 1978;299:999–1001.
3. Yong E. How reliable are cancer studies? The Atlantic. 18 January 2017.
4. Begley CG, Ioannidis JP (2015) Reproducibility in science: improving the standard for basic and preclinical research. Circ Res 116(1):116–126
5. Levine M, Ensom M (2001) Post hoc power analysis: an idea whose time has passed? Pharmacotherapy 21(4):405–409
6. Goodman SNB, Jesse A (1994) The use of predicted confidence intervals when planning experiments and the misuse of power when interpreting results. Ann Intern Med 121(3):200–206
7. Maca J, Gallo P, Branson M, Maurer W (2002) Reconsidering some aspects of the two-trials paradigm. J Biopharm Stat 12(2):107–119
8. Ruberg SC. Victoria. Providing evidence of efficacy for a new drug. Stat Med. 1998;17:1813–23.
9. Senn S. Statistical issues in drug development. 2nd ed. Wiley; 2007.
10. Wacholder S, Chanock S, Garcia-Closas M, El Ghormli L, Rothman N. Assessing the probability that a positive report is false: an approach for molecular epidemiology studies. J Natl Cancer Inst. 2004;96(6):432–42.
11. Gao J. P-values: a chronic connundrum. BMC Med Res Methodol. 2020;20.
12. Wasserstein R, Lazar NA (2016) The ASA statement on p-values: context, process, and purpose. Am Stat 70(2):129–133
13. Ledley RLL (1959) Reasoning foundations of medical diagnosis. Science 130:9–21
14. Senn S. LinkedIn Corporation. 2021. [cited 2021]. https://www.linkedin.com/pulse/screen-test-stephen-senn/.
15. Dawid A (1976) Properties of diagnostic distributions. Biometrics 32(3):647–658
16. Miettinen O, Caro JJ (1994) Foundations of medical diagnosis: what actually are the parameters involved in Bayes' theorem? Stat Med 13:201–209
17. Leeflang M, Bossuyt P, Irwig L (2009) Diagnostic test accuracy may vary with prevalence: Implications for evidence-based diagnosis. J Clin Epidemiol 62(1):5–12
18. Hlatky M, Pryor D, Harrell F, Califf R, Mark D, Rosati R. Factors affecting sensitivity and specificity of exercise electrocardiography. Am J Med. 1984;77.

19. Chikere C, Wilson K, Graziadio S, Vale L, Allen A (2019) Diagnostic test evaluation methodology: a systematic review of methods employed to evaluate diagnostic tests in the absence of gold standard—an update. PLoS ONE 14(10)
20. Black M, Craig B (2002) Estimating disease prevalence in the absence of a gold standard. Stat Med 21(18):2653–2669
21. Joseph L, Gyorkos T, Coupal L (1995) Bayesian estimation of disease prevalence and the parameters of diagnostic tests in the absence of a gold standard. Am J Epidemiol 141(3):263–272

Margaret MacDougall is a senior lecturer in medical statistics and researcher in education at the Usher Institute, Edinburgh Medical School. She developed an interest in philosophy early on in her career through her Ph.D. research in mathematical logic, constructive set theory and the philosophy of mathematics. She has conceived and led eleven funded projects in medical education and statistics and supervised research staff and students at undergraduate and postgraduate levels through various spin-off projects. Margaret seeks to use her research to promote curriculum innovation and enhance student learning in statistics at undergraduate and postgraduate levels and has published on an international scale in the fields of education, medicine, philosophy and statistics. She has served as an invited speaker, chair and conference organiser at various local, national and international events relating to statistics, and learning and teaching in higher education. Currently, she serves as Vice Chair of the Teaching Statistics section of the Royal Statistical Society.

Chapter 4
Increasing the Positive Predictive Value in RCTs: Lessons from Bayes' Rule

Margaret MacDougall

Abstract In this chapter, we shall use Bayes' rule to investigate the parameterisation of the positive predictive value (PPV) as a means of reviewing the possible management of this statistic within the context of null hypothesis significance testing (NHST). We shall also see how this rule can be supportive in seeking to circumvent the influence of the base rate fallacy and the fallacy of confusion of the inverse, through giving precedence to the PPV. This will enable us to consider the Bayesian approach to hypothesis testing, together with a few of the challenges which it presents. As we shall see, while reduction of the significance level, α, may seem to be a particularly promising approach to managing the PPV, this approach should not be perceived *ipso facto* as a remedy for a so-called replication crisis. Furthermore, in our interpretation of the PPV, we need to remain sensitive to the role of the statistical effect size and clinical significance in deciding whether to accept a statistically significant outcome in favour of an intervention as a positive result. While we shall acknowledge the limitations of the Bayesian approach to NHST, the gentle introduction to Bayesian statistics provided here may be of interest to educators seeking a primer for use with undergraduate medical students and students of other health professions within the constraints of a crowded curriculum.

4.1 Overview

In Sect. 4.2, we shall focus on the tendency for the positive predictive value (PPV) to be overestimated, and the accompanying need to respond by ensuring that the problem of a low PPV is acknowledged and possible remedies explored by consideration of its parameterisation, here within the context of null hypothesis significance testing (NHST). For this purpose, we will use Bayes' rule, which we previously met in Chap. 3 within the context of diagnostic testing. This will lead naturally in

M. MacDougall (✉)
Usher Institute, Edinburgh Medical School, Edinburgh, UK
e-mail: margaret.macdougall@ed.ac.uk

M. MacDougall (ed.), *A Medical Educator's Guide to Thinking Critically about Randomised Controlled Trials: Deconstructing the "Gold Standard"*,
https://doi.org/10.1007/978-3-031-25859-6_4

Sect. 4.3 to a consideration of the Bayesian approach to NHST together with its role in explicitly recognising the prior (or base rate) probability and its influence on the PPV, thus taking the focus away from the achievement of statistical significance per se towards optimising the probability that such a result reflects the truth about the effectiveness of an intervention. As well as highlighting that Bayesian statistics has its own limitations, the above introduction to Bayesian statistics will enable us to explore the essential role of α in managing the PPV and the importance of avoiding a one-size-fits-all approach in choosing a new threshold for statistical significance. It is hoped that this focus on the PPV will prove useful to educators in dissuading trainee health professionals from viewing statistical significance as an end in itself and relatedly, relying on statistically significant findings from a single study—even a randomised controlled trial (RCT)—to inform patient care.

4.2 Managing the PPV

As we have seen in Chap. 3, the p-value fallacy, the base rate fallacy and confusion of the inverse each have the potential to mislead the uninitiated into attributing too much weight to the achievement of statistical significance, with the effect that the false discovery rate (FDR) is neglected or confused with α, and hence potentially underestimated, with its complement, the PPV, being overestimated.

We have also noted that the PPV is directly proportional to the base rate, which, in NHST, pertains to the probability, $P(\boldsymbol{H_a})$, of the study hypothesis, H_a, being true. As a reminder from Sect. 3.5.2.2, using classical logic, we can take this probability to be equivalent to the probability, $P(not\ \boldsymbol{H_0})$, of the null hypothesis, H_0, being false. We shall now use Bayes' rule for conditional probabilities to examine the parameterisation of the PPV more closely.

In Chap. 3, for events A and B, Bayes' rule was defined as follows:

$$P(A|B) = \frac{P(B|A)P(A)}{P(B)}. \tag{4.1}$$

Also, conditional probability *Pr10* was used to represent the PPV for NHST in prose. Using more formal language, the PPV can be abbreviated as $P(not\ \boldsymbol{H_0}|p \leq \alpha)$, where H_0 is the null hypothesis, | is the conditional sign introduced in Chap. 3 and p is the p-value. Using Eq. 4.1, above, this leads to the following formula:

$$P(not\ \boldsymbol{H_0}|p \leq \alpha) = \frac{P(p \leq \alpha|not\ \boldsymbol{H_0})P(not\ \boldsymbol{H_0})}{P(p \leq \alpha)}. \tag{4.2}$$

Let's consider a few of the terms in Eq. 4.2 separately. From Chap. 3 we should see that $P(p \leq \alpha|not\ \boldsymbol{H_0})$ matches the statistical power, $1-\beta$, and, as noted above, $P(not\ \boldsymbol{H_0})$ matches the base rate.

Also, we can expand the expression '$P(p \leq \alpha)$' as follows:

$$P(p \leq \alpha) = P(p \leq \alpha | not\, \boldsymbol{H_0}) \times P(not\, \boldsymbol{H_0}) + P(p \leq \alpha | \boldsymbol{H_0}) \times P(\boldsymbol{H_0}). \tag{4.3}$$

Equation 4.3 involves an application of the law of total probability (Chap. 3), which expresses the total probability of an outcome (here, the outcome $p \leq \alpha$) in terms of the distinct events whereby this outcome can be achieved (here, in terms of H_0 being either false or true). From our definitions of α and 1−β as conditional probabilities in Chap. 3 (see *Pr1* and *Pr9*) and using 'π' to denote the base rate, we can rewrite this equation as follows:

$$P(p \leq \alpha) = (1 - \beta)\pi + \alpha(1 - \pi). \tag{4.4}$$

We can use the above observations to rewrite Eq. 4.2 as follows:

$$P(not\, \boldsymbol{H_0} | p \leq \alpha) = \frac{(1 - \beta)\pi}{(1 - \beta)\pi + \alpha(1 - \pi)}. \tag{4.5}$$

In keeping with how the PPV was introduced earlier in this section, let's replace '$P(not\, \boldsymbol{H_0} | p \leq \alpha)$' by '*PPV*' in Eq. 4.5 to obtain[1]:

$$PPV = \frac{(1 - \beta)\pi}{(1 - \beta)\pi + \alpha(1 - \pi)}. \tag{4.6}$$

The reader might like to invite students to confirm that, within the context of diagnostic statistics, the PPV, *P(individual has the disorder (according to the gold standard) if they have tested positive for that disorder)*, can be expressed in a similar way, as previously shown in Eq. 3.7 of Chap. 3, where, in this case, the prevalence of the disorder is the special case of π for in diagnostic setting.

From Eq. 4.6, it is possible to see that the PPV is inversely proportional to α. In order to see how, based on this equation, the PPV is influenced by π and 1−β, we are going to apply a useful algebraic principle that if we divide the numerator and denominator of a fraction by the same (non-zero) term, the resultant fraction is mathematically equivalent to the original one. In what follows, we shall correspondingly assume that both the base rate and statistical power are greater than zero.

Firstly, then, let's divide the numerator and denominator of the right-hand side (RHS) of Eq. 4.6 by π to obtain:

---

[1] Our journey into Bayesian statistics in this chapter is intentionally at an introductory level to respect the more general learning needs of medical students and students of other health professions. However, to ensure clarity in supporting previous work, it is noted here on passing that alternative routes to deriving the formula for the PPV are provided by Ioannadis [1]. These approaches involve the prior odds, R (the ratio of the number of true study hypotheses to the number of false study hypotheses for those hypotheses tested for a given field of study). Ioannadis' most basic formula using R can be translated into Eq. 4.6 by noting that $R = \frac{\pi}{1-\pi}$.

$$PPV = \frac{1-\beta}{(1-\beta)+\alpha\left(\frac{1}{\pi}-1\right)}. \tag{4.7}$$

Note from this equation that whenever the probability, π, of having a true study hypothesis is exactly 1, the PPV is also exactly 1. In diagnostic testing, the equivalent relationship is that when the prevalence of a condition is 100% in the study population, the percentage of positive diagnoses based on the diagnostic test which are true positives (relative to the chosen gold standard) is, unsurprisingly, also 100%.

Typically, π will be less than 1, meaning that $\frac{1}{\pi}-1$ will be positive. Notice that as π increases, $\frac{1}{\pi}-1$ decreases and the denominator on the RHS of Eq. 4.7 also decreases, causing the PPV to increase. This is consistent with what we observed using Eq. 3.10 in Chap. 3 concerning the relationship between the PPV and the base rate, prevalence, within the context of diagnostic testing. Notice, also, that as the significance level, α, increases, this time the above denominator increases, and in turn, the PPV decreases. Thus, the PPV is inversely proportional to α.

Now let's divide the numerator and denominator of the RHS of Eq. 4.6 by the statistical power, $1-\beta$, to obtain:

$$PPV = \frac{\pi}{\pi + \frac{\alpha}{(1-\beta)}(1-\pi)}. \tag{4.8}$$

We can see that as $1-\beta$ increases, the denominator on the RHS of Eq. 4.8 decreases and in turn, the PPV increases. Thus, the PPV is also directly proportional to $1-\beta$. Thus, we can see that the PPV is increased both by decreasing the statistical significance level, α, and by increasing the statistical power, $1-\beta$, respectively or, equivalently, both by decreasing the probability of a Type I error and by decreasing the probability of a Type II error, respectively. Note, however, that, for a fixed value of α, an increase in $1-\beta$ requires an increase in the required minimum sample size, which helps to explain why the absence of positive results from small studies ought to be interpreted with caution.

While, in attempting to manage the PPV, we may be able to control each of α and β, π can continue to influence the PPV unnoticed, largely due to the base rate fallacy (see Sect. 3.5.2.2). This being the case, questionable research practices (QRPs), including scientific fraud, are a force to be reckoned with in managing the PPV, with QRPs having the potential to make false hypotheses seem plausible. Publications offering guidelines on the conduct or reporting of clinical research have proliferated over recent years and have been well publicised through bodies such as the World Health Organisation and the Equator Network. Nevertheless, legitimate concerns of a political nature remain, including in terms of strict adherence to guidelines.

Even in such contexts, it is important to get to the heart of the matter in recognising the importance of proactively exploring meaningful prior estimates of π as a means of estimating the PPV more accurately. The question of how this can best be achieved is itself open to philosophical debate, knowing that what we consider to be a strong rationale for or against a hypothesis today, possibly based on prior pre-clinical (laboratory) research, may be overruled later by fresh insights that need to be gleaned on the basis of cumulative scientific evidence over time. In opening up this challenging area of need and uncertainty in a transparent way, this chapter will hopefully stimulate growth in working towards a better understanding of how to estimate π, possibly through the intervention of Bayesian approaches. Addressing this challenge is, of course, beyond the scope of this chapter.

Key Learning Point 4.1

In seeking to manage the PPV, it is instructive to consider the parameterisation of this statistic as a special case of Bayes' rule. Based on simple algebra, it is evident that the PPV is directly proportional to π and $1-\beta$ and inversely proportional to α. This shows that efforts to manage the PPV could include increasing statistical power and decreasing the statistical significance level, although a low probability of the study hypothesis being true–a low π–would still attenuate the PPV.

The observations under Key Learning Point (KLP) 4.1 are important in managing the PPV, a strong incentive being that of reducing physician use of reported false positive results, including those of RCTs, in clinical decision making. For example, through misplaced trust exclusively in the quality of the design, conduct and reporting of the study and conformity to standard drug regulatory procedures, a couple of Phase III RCTs designed within a context where the base rate is low may prove to be a liability. Such trials involve hundreds, or even, thousands, of patients and are typically focused on assessing the effectiveness of an intervention against a standard treatment. As we noted in Chap. 3, use of the two-trials rule to justify the release of a treatment intervention is common. Nevertheless, in the application of this rule, a low base rate can still result in a modest PPV for the trials and thus a considerable risk that statistically significant results in favour of the intervention are misleading. In a clinical setting, use of such results unchecked may in turn lead to a patient foregoing a more beneficial existing treatment option or to scientific progress in the development of such an option being hindered [2].

In response to this concern, with the benefits of available time and funding, an emphasis can be placed upon the need to ensure that scientific rationales for choice of hypotheses in RCT studies have themselves been evaluated and approved within scientific communities through preclinical laboratory testing. That is not to say that such testing *guarantees* a high π. False study hypotheses can be conceived on the basis of small laboratory studies, and if the clinical studies are in turn themselves small, combining the results in a meta-analysis may strengthen the degree of deception, particularly if the aggregate effect size turns out to be rather large. Assuming random sampling, for underpowered studies, the smaller the sample size, the more likely the effect size estimate from the sample is to overestimate the true effect size and that in turn, a combination of multiple wrongs in a meta-analysis will be misconstrued as a statistical right. More generally, the way researchers respond to positive statistical findings ought to be characterised by a healthy scepticism rather than, as is more common, by euphoria.

The need to adopt this perspective is illustrated by published findings in response to two meta-analyses, involving RCTs only, for assessing the effectiveness of magnesium in the treatment of myocardial infarction (MI) [3]. The scientific rationale for the meta-analyses followed that of a number of earlier small RCTS, which relied on previous reports of the antiarrhythmic action of magnesium [3, 4]. Therefore, it would be unfair to suppose that concerns about the misleading conclusions from this study can be explained in terms of a poor scientific foundation. In an editorial in the journal *Circulation*, the highly significant results and large reductions in incidence rates represented by the effect size estimates from these meta-analyses were used to lend weight to the claim that "the totality of the evidence probably would persuade the majority of physicians to use the intervention" [4]. Nevertheless, in the same article the authors recommended awaiting the results of a mega-trial involving the same intervention "before routinely using magnesium" [4]. This was based on concerns about clarity relating to the interpretation of the results, including in relation to the type of patient, the sustainability of the outcomes and the precision of the estimation of the level of short-term mortality reduction.

The mega-trial, the fourth International Study of Infarct Survival (ISIS-4), was later to involve 58,050 patients [5]. It is beyond the scope of this chapter to cross-evaluate the research and reporting in the primary studies for the meta-analyses against those in the mega-trial, except to note that the mega-trial involved a consistent large-scale approach which could not be emulated across the different studies included in each of the meta-analyses, although the results of the meta-analyses were key drivers in designing ISIS-4 [6]. Importantly, the mega-trial reported that there was no statistically significant reduction in short-term mortality either overall or through consideration of well-defined patient groups. This trial also reports that for several adverse secondary outcomes, a statistically significant increase in rate was found to be associated with use of intravenous magnesium, thus undermining the general safety of the intervention.

With reference to the above meta-analyses and other meta-analyses, the results of which had stimulated large-scale "more definitive studies", such as ISIS-4, Yusuf later offers the following perspective on the finding that the large-scale studies did not support the 'positive' findings from the corresponding meta-analyses:

> [I]f one assumed that only moderate sized effects were possible, the apparent large effects observed in the meta-analyses of small trials with magnesium … should perhaps have been tempered by this general judgment. If a result appears to be too good to be true, it probably is. If such a judgment were applied, then one would not have accepted these meta-analyses based on small numbers of events as being conclusive, especially when they suggested very large effect sizes that were unlikely to be plausible [6].

These observations are important, generally speaking, but also in relation to having a better understanding of the efficacy of intravenous magnesium in reducing short-term mortality rates, noting that the authors of one of the meta-analyses discussed in the Circulation editorial had previously reported that, "Our study provides strong evidence that intravenous magnesium may have a beneficial effect on mortality" [7]. Also, it is implicit from Yusuf's remarks and from the examination of funnel plots for detecting publication bias [3] that the criteria assumed for choice of studies to include in the meta-analysis were, among other design issues, instrumental in producing inflated results for the effectiveness of intravenous magnesium [7]. Arguably many of these issues could have been managed had the existing Preferred Reporting Items for Systematic Review and Meta-Analysis Protocols (PRISMA-P) [8] been available and implemented early on. Nevertheless, they do alert us to the need to be cautious in responding to the traditional evidence-based medicine (EBM) pyramid, which places systematic reviews, including meta-analyses, at the apex of the pyramid, corresponding to best evidence, while bypassing the potential range in quality of the individual studies which contribute to such reviews and which are represented by the lower layers of the pyramid [9].

The lesson from this case study is summarised in KLP 4.2.

Key Learning Point 4.2

Meta-analyses, including those exclusively involving RCTs, inflate the importance of positive results arising from multiple misleading small studies. As such, they have the potential to contribute to the problem of low PPVs. They ought, therefore, to be included among the types of study requiring remedial attention in programmes aimed at managing the PPV, irrespective of their position at the apex of the traditional EBM pyramid.

Having noted that concerns about the PPV cannot be generally resolved through use of meta-analyses, let's now consider the behaviour of the PPV more carefully in relation to the parameters in Eq. 4.6. Figure 4.1 illustrates the nature of the relationships between the PPV and each of π and $1-\beta$ for different values of α.

As is well known, statistical power is a monotonic increasing function of sample size and is also influenced by other parameters, the identity of which is determined by the choice of statistical hypothesis test and the values of which are estimated from prior knowledge relating to the study population. Therefore, in seeking to increase statistical power, it makes sense, theoretically speaking, to seek to increase the sample size, although in designing an RCT, this may be challenging, due to constraints on funding, resources and time for conducting the trial. We shall return to this point shortly. Note also from Fig. 4.1b that, even were it possible to increase $1-\beta$ to 0.95, if the commonly used value of 0.05 for α were assumed, the PPV would remain below 70% for a low π of 0.1. This illustrates the general point that for $\alpha = 0.05$, where π is low, increasing $1-\beta$ will deliver a modest return and is therefore not cost efficient.

From the perspective of managing the PPV, it is evident from Fig. 4.1 that α has considerable potential, as expressed in KLP 4.3. However, in balance, it is also important to bear in mind that for a given sample size, decreasing α (the FPR) will lead to an increase in the Type II error rate (equivalently, a decrease in $1-\beta$), although, based on our above findings, the influence of these two competing effects on the value of the PPV is likely to be stronger in the case of α.

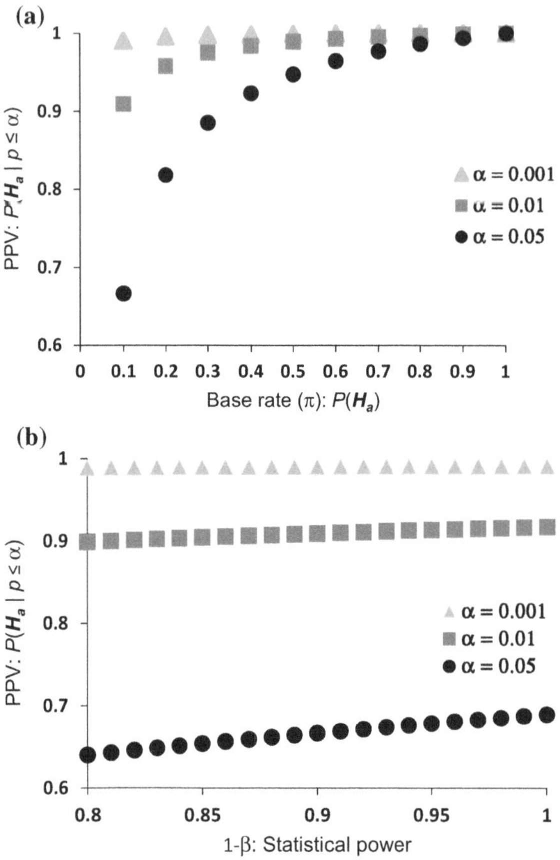

Fig. 4.1 Relationship in NHST between the positive predictive value (PPV) and a) the base rate (π), with $\beta = 0.1$ and b) the statistical power, $1-\beta$, with $\pi = 0.1$, for different values of α

Key Learning Point 4.3

For particularly small values of α (such as 0.001), the PPV is close to 1 and is robust to changes in π and $1-\beta$. This suggests that minimising α is a worthwhile strategy to consider in aiming to maximise the PPV.

Options to control the PPV by reducing α have been considered elsewhere. In particular, for those intent on minimising reputational damage based on false reliance on postive results from NHST, Colquhoun recommends reducing α to 0.001 as a means of keeping the PPV across studies at an acceptable level (above 0.95) [10]. For the study of new effects, others recommend a significance level of 0.005 as a means of reducing the FDR to "reasonable" levels, while recommending that cases where statistical significance is achieved with $\alpha = 0.05$, but not with $\alpha = 0.005$, be reported as providing "suggestive evidence" [11].

With respect to this advice, care should be taken to contrast the behaviours of the p-value, as a random variable, depending on whether H_0 or H_a is true. Under H_0, the p-value has a uniform distribution across the interval [0, 1], irrespective of the sample size. By contrast, under H_a, the p-value is inversely proportional to the sample size. Correspondingly, for a well-powered study, we ought to be fairly sceptical about the truth of H_a in those cases where the p-value is equal to 0.05 or marginally less than 0.05. This truth is commonly overlooked in medical publications, with an emphasis being placed on what might have been achieved given more data, to the neglect of concern about what ought to have been achieved, given the observed data, if H_a were true. Combined with infatuation with achievement of statistical significance as an end in itself, such bias toward borderline significant cases can serve as a breeding ground for false positives within the context of NHST, not to mention publication bias.

These concerns in turn strengthen the case for reducing α as a strategy for managing the PPV at the level of the individual study hypothesis. We should be careful, however, to avoid misconstruing this recommendation as a strategy for addressing the replication crisis that is reputed to particularly affect psychology [12] and medicine. Here 'replicability' can be understood to mean the capacity to obtain consistent results *across studies* aimed at answering the same scientific question with similar methods but with new data collection [13]. In turn, the replication crisis is commonly depicted as the recent escalation of difficulties in replicating findings of published studies into a crisis. Within the context of clinical research and elsewhere, the above difficulty refers, more precisely, to the age-old problem [14] that hypotheses that are validated by one study and then published are frequently not validated by subsequent studies. The latter studies typically find no evidence of an effect or produce effect size estimates markedly smaller than the original study, thus 'failing to replicate findings'.

If adopting an α as low as 0.005 were to present an attractive potential trade-off between the PPV and statistical power as a response to the replication crisis so-defined, we would expect to already see a convincing trend towards particularly low p-values in studies pertaining to long-established positive findings based on the most commonly assumed values for statistical power (within the range 0.80 – 0.90). The absence of this trend points to the irreconcilable tension between achieving replication, as defined above, as an end in itself, and advancing the cause of ground-breaking research from a patient-centred perspective. Instead, consideration of a plausible estimate or range of estimates for π on a per-hypothesis basis ought to inform the level of stringency for α with the support of Eq. 4.6, above.

Even where the p-value for an initial study is as low as 0.005, for $\alpha = 0.05$, the probability of the duplicate study not achieving statistical significance can be as high as about 30%, due to the underlying distributional properties of the chosen hypothesis test [15]. This finding should help to moderate the following claims by Benjamin et al.:

> For fields where the threshold for defining statistical significance for new discoveries is $P < 0.05$, we propose a change to $P < 0.005$. This simple step would immediately improve the reproducibility of scientific research in many fields [11].

It should also serve to illustrate the relevance of the following remarks by Devezer et al. to attempts to address concerns about irreplicability:

> We expect methodological reforms to be held to standards that are at least as rigorous as those we expect of empirical scientists [16].

The idea, however, that irreplicability is a problem to be solved ought to be handled with care, as is illustrated by the following general result for an original and duplicate trial involving a two-tailed statistical test of significance across two groups under particular distributional assumptions[2] [15]:

> If the first trial yields $p = \alpha$ then the probability that a second trial will be significant at significance level α (and in the same direction as the first trial) is 0.5 [17].

This result reflects the fact that, for *any* non-zero value of α, conditional on the outcome of the first trial involving borderline significance, there are cases where, based on the second trial, irreplicability is as likely as replicability, purely on theoretical grounds.

Also of interest is improved rigour in the interpretation of the experimental design and in turn, the appropriate choice of model for hypothesis testing. The latter choice influences the accuracy of the standard error, which is designed to tell us how well we can expect a sample effect size to estimate the parent population effect size. A poorly estimated standard error can lead to false conclusions from hypothesis testing, including through leading to over-estimation or, more typically in research, under-estimation, of the p-value in NHST. When considering RCTs, a useful example in this respect is the problem of misspecification of cluster membership in the design of cluster-randomised trials. Cluster membership of patients in a trial can vary during the period that the trial is conducted, through, for example, patient relocation to a different treatment centre. Alternatively, cluster membership may not have been consistently recorded with appropriate future analyses in view, leading to problems with missing data or misallocation of patients to clusters. This inevitably leads to invalid estimates of the standard error, which take into consideration the assumed, rather than the actual, correlation structure of the data, and in turn Type I and Type II errors, the first of which is the more common, due to the extent of clustering being underestimated or misclassification of patients across existing clusters [18]. Use of individually randomised, parallel group RCTs in lieu of cluster randomised trials, where the latter are required, is a more basic means of neglecting clustering, with data being analysed at the level of the individual patient rather than, for example, the GP catchment area. Misappropriation of the type of experimental design in this case is well-recognised as a potential source of false positives, through underestimating standard errors for effect size estimates and, in turn, the corresponding p-values [19].

---

[2] These conditions include lack of an informative prior probability distribution for the true treatment effect and that the observed difference is the same as the true (or, population difference).

Use of a more stringent α could certainly serve as a gatekeeper for such studies in reducing the number of falsely reported positive outcomes. However, with accurate reporting and future meta-analyses in view, it would be preferable to ensure at the stage of study protocol design, that data collection can support accurate monitoring and recording of cluster membership.

It is important to bear in mind, though, that if the standard for α is raised (by decreasing the value of this parameter), then in order to maintain the existing level of statistical power, a considerable increase in minimum required sample size may be needed. This is the case because the value of $1-\beta$ for a given sample size falls with decreasing values of α [20]. As we have noted earlier, increasing the sample size can prove challenging. Depending on the nature of the case scenario, the availability of eligible patients within a given time period may be too limited to design a study to meet this standard. This might be the case, for example, in an RCT assessing the effectiveness of exercise training on outcomes in solid organ transplant recipients, if the number of eligible patients having undergone solid organ transplantation within the time period of interest is rather modest. Despite the onset of the big data revolution, real clinical practice must continue to take its course and there are clinical contexts within which the suggestion in KLP 4.3 can place unreasonable demands on researchers.

For completeness, it is also helpful to note that the (statistical) effect size is an important parameter which is positively correlated with the PPV, and that an increase in this parameter increases the rate at which statistical power, and in turn, the PPV, increases with increasing sample size. As a point of clarity, it is important to contrast the latter behaviour with the property highlighted in Chap. 1 that α is invariant to changes in sample size.

Thus far, we have considered parameters which influence the PPV, which, as such, are equally capable of influencing the FDR, including among well-conducted RCTs. This is important in helping us to see how we may have some level of control over the PPV. We have also noted, however, that there are limitations in the extent to which α, $1-\beta$ and π can be controlled in order to manage the PPV. This ought to be of interest to practicing health professionals in being more sensitive to the importance of π in making informed decisions about use of RCT findings in clinical decision making. This can be supported by making estimates of the PPV more readily available within the context of hypothesis testing and the reporting of the results thereof, including in academic journals.

We have also noted that strategies aimed at optimising the PPV should not be misconstrued as addressing study irreplicability in a general sense. With Chap. 7 in view, it should also be recognised that study replicability need not serve as an indicator of truth and as such, should not be viewed as an end in itself. Indeed, as is indicated elsewhere [16], depending on the stringency and correctness of the model assumptions assigned to a procedure for testing a hypothesis, some true results are highly unlikely to be replicable while some false results are highly likely to be replicable. Correspondingly, we are wisely warned by Devezer et al. against adopting the rationale that "if a result is independently reproduced many times, it must be a true result." and thus, the common belief that "Replication is a means of increasing the confidence in the truth value of a claim" [16]. Use of incorrect models, and methodologies more generally across studies, can make all the difference in this respect. Relatedly, the need to increase the NPV and as such, reduce the likelihood of statistical testing giving rise to false negatives, should be recognised. All to readily, comfort can be derived from the replicability of negative results from NHST, the falseness of which remains undetected. Such considerations are, however beyond the scope of this chapter.

We shall now briefly consider the Bayesian approach to hypothesis testing as a means of explicitly recognising the PPV in hypothesis testing and as such, avoiding being misled by two of the three logical paradoxes that we considered in Chap. 3.

4.3 The Bayesian Approach to Hypothesis Testing

4.3.1 The Basic Bayesian Model

An important difference between NHST and Bayesian statistics is that while, through focusing on statistical significance, the first of these places an emphasis on the probability of the data conditional on the null hypothesis being true, the second is centred on the plausibility of a study hypothesis conditional on the data. As we shall see shortly, calculation of the PPV is a natural product of the latter approach. Thus, at the very least, as long as NHST continues in medicine as an accepted approach for publication, an awareness of the Bayesian approach is important in ensuring that the PPV is properly understood and used within a context where an important new scientific result is found to be statistically significant.

To appreciate the Bayesian approach to hypothesis testing more fully, it is helpful first to think of Bayesian inference more generally as responding to a research question within a context where there is at least one unknown parameter of interest, together with relevant observed data [21]. Bayesian inference can then be formulated in terms of the following steps (ibid.):

1. Specify a prior probability distribution (the *prior*) on the unknown parameter(s). (This often reflects accumulated knowledge about the research question.)
2. Summarise the observed data using a likelihood function.
3. Condition the likelihood function on the prior.
4. Construct the posterior distribution (the *posterior*) to produce updated knowledge concerning the unknown parameter(s) through multiple simulations and hence answer the research question.

While the above model relates to a more general notion of distribution, in the current chapter, as in Chap. 3, we are focusing on the case where the prior pertains to a single probability. Let H be the study hypothesis under investigation, $\boldsymbol{H}$ the event that H is true and D some pertinent data, such as the outcome of a trial. (Witin the context of Bayesian statistics, the pertinent data, D, can take a variety of forms).

The probability *P(**H**)* is the 'prior probability' of ***H***—that is to say, our degree of belief (or certainty) that ***H*** is true prior to observing *D*. The probability, *P(D)*, is our degree of belief concerning the data—for example, how likely we think the trial is to produce *D*. Thus, *P(D* | ***H****)* is the conditional probability of the data given ***H*** (the likelihood function, or likehood) and P(***H*** | *D)* is the conditional probability of ***H*** given the data (the posterior probability, or posterior).

Bayesianism is anchored on Bayes' rule (Chap. 3). We can use the above probabilities to present a particular instance of this rule as defined in Eq. 4.1 as follows:

$$P(\boldsymbol{H}|D) = \frac{P(D|\boldsymbol{H})P(\boldsymbol{H})}{P(D)}. \tag{4.9}$$

To make the application of steps 1–4 clearer in the formation of Eq. 4.9, this equation may be rewritten as follows, where θ may be interpreted as a vector containing the parameters used to define ***H***:

$$\underbrace{P(\boldsymbol{H} \mid D)}_{\substack{\text{Posterior}\\ \text{probability}\\ \text{concerning } \theta}} = \underbrace{P(\boldsymbol{H})}_{\substack{\text{Prior}\\ \text{probability}\\ \text{concerning } \theta}} \times \underbrace{\frac{P(D \mid \boldsymbol{H}))}{P(D)}}_{\substack{\text{Updating factor}\\ \text{concerning } \theta\\ \text{for } P(\boldsymbol{H} \mid D)}} \tag{4.10}$$

Equation 4.9 tells us that posterior probability *P(**H*** | *D)* is directly proportional to both the prior probability *P(**H**)* and the conditional probability *P(D* | ***H****)* and inversely proportional to the marginal probability *P(D)*. Equation 4.10 tells us that if the experiment produces outcome *D*, then we should update the probability we attach to the truth of *H* from the value of *P(**H**)* to the value of *P(**H*** | *D)* by means of the updating factor $\frac{P(D \mid \boldsymbol{H})}{P(D)}$. Here, *P(D* | ***H****)* plays the role of the likelihood function in item 2 of the list characterising Bayesian inference which we considered earlier.

Bayes' rule is explicit in making the probability of the truth of *H* after the evidential data have been obtained dependent on the probability attached to the truth of *H* before the experiment. This is unlike the standard use of NHST we considered earlier, where the evidence from the trial alone, as determined by hypothesis testing, determines whether we should reject the null hypothesis in favour of the alternative hypothesis—the antecedent plausibility of the latter hypothesis is irrelevant.

In the Bayesian approach to hypothesis testing, information, the base rate, which the unwary might otherwise overlook, is explicitly represented by the prior probability, *P(**H**)*, from the outset. In this setting, the terms, 'base rate', 'prior probability' and 'prior' are interchangeable.

Now let's revisit the PPV in diagnostic testing, the parameterisation of which, as we saw in Chap. 3, has traditionally been modelled according to Bayes' rule. For our convenience, Eq. 3.3, which arises from this application of Bayes' rule can be expressed as follows:

$$PPV = prevalence \times \frac{sensitivity}{P(D+)}, \tag{4.11}$$

with $P(D+)$ representing the probability of a positive diagnostic result.

This helps us to see more clearly that, in this case, *prevalence* is the prior probability and $\frac{sensitivity}{P(D+)}$ is the updating factor, thus forming an analogy with the properties of Eq. 4.10, above, used to express the PPV within the context of NHST.

4.3.2 The Bayesian Approach in Diagnostic Testing Revisited

Consider the Harvard University problem presented in Chap. 3. Here, we took ***H*** to be the event that the hypothesis, *H,* that an individual has the disease is true. Thus, $P(\boldsymbol{H})$ was the prevalence, or base rate, known in more formal Bayesian terminology as the *prior*. Within this setting, for the rare disease in the Harvard University problem introduced in the same chapter, $P(\boldsymbol{H}) = 0.001$. Also, the data *D* is the outcome that an individual tests positive for the disease, which, for consistency with the notation in Chap. 3, we can denote by *D*+. In turn, $P(D+ \mid \boldsymbol{H})$ and $P(not\ D+ \mid not\ \boldsymbol{H})$ are the sensitivity and specificity of the test, respectively, which we know from the reliability to be 0.95. Further, using classical logic, *not D*+ is equivalent to the outcome that an individual tests negative test, which, as is in Chap. 3, we denote here by *D*– . Using the above terminology, we are going to use Bayes' rule to solve the above problem and compare the answer we get with that which we obtained in Chap. 3 using Table 3.1.

$$P(D+ \mid not\,\boldsymbol{H}) = 1 - specificity = 1 - P(D- \mid not\,\boldsymbol{H}) = 0.05.$$

In Chap. 3, we also saw that

$$P(D+) = P(D+ \mid \boldsymbol{H}) \times P(\boldsymbol{H}) + P(D+ \mid not\,\boldsymbol{H}) \times P(not\,\boldsymbol{H}). \tag{4.12}$$

By substituting the relevant values pertaining to the Harvard University problem into the right-hand side of Eq. 4.12, we obtain:

$$P(D+) = 0.95 \times 0.001 + 0.05 \times 0.999 = 0.0509.$$

We have $P(\boldsymbol{H})$, $P(D+ \mid \boldsymbol{H})$ and $\mathrm{P}(D+)$, so we can now use the special case of Bayes' rule represented by Eq. 4.9 to obtain the PPV, $P(\boldsymbol{H} \mid D+)$ as follows :

$$P(\boldsymbol{H} \mid D+) = \frac{0.95 \times 0.001}{0.0509} = 0.019,$$

which, with the same degree of accuracy, is the answer we obtained in Chap. 3 from the frequencies in Table 3.1.

4.3.3 The Bayesian Approach to Evaluating Interventions

We have already noted in Chap. 3 that there are limitations in the use of Bayes' rule for representing the parameterisation of the PPV in diagnostic testing (see KLPs 3.4 and 3.5 for a summary). Nevertheless, familiarity with the use of Bayes' rule in this way and the corresponding application in the previous section should provide insight into its alternative use in testing the study hypothesis relating to the effectiveness of an intervention, as there is a direct analogy.

Here, hypothesis H is a function of one or more parameters pertaining to the effect size. For simplicity, in applying steps 1–4, above, we shall restrict ourselves, as in Sect. 4.2, to the case where H can be replaced by either of the competing hypotheses H_0 and H_a from NHST. We shall also assume here that these hypotheses take the form $\theta_1 = \theta_2$ and $\theta_1 \neq \theta_2$, respectively for parameters θ_1 and θ_2.

To illustrate the application of Bayes' rule in the case of H_a, let us consider a single placebo-controlled trial of the antiviral medication Remdesivir for treatment of COVID-19. If, taking positives to be statistically significant outcomes in favour of Remdesivir, we assume the proportion of true positives in Table 3.2 to be the base rate, then this base rate serves as the prior scientific knowledge we require for estimating $P(\boldsymbol{H_a})$, where H_a is the study hypothesis that at 14 days, there is a difference in the proportions, θ_1 and θ_2, of COVID-19 cases in the Remdesivir and placebo groups, respectively.

Based on our observations in Chap. 3, we could estimate that $P(\boldsymbol{H_a})$ is of value 0.1. This is, of course, a rather simplistic approach to estimating $P(\boldsymbol{H_a})$, as it is based on the idea that the prior is a constant across an unspecified family of clinical hypotheses, including the current one. In this context, the data, D, serve as evidence in support of H_a and include the observed outcomes corresponding to the population parameters θ_1 and θ_2. Let's take D to constitute a statistically significant difference between the corresponding sample proportions, $\hat{\theta}_1$ and $\hat{\theta}_2$, of COVID-19 cases at 14 days, respectively. We will now explore NHST within the framework of Bayesian statistics, with the bonus that, as we shall see, the base rate is explicitly recognised.

For the special case of NHST, $P(D \mid \boldsymbol{H_a})$ and $P(D \mid not\ \boldsymbol{H_a})$ (equivalently here, $P(D \mid \boldsymbol{H_0})$) represent the statistical power of the hypothesis test and the statistical significance level respectively, which, in keeping with common practice, we can choose to be 0.90 and 0.05, respectively.

Adapting Eq. 4.12 to our current context, we obtain,

$P(D) = (0.9 \times 0.1) + (0.05 \times 0.9) = 0.135$. Thus, based on Eq. 4.9, we can calculate $P(\boldsymbol{H_a} \mid D)$ as follows:

$P(\boldsymbol{H_a} \mid D) = \frac{0.9 \times 0.1}{0.135} = 0.67$. Using the same level of accuracy, this value is in agreement with the corresponding value obtained earlier from the frequencies in Table 3.2.

A similar approach could be taken to estimating the posterior probability $P(\boldsymbol{H_0} \mid D)$.

Use of Bayes' rule in this way, not only explicitly acknowledges the base rate, $P(\boldsymbol{H_a})$, but also, leads to the generation of the PPV in the form of $P(\boldsymbol{H_a} \mid D)$, thus avoiding the pitfalls of the base rate fallacy and the confusion of the inverse. This leads to our next KLP.

Key Learning Point 4.4
The Bayesian approach to evaluating conditional probabilities avoids the pitfalls of the various fallacies because it has a place for explicitly acknowledging the prior (or, base rate) probability, $P(\boldsymbol{H}_a)$, for the study hypothesis, H_a, under consideration. This property could prove useful in the identification of fields of research where there is a need to better manage the PPV.

Note, however, that use of the Bayesian approach in the above way to better understand the results of studies involving NHST does not involve dispensing with the p-value, and therefore, the p-value fallacy remains at large. For example, the estimate for the base rate (equivalently, $P(\boldsymbol{H}_a)$) of 0.1 that we have assumed in this chapter is a product of the prevailing use of the p-value from the perspective of NHST in medicine, and in particular, pre-clinical medicine. This is because the trials upon which this estimate relies themselves used NHST. Thus, it is possible for a Bayesian approach to hypothesis testing to circumvent both the base rate fallacy and the fallacy of confusion of the inverse while depending on an estimate for the base rate arising from previous applications of NHST which are subject to the p-value fallacy.

Our calculations have helped us to see that in a variety of contexts, statistical significance may not be a strong indicator of the truth of the study hypothesis, with about a third of cases being false positives in the sense that the PPV for statistical significance is approximately 0.67 and the corresponding FDR is therefore approximately 0.33. More research in estimating the PPV in a range of clinical contexts would strengthen the case for managing the PPV. This ought to be complemented, however, with thorough communication between statisticians and clinicians at the stage of protocol design in pre-specifying thresholds for clinical significance based on statistical effect size estimates. (Traditional thresholds for capturing small, medium and large statistical effect sizes, for example, need not capture the clinical equivalents of these categories on assessing the same effect size estimates in clinical terms [22].)

Given the above findings and our previous estimate based on approximations to the base rate from pre-clinical medicine, it is possible that in a variety of clinical contexts, the PPV for NHST is rather low. There is a resultant need for a greater focus on estimating the base rate at the study design stage in clinical research. Where appropriate, the quality of this research can be improved through in turn seeking to optimise the PPV based on the principles considered in this chapter.

4.3.4 A Note on the Bayes Factor

Use of the Bayes factor is also a well-known approach to summarising the evidence in favour of one scientific hypothesis relative to the other. In the terminology used above, the Bayes factor may be interpreted as the ratio $B_{a0} = \frac{P(D\,|\,\boldsymbol{H}_a)}{P(D\,|\,\boldsymbol{H}_0)}$, which, in the language of the basic Bayesian model (Sect. 4.3.1), is a ratio of likelihoods. Using pre-specified criteria, the weight of the evidence in favour of H_a rather than H_0 can then be assessed in terms of the value of B_{a0} [23]. More generally, the Bayes factor is frequently considered as an alternative to the p-value in providing a decision rule in relation to a null and alternative hypothesis. While this and other alternatives to NHST may go some way to tackling the p-value fallacy, debates about their suitability remain. The choice of threshold for B_{a0} in support of choosing H_a in favour of $\boldsymbol{H_0}$ is contentious, as is the idea that this threshold value should justify transitioning from a ratio of conditional probabilities to choice between two marginal probabilities. Indeed, if the extent to which one hypothesis is more likely to explain the data than the other is very high, this does not necessarily indicate that the first of the two hypotheses is very likely to be true [24]. In understanding this point, the concept of relative risk provides a useful analogy for students to consider; notably, a high relative risk can arise for a very low baseline risk in one group without requiring the risk for the comparator group to be high. For example, if the relative risk of hospital admission in COVID-19 patients, involving the risk of admission without prior intake of honey and lemon with hot water divided by the corresponding risk with prior intake of this home remedy, turns out to be two, that does not mean to say that the absolute risk of admission without prior intake of the home remedy is itself high!

Despite existing scepticism concerning use of the Bayes factor, it is of interest to note that the value of 0.005 for α recommended by Colquhoun as having the potential to contribute to acceptable levels for the FDR, and hence the PPV, is also recommended indirectly elsewhere with reference to the Bayes factor. In particular, Benjamin et al. note that in assuming the value 0.005, for a point null hypothesis, α can be translated into a Bayes factor "between 14 and 26 in favour of" the study hypothesis, which is "representative of 'substantial' to 'strong' evidence according to conventional Bayes factor classifications" [11].

4.3.5 Moving Beyond NHST in Statistical Hypothesis Testing

Evidence, as presented by data D in a more general sense need not favour H_a, as D need not constitute a statistically significant difference in favour of H_a. Correspondingly, in moving beyond NHST, the interpretations of $P(\boldsymbol{H_a} \mid D)$ and $P(\boldsymbol{H_0} \mid D)$ are not restricted to the PPV and the FDR, respectively. In a general sense, the balance in favour of either of the two hypotheses H_0 and H_a is recognised according to which of the two posterior probabilities, $P(\boldsymbol{H_a} \mid D)$ and $P(\boldsymbol{H_0} \mid D)$, are the higher. The procedure which draws this comparison, by asking, given data, D, which of the two hypotheses, H_a and H_0, is the more likely, is referred to as the *maximum* a posteriori (*MAP*) test, which, in its more general form, can accommodate more than two conditional probabilities, not necessarily involving null and alternative hypotheses.

A more sophisticated approach is to adapt the MAP test to take into consideration the respective costs, C_{a0} and C_{0a}, of favouring H_a when H_a is false and favouring

H_0 when H_0 is false, respectively. The decision process then involves favouring H_a if and only if $P(\boldsymbol{H_a} \mid D)\ C_{0a} > P(\boldsymbol{H_0} \mid D)\ C_{a0}$.

The latter approach rests on the assumption that the two costs are unlikely to be the same. For example, in response to the relative evidence in favour of each of H_0 and H_a in a placebo-controlled trial, the cost of wrongly withholding a new intervention may be greater than that of wrongly administering the intervention.

The more general philosophical question of where the prior, should come from is an important one. Here, for a given hypothesis, H_{a}, the prior, $P(\boldsymbol{H_a})$, has to be the scientific plausibility of H, taking into consideration one's background knowledge, which may not include an established estimate for $P(\boldsymbol{H})$ within a given specialty. In this context, the base rate is estimated, to varying extents, according to subjective criteria. Therefore, in practice, a Bayesian analysis of a test of a scientific hypothesis, such as a claim about the efficacy of a novel treatment, will typically have to use a range of priors for $P(\boldsymbol{H})$, so that any 'reasonable' prior a scientist might recommend is included in that range.

But what is a 'reasonable' prior? In attempting to respond to the above question, considerable progress has been made in Bayesian inference more generally through reliance on distributions arising from additional data to *D*, subjective knowledge and clinical expertise. Nevertheless, on account of the need for choices that generate reliable probabilistic predictions, including when considering novel scientific hypotheses, the search for a suitable estimate for $P(\boldsymbol{H})$ may, in many contexts, prove inconclusive [23]. The Bayesian approach therefore has its own limitations.

It is also important to acknowledge that in a more purist Bayesian approach than we have the scope to cover in this chapter, priors take the form of probability density functions (pdfs). Pdfs are typically continuous functions and allow probabilities to be estimated within a range of values of a variable as an area under the curve for the pdf, and mathematically, as an integral.

In this Bayesian context, probability, as viewed in terms of frequencies, is superseded by plausibility, which is subjective, since dependent on the choice of prior. Where plausibility relies on evidence from one or two previous studies, it is arguably contaminated with the same sort of flaw as we considered earlier with reference to the p-value fallacy, namely that evidence from a single study is being over-stretched in the interpretation of new studies. Furthermore, statistical power falls with the decreasing usefulness of the chosen prior. Therefore, there can be instances where meta-analyses arising from well-designed primary studies of moderate size involving NHST are truer to the data than Bayesian analyses.

4.4 Conclusion

As we have seen in this chapter, RCTs are not guaranteed to provide an elite passageway out of the problem of having a low PPV in statistical hypothesis testing. While this does not undermine the RCT study design per se as a gold standard, our above analysis does confirm that where parameter values for the base rate, statistical power and especially, statistical significance, are sub-optimal, choice of the RCT study design and absence of QRPs cannot compensate for these deficiencies. Thus, for example, there may be instances where a better-powered non-RCT study involving a lower value of α by comparison with an RCT counterpart generates a more valid finding, assuming that the study hypotheses are the same. While minimising α is a viable approach to maximising the PPV, this approach needs to be handled with care, given that a universal value for α can lead to false negatives in cases where the PPV is already high on using a less stringent value for α.

Statistical significance is not an end in itself, even where a particularly small value is assumed for α. Recognition of Bayes' rule within the context of NHST may go some way to attributing due importance to PPV when considering a study hypothesis, H. Nevertheless, estimation of the prior, $P(\boldsymbol{H})$, equivalently, the base rate (π), can appear arbitrary in the parameterisation of this rule, there being complexities in some cases in identifying reasonable priors. This can make retention of NHST a more attractive option than a purely Bayesian approach.

Nevertheless, the application of Bayesian statistics in medicine has been popularised in recent years, including through the rise of machine learning and artificial intelligence techniques to enhance the quality of RCTs. Thus, arguably, in future work, there is a nettle to be grasped in preparing undergraduate medical students to make responsible choices regarding the value added from involving Bayesian approaches in their future research. In more recent years, within undergraduate medicine, students have been gaining exposure to the basic principles of Bayesian statistics and their application. Emerging routes for curriculum development in Bayesian statistics at a more advanced level include EBM programmes and options within honours degree programmes for medical students. Such opportunities require to be balanced with prevailing entry requirements in the mathematical sciences on entry to medical school, including at graduate level, and the widening of access to medical degree programmes in recent years, a general topic which itself invites considerable discussion!

References

1. Ioannidis JP. Why most published research findings are false. PLoS Med. 2005;2(8):0696–701.
2. Bird A. Understanding the replication crisis as a base rate fallacy. Br J Philos Sci. 2020;Ahead of print.
3. Higgins J, Spiegelhalter D. Being sceptical about meta-analyses: a Bayesian perspective on magnesium trials in myocardial infarction. Int J Epidemiol. 2002;31(1):96–104.
4. Yusuf S, Teo K, Woods K. Intravenous magnesium in acute myocardial infarction: an effective, safe, simple and inexpensive treatment. Circulation. 1993;87:2043–6.
5. ISIS-4 (Fourth International Study of Infarct Survival) Collaborative Group. ISIS-4: a randomised factorial trial assessing early oral captopril, oral mononitrate, and intravenous magnesium sulphate in 58,050 patients with suspected acute myocardial infarction. Lancet. 1995;345:669–85.
6. Yusuf S. Meta-analysis of randomised trials: looking back and looking ahead. Control Clin Trials. 1997;18:594–601.
7. Teo K, Yusuf S, Collins R, Held P, Peto R. Effects of intravenous magnesium in suspected acute myocardial infarction: overview of randomised trials. BMJ. 1991;303:1499–503.

8. Shamseer L, Moher D, Clarke M, Ghersi D, Liberati A, Petticrew M, Shekelle P, et al. Preferred reporting items for systematic review and meta-analysis protocols (PRISMA-P) 2015: elaboration and explanation. BMJ. 2015;349(g7647).
9. Murad M, Asi N, Alsawas M, Alahdab F, Asi N, Alsawas M, et al. New evidence pyramid. BMJ Evid-Based Med. 2016;21(4):125–7.
10. Colquhoun D. An investigation of the false discovery rate and the misinterpretation of p-values. R Soc Open Sci. 2014;1.
11. Benjamin DJ, Berger JO, Johannesson M, Nosek BA, Wagenmakers EJ, Berk R, Bollen KA, Brembs B, Brown L, et al. Redefine statistical significance. Nat Hum Behav. 2017;2.
12. Diener E, Biswas-Diener R. The replication crisis in psychology. In: Biswas-Diener R, Diener E, editors. Nova textbook series: Psychology. Champaign, IL: DEF Publishers; 2020.
13. Committee on Reproducibility and Replicability in Science et al. Reproducibility and replicability in science. Washington, DC; 2019.
14. Fanelli D. Is science really facing a reproducibility crisis, and do we need it to? PNAS. 2018;115(1):2628–31.
15. Goodman S. A comment on replication, p-values and evidence. Stat Med. 1992;11:875–9.
16. Devezer B, Navarro DJ, Vandekerckhove J, Ozge Buzbas E. The case for formal methodology in scientific reform. R Soc Open Sci. 2021;8(3).
17. Senn S. A comment on replication, p-values and evidence. Goodman SN. Stat Med. 1992;11:875–9. Stat Med. 2002;21:2437–44.
18. Desai MB, Robinson T. On the use of robust estimators for standard errors in the presence of clustering when clustering membership is misspecified. Contemp Clin Trials. 2013;34(2):248–56.
19. Bland J. Cluster randomised trials in the medical literature: two bibliometric surveys. BMC Med Res Methodol. 2004;4(21).
20. Lieber RL. Statistical significance and statistical power in hypothesis testing. J Orthop Res. 1990;8:304–9.
21. Baig SA. Bayesian inference: an introduction to hypothesis testing using Bayes factors. Nicotine Tob Res. 2020;22(7):1244–6.
22. Griffiths PN, Needleman J. Statistical significance testing and p-values: defending the indefensible? A discussion paper and position statement. Int J Nurs Stud. 2019;99.
23. Kass RER, Adrian E. Bayes factors. J Am Stat Assoc. 1995;90(430):773–95.
24. Tendeiro JN, Kiers HA. A review of issues about null hypothesis Bayesian testing. Psychol Methods. 2019;24(6):774–95.

Margaret MacDougall is a senior lecturer in medical statistics and researcher in education at the Usher Institute, Edinburgh Medical School. She developed an interest in philosophy early on in her career through her Ph.D. research in mathematical logic, constructive set theory and the philosophy of mathematics. She has conceived and led eleven funded projects in medical education and statistics and supervised research staff and students at undergraduate and postgraduate levels through various spin-off projects. Margaret seeks to use her research to promote curriculum innovation and enhance student learning in statistics at undergraduate and postgraduate levels and has published on an international scale in the fields of education, medicine, philosophy and statistics. She has served as an invited speaker, chair and conference organiser at various local, national and international events relating to statistics, and learning and teaching in higher education. Currently, she serves as Vice Chair of the Teaching Statistics section of the Royal Statistical Society.

Chapter 5
What Do RCTs Tell Us, and Could They Tell Us More? Looking Within and Beyond the Study Sample

Julius Sim, Gillian Lancaster, and Martyn Lewis

Abstract Randomised controlled trials (RCTs) are generally regarded as the gold-standard method to answer questions of treatment effectiveness. Carefully designed and rigorously implemented, they can provide unbiased estimates of average treatment effectiveness in a wide range of clinical presentations. However, they have two potential drawbacks. First, owing to the variability of treatment effects at the level of individual patients, the aggregate findings of an RCT may be difficult to apply to a particular patient or type of patient. Subgroup analyses can produce more specific findings in this regard but have some shortcomings of their own and must be conducted, and reported, carefully. Second, while RCTs can normally claim a high degree of internal validity, their external validity can be more questionable. This results from the fact that participants in an RCT may not be representative of the target population to which their findings are intended to apply, and from differences between the experimental context in which RCTs are conducted and the world of everyday clinical practice. This chapter will discuss these difficulties and suggest ways in which they may be mitigated.

5.1 Introduction and Overview

The randomised controlled trial (RCT) occupies a prominent position in the hierarchy of designs for determining clinical effectiveness [1]. It is generally regarded as the gold-standard design for establishing a cause-and-effect relationship between treatment and outcome, if carefully designed, adequately powered and, where necessary, informed by appropriate preliminary pilot or feasibility studies [2]. However, whilst much can be learned from RCTs, there are several limitations to the validity and usefulness of the information that they provide. Drawing upon the statistical properties of RCT designs, as well as broader methodological and epistemological

J. Sim (✉) · G. Lancaster · M. Lewis
Keele University, Keele, UK
e-mail: j.sim@keele.ac.uk

M. MacDougall (ed.), *A Medical Educator's Guide to Thinking Critically about Randomised Controlled Trials: Deconstructing the "Gold Standard"*,
https://doi.org/10.1007/978-3-031-25859-6_5

considerations, in this chapter, we shall explore some of these limitations in terms of seeking to look, not only 'beyond' the study sample to the population from which it was drawn but also, 'within' the study sample, such as by drawing conclusions that relate to strict subgroups of the study population.[1]

First, we shall explore the variability of treatment effects and consider the extent to which one can draw more specific conclusions regarding treatment effectiveness than those expressed by the aggregate conclusions of the RCT; we shall focus in particular on the use, and misuse, of subgroup analyses. Our focus will then shift to the generalisability of the results of RCTs: the extent to which conclusions drawn from the data can be applied more broadly to other samples and other contexts. In the process, we shall outline some respects in which the generalisability of these conclusions can be evaluated. Unless indicated otherwise, in this chapter, we shall focus on parallel group, individually randomised controlled trials.

In what follows, we shall refer in places to two fundamental forms of validity: internal and external, both of which were defined in Chap. 2 (see Key Learning Point (KLP) 2.3). From the first of these definitions, it is implicit that the degree of internal validity can be represented by the extent to which explanations of the treatment effect can be restricted to the interventions being tested. By contrast, based on the second definition, the degree of external validity can be represented by the extent to which the conclusions of a study can be applied beyond the sample and the specific context in which the study was conducted—to other similar patients and to a clinical rather than a research context. Later in this chapter, we shall revisit internal and external validity and consider the relationship between them.

5.2 Variability of Treatment Effects

When two treatments are compared in an RCT, the difference between the two groups estimates the average treatment effect for the population from which the sample was drawn. In such a study, we cannot obtain individual treatment effects because we do not know what a particular individual's response would have been to the treatment that he or she did not receive. As already noted in Chap. 2, this unknowable, and hence unmeasurable, response to the treatment not received is called a counterfactual outcome—*counterfactual*, for short [3, 4].

Thus, RCTs produce aggregate conclusions; provided certain distributional assumptions are in place, they estimate, *on average*, how much more or less effective two interventions (or, for more complex designs, more than two interventions) are in comparison to one another based on the study sample. Any one such conclusion can in turn be applied to the population from which the sample was drawn (assuming

[1] The term 'strict subgroups' pertains to subgroups of a parent group, non-inclusive of the parent group—here the entire study population. In what follows, the term 'subgroups' and its derivatives assume this strict interpretation, although the term 'strict' is dropped for convenience.

representativeness). However, understanding the average effect of an intervention only takes us so far. As Yusuf et al. state,

> If all patients shared identical characteristics and equivalent risks, the overall analysis of a clinical trial would be sufficient to provide medical practitioners with an unambiguous guide to treatment. In practice, however, patients with a specific disease may differ greatly one from another [5, p. 93].

This is the issue of *heterogeneity of treatment effects* [6]. Aggregate estimates from RCTs are our best estimates of treatment effectiveness, but they are based on a distribution of individual effects within the treatment groups, and these distributions are likely to overlap. Information on the response of a future individual patient is limited, and some future patients may do better on the overall worse-performing treatment or do worse on the overall better-performing treatment. This is illustrated in Fig. 5.1, which displays four hypothetical participants in a trial of two interventions.

In this figure, Participant 1's outcome after receiving Intervention A is considerably worse than if they had received Intervention B, whilst Participant 2's outcome under Intervention A is moderately better than under Intervention B. Participant 3 has a somewhat worse outcome under Intervention B than under Intervention A, whilst Participant 4 has a very much better outcome under Intervention B than under Intervention A. Clearly, the overall treatment effect gives limited information on the way in which individual participants may respond to the trial interventions. Hence, to assume that aggregate conclusions can be straightforwardly applied to individual patients is to commit the logical fallacy of division, whereby what is true for the

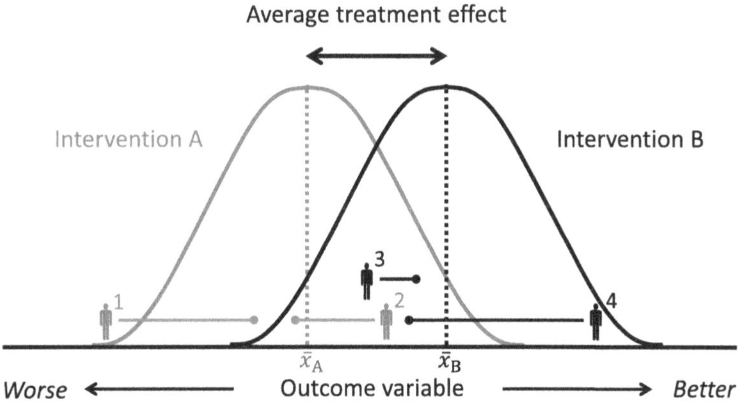

Fig. 5.1 Schematic representation of heterogeneity of treatment effects. *Note* The distribution of individual treatment effects is represented by the two curves—grey for Intervention A and black for Intervention B—and the round-headed arrows represent hypothetical individual treatment effects. The average (mean) treatment effect is the difference between the mean treatment responses, $\bar{x}_A$ and $\bar{x}_B$, for the two groups and shows Intervention B to be on average more effective. Four participants (coloured grey or black in relation to their treatment assignment) are marked. For each of them, the associated round-headed arrow points to their unobserved outcome under the other treatment (i.e., their score on the outcome variable had they received the other intervention)—the counterfactual outcome (Chap. 2)

whole is inappropriately assumed to be true for its constituent parts [7].[2] It is clear from the figure that the average treatment effect in the trial may not be informative as to the treatment effects that would be experienced by individual participants if they had received one intervention rather than the other.

Key Learning Point 5.1 Randomised controlled trials provide estimates of the aggregate treatment effect in respect of two or more interventions. However, individual treatment effects may differ in both direction and magnitude from this aggregate effect, and the conclusions of a trial may therefore not be informative as to the responses of individual participants.

Hence, our aggregate conclusion regarding a trial cannot necessarily be applied to any individual participant or to a particular type of patient [10]. Moreover, owing to this heterogeneity of treatment effects, showing the *average* superiority of one treatment over another is not informative as to the *proportion* of participants receiving the superior treatment whom one might expect to have a positive treatment response [11].[3] Superior clinical effectiveness of one treatment over another at the aggregate level could result from a comparatively small number of participants exhibiting a very large effect or, equally, a larger number of participants exhibiting a more modest effect, and knowing which was the case would be clinically useful information. One approach to attempting to dig beneath the average treatment effect is the use of subgroup analyses.

5.3 The Application of Subgroup Analyses in RCTs

5.3.1 Statistical Considerations

Subgroup analyses may take two forms. First, they may seek to determine whether the treatment effect observed in the trial is consistent, in terms of its magnitude and direction, across categories of patients defined by a subgrouping variable [12]. Second, subgroup analyses may additionally involve calculating separate estimates for such categories (what we refer to below as subgroup effects). These categories

[2] A specific example of the fallacy of division, common in epidemiological studies, is Simpson's paradox, whereby the direction of an effect may be reversed when subgroups are combined; for example, an odds ratio in each of the subgroups may be greater than unity, but less than unity when the subgroups are pooled [8]. This paradox is related to the weighting of the subgroups when combined to form a pooled estimate, which is unlikely to occur in an RCT in which equal numbers of participants are allocated to each arm of the study. See Tu et al. [9] for a discussion of the reversal paradox, of which Simpson's paradox is a particular manifestation.

[3] This assumes that treatment effect is measured on some type of scale—if measured on a dichotomy, the 'average' would be the proportion.

of a subgrouping variable may represent demographic or clinical characteristics of the study participants (such as sex, disease severity, or the presence or absence of comorbidities), or may represent the influence of other interventions (e.g., whether or not participants receiving one of two types of anxiolytic medication are also undergoing cognitive behavioural therapy).[4] Subgrouping variables are thus potential *moderators* of a treatment effect [14]. Moderation in this context is when the magnitude and/or direction of the relationship between two variables (here, the treatment received and clinical outcome) depend in part on a third variable (here, the subgrouping variable) [15].

The nature of subgroup analyses is presented diagrammatically in Fig. 5.2. Each level of the *subgrouping variable* (sex) represents a *subgroup* (either men or women). Following Yusuf et al. [5], we use the term *interaction* for a treatment effect that differs *across* subgroups (also referred to as effect modification), and the term *subgroup*

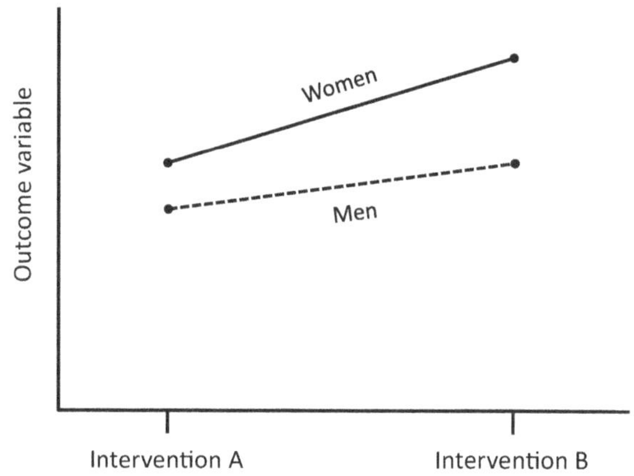

Fig. 5.2 An interaction plot for a hypothetical study involving a subgroup analysis based on sex. *Note* Sex is the subgrouping variable. The vertical axis represents the outcome variable for the trial (with higher scores being better) and the horizontal axis shows the two interventions (A and B). The lines in the plot represent the *subgroup effects*, namely the treatment effect for each of the two subgroups: women (solid line) and men (dashed line). The slope of each of these lines shows the magnitude of the subgroup effect. The extent to which the two lines are not parallel (i.e., they have different slopes) indicates the degree of *interaction*—i.e., different treatment effects in the subgroups—which can be assessed visually and/or through a statistical test. The circular markers at the end of each line indicate the effect of a particular intervention for a specific *level* of the subgrouping variable. For example, the marker at the left end of the dashed line indicates the effect of Intervention A specifically among men

[4] If an additional treatment is being tested in the study by *design*, this would be a factorial trial, with hypotheses relating to both interventions [13]. However, the presence of an additional treatment may pre-date the study and not therefore represent a prior research hypothesis. The additional treatment would not be of interest in its own right but in terms of the extent that it may modify the effect of the primary treatment that is being tested.

effect to refer to a treatment effect *within* a specific subgroup. These different types of effect correspond to the two different forms of subgroup analysis outlined above.[5]

We can classify subgroups as either *proper* subgroups (where subgrouping is defined in terms of information available at baseline) or *improper* subgroups (where subgrouping is defined in terms of information measured during follow-up, including information that has changed from baseline at follow-up due to loss of patients) [5, 17]. A particular danger with improper subgrouping is that the treatments tested in the trial may influence how the subgroups are created and as such contaminate the perceived influence of the subgrouping per se in explaining differential treatment effects across these subgroups. Thus, while the aim is to explore the explanatory role of the subgrouping on treatment effectiveness, the treatment itself may have already influenced how the subgroups are defined and the characteristics of the patients within these groups, thus presenting a circularity in the reported relationship between the improper subgroups and the study intervention. As an example, in a follow-up investigation at 12 months for a trial involving a new treatment for chronic neck pain, we might wish to explore the influence of age at baseline on the effectiveness of the new treatment. While baseline ages will, of course, have remained unchanged at the level of the individual patient, their distribution may have been affected by older patients being more susceptible to experiencing adverse effects and having dropped out of the study prior to follow-up. The loss of these patients would possibly in turn influence the choice of cut-offs for the age-based (improper) sub-groups and inflate the treatment effectiveness for younger patients when comparing subgroups of older and younger patients. As a simpler example, consider a trial of two types of rehabilitation for chronic neck pain in the same patient population, with physical function as the primary outcome. In this trial, at 12 months, participants are subgrouped in terms of their scores at this time point on a scale measuring self-efficacy (i.e., confidence in one's ability to achieve a goal), to assess whether treatment effectiveness might differ between those with high and low self-efficacy. However, it is possible that participants' self-efficacy scores, and thereby the creation of the subgroups, have already been influenced by the type of rehabilitation received. As the subgrouping may as such have been contaminated by the effect of the type of rehabilitation received, a greater treatment effect estimate in one subgroup may arise at least partly because a positive treatment response has already played a role in defining that subgroup. Subgrouping based on baseline values of self-efficacy would have avoided this problem. Key terms relating to subgroup analyses more generally are explained in Table 5.1, including a distinction between the concepts of interaction and confounding.

As has been repeatedly emphasised in the literature, a subgroup analysis should normally take the form of a single test for interaction, rather than a series of separate tests relating to subgroup effects within each level of a subgrouping variable [18–22]. The latter approach produces subgroup-specific estimates of the treatment effect (e.g., just for men or just for women), but does not directly address the question of

[5] Some, such as Sun et al. [16], use the terms subgroup effect and interaction synonymously, but we believe it is preferable to distinguish them in the way that Yusuf et al. [5] do.

whether there is evidence for inconsistency in the treatment effect across subgroups (e.g., through exploring whether there is evidence to suggest that the treatment effect for men differs in magnitude and/or direction from that for women). Interaction tests are, moreover, less prone to Type I errors (false positives) than tests within specific subgroups [23].

Subgroup analyses are commonly distrusted unless they are strictly a part of the study protocol and the study has been statistically powered to determine their statistical significance. Accordingly, subgroup analyses that are not prespecified but undertaken during or after the previously planned study are criticised for being "exercises in pure data dredging" [24]. Greenhalgh highlights the issue in her list of ways to cheat on statistical tests:

> If your results prove uninteresting, ask the computer to go back and see if any particular subgroups behaved differently. You may find that your interventions worked after all in Chinese women aged 52–61 [25, p. 63].

Table 5.1 Terminology relating to subgroup analyses

Term	Definition
Overall treatment effect	The overall treatment effect, without regard to any subgroups; in the terminology of experimental design, this is a *main effect.*
Subgroup analysis	An analysis that examines treatment effects in relation to categories (subgroups) of participants defined by a subgrouping variable. This may involve an estimate of the extent to which a treatment effect differs *across* categories of the subgrouping variable (an interaction effect estimate) or an estimate of the treatment effect *within* an individual category (a subgroup effect estimate). The former is the preferred approach [18–22].
Interaction	A situation in which the treatment effect is inconsistent across two or more subgroups, differing in magnitude (known as an ordinal, or quantitative, interaction) and/or direction (known as a disordinal, or qualitative, interaction). An interaction can be formally tested for statistical significance as part of the statistical model or can be gauged descriptively in terms of its magnitude, without recourse to statistical testing.
Subgroup effect	A treatment effect estimated within a specific subgroup; in the terminology of experimental design, this is a *simple effect*. With respect to a particular subgrouping variable (e.g., sex or the presence/absence of a concomitant intervention), a subgroup effect relates to a particular level of such a variable (e.g., female, no concomitant intervention).
Proper subgroup	A subgroup defined in terms of a baseline variable or other information taken account of prior to randomisation.

(continued)

Table 5.1 (continued)

Term	Definition
Improper subgroup	A subgroup defined in terms of a variable at a stage of measurement that occurs after randomisation. A particular danger here is that the treatments tested in the trial may influence how the subgroups are determined. For example, if, at follow-up, age is used to subgroup participants in a drug trial, older patients less tolerant of side-effects who have dropped out of the study will not be included, possibly creating the impression that the treatment is more effective in younger patients when, in fact, a difference in tolerance levels has created imbalance across older and younger patients *after* randomisation, leading to an unfair comparison across these subgroups.[17].
Confounding	Whereas interaction refers to the modification of a treatment effect by a subgrouping variable, leading to different estimates of this effect in the subgroups defined by this variable, confounding refers to a distortion or biasing of the treatment effect by another variable, thereby representing a threat to internal validity. In an RCT, this can occur if this other variable satisfies each of the following conditions, and as such, *is a confounder*: (i) it is associated with the treatment groups (i.e., has different mean values in each group or is represented in different proportions in each group); (ii) it is associated with the outcome variable; and (iii) it is not a component or a result of the intervention being tested [26, 27]. For example, a trial of two different drugs for hypertension might show a different rate of deaths between the treatment groups. However, if the proportion of participants with lung cancer was higher in one group, lung cancer would be associated with the treatment groups. As lung cancer is also known to be a cause of death, it could also be associated with the outcome. As such, imbalance in lung cancer status could serve as an alternative explanation of the observed treatment effect; that is, lung cancer status could serve as a confounder [28]. Suppose, however, one of two interventions aimed at controlling blood glucose had a side-effect in relation to symptoms of nausea, such that levels of nausea differed across the treatment arms. Even if nausea were associated with the outcome variable (blood glucose), its imbalance across treatment arms would not be a confounder, because nausea is a result of the intervention, thereby failing to satisfy condition iii, above. Randomisation helps to reduce confounding through balancing of confounders, and this is more effective with increasing sample size, mitigating the influence of condition i, above.

Subgroup analyses should therefore normally represent prior hypotheses—based on previous pathophysiological, clinical or epidemiological knowledge—and are best applied only to the primary outcome variable in the study [19, 29]. If subgroups are determined in relation to baseline values of a continuous subgrouping variable that is liable to change over the course of the trial, and three or more such subgroups are defined, this can be problematic. The baseline values in relation to which the extreme subgroups are defined (i.e., those determined in terms of the highest and lowest values of the subgrouping variable) are likely to change in the direction of less extreme values (towards the average) during the study, thereby creating 'unstable' subgroups. This is on account of the phenomenon of regression to the mean [30], as defined in Chap. 1. In such circumstances, the definition of these subgroups at follow-up, when they are analysed, will differ from when they were originally defined at baseline. For example, subgroups might have been created at baseline in terms of mild, moderate and severe levels of anxiety based on scores on an anxiety scale. Over the course of the trial, many of the more extreme values, in either direction, may regress towards the overall mean level of anxiety of the participants in the trial. Accordingly, the initial values that were used to define the subgroups may now differ, causing a change in the 'meaning' of these subgroups when the data are analysed. Caution should therefore be exercised when creating subgroups from a continuous variable that is liable to change over time, especially if there are a number of values at one or both extremes of the scale.

On a technical note, it is the authors' viewpoint that if an interaction is to be formally tested within the statistical model, it is advisable to fit an initial model with just the main effects and then introduce the interaction term in a second model. This is because the introduction of an interaction term will influence the magnitude and associated p-value of a main effect, and one usually wants an estimate of the main effect that is free from this influence (i.e., before adding the interaction term). Additionally, a comparison, or meta-analysis, of overall treatment effects presented in different studies would be difficult if some of these had been affected by (possibly a variety of) interaction terms in the statistical model and others had not. An initial model without the interaction terms avoids this problem. It should be noted that subgrouping variables are likely to be included in this initial model, as variables for which adjustment is needed, but only in the form of main effects, not in the form of an interaction term.

Whereas the overall (main) effect of an intervention expresses its effect without regard to any subgroups, an interaction expresses the extent to which the effect of the intervention depends upon which subgroup is being considered. Such interactions may be either ordinal (or quantitative) or, less frequently, disordinal (or qualitative). In an ordinal interaction, the treatment effect is in the same direction in different subgroups but is of a different magnitude. For example, one treatment may be more effective than another for all subgroups defined by a subgrouping variable, but to a greater extent in one subgroup than in one or more others. This is illustrated in plot A in Fig. 5.3, where the somewhat steeper slope of the line for women suggests that the superiority of Intervention B over Intervention A is more pronounced than it is for men. In a disordinal interaction, the direction of the treatment effect changes between subgroups (and may or may not differ in its magnitude). Here, one treatment may be more effective than its comparator in one subgroup, but less effective than its comparator in another. In plot B in Fig. 5.3, this is represented by a crossing of the lines for women and men, indicating that the treatment effect would be reversed across the subgroups. A disordinal interaction is in most cases less likely than an ordinal interaction. However, it may occur if a treatment has the potential to be harmful

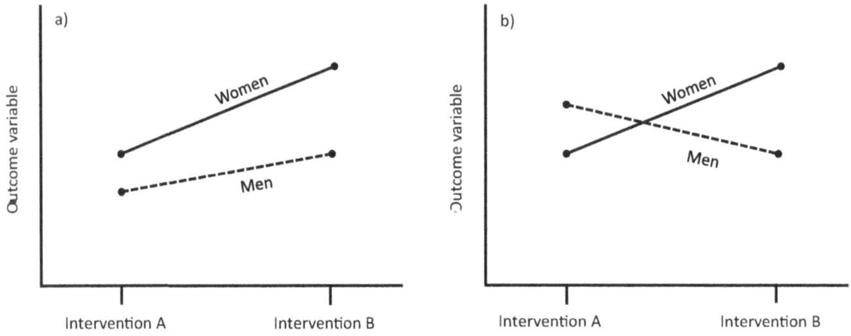

Fig. 5.3 An ordinal (plot **a**) and a disordinal (plot **b**) interaction. *Note* As higher values on the outcome variable are better, it can be seen from the ordinal interaction in plot (**a**) that Intervention B is superior to Intervention A in both subgroups, but to a greater extent among women. Plot (**b**) displays a disordinal interaction whereby Intervention B is superior to Intervention A in women, but the opposite is the case for men

in a certain subgroup of participant (e.g., patients exhibiting high chronicity), but beneficial in one or more other subgroups (e.g., patients exhibiting moderate or low chronicity). This would cause the reversal of the direction of the treatment effect in each of these subgroups.[6] An RCT of two forms of short-term psychotherapy (interpretive versus supportive) by Ogrodniczuk et al. [31] provides an example. Male participants in the trial showed a better response to interpretive than to supportive psychotherapy, whereas female participants exhibited a better response to supportive than to interpretive therapy. A finding such as this is important, as it suggests that a change in treatment may not just be ineffective but contraindicated in one group of patients.

Statistical testing of interactions within a subgroup analysis raises a problem in relation to statistical power; this stems from the fact that the variance associated with an interaction term in a statistical model is larger than that associated with a main effect. Hence, a study whose sample size has been set at the minimum required to detect a main effect as statistically significant will have insufficient power to detect an associated interaction of equivalent magnitude. Typically, four times as many participants are required to detect as significant an interaction as to detect a main effect of the same magnitude, at the same level of statistical significance [32–34]. If the magnitude of the interaction is expected to be smaller than that of the main effect, the required sample size will be even greater, and may well be prohibitive. For a trial powered to detect only a main effect, therefore, any subgroup analyses

[6] It may occur that there is an effect in a particular direction in one or more categories and no effect at all in one or more others. This difference in magnitude across categories is probably best regarded as an ordinal interaction even though it does not satisfy the strict definition of this term—the important issue is that there is no reversal of direction of effect across subgroups.

are subject to a high probability of Type II error (false negative) and the associated estimates will be inaccurate, with wide confidence intervals (CIs).

Even if a trial has been powered in relation to a specific interaction, if many such interactions are tested, this raises equally problematic issues of multiplicity and Type I error inflation. In this connection, Lee et al. [35] report an interesting simulation study. For this purpose, they used a population of 1073 patients with coronary artery disease, all of whom had received the same treatment. Although these patients were randomised in the simulation to two groups, as these two groups had not received different interventions, there was, as expected, no significant difference in overall survival. Notwithstanding this result, a significant subgroup effect was found between groups in respect of the subgroup of patients with three-vessel coronary disease and abnormal left ventricular contraction. The authors explained this by the presence of several small differences in certain prognostically important baseline variables with respect to this subgroup. Thus, despite the good baseline comparability of the arms of the study in their entirety, there was imbalance in baseline characteristics across the study arms in terms of this subgrouping variable. Stratifying randomisation by the subgrouping variable(s) would offset this tendency by causing the baseline characteristics of subgroups to be balanced across the arms of the trial [36] (see also, Chap. 2).

Additional caution regarding statistical testing of subgroup effects is provided by Wittes [35]. She points out that, for two treatments that are equal in their effectiveness, creating 20 mutually exclusive subgroups and testing each of these at $p \leq 0.05$ will result in a 64% probability of at least one Type I error.[7] She further points out, however, that for any such subgroup, the probability of a Type I error is the same, regardless of its size. As a result:

> ...the smaller the subgroup, the larger the observed effect size must be to show statistical significance and because each subgroup, no matter what its size, is equally likely to show statistical significance when absolutely no effect of treatment exists, a false-positive effect in a small subgroup will appear larger than a false-positive effect in a large subgroup... Thus, if a statistically significant result is a false positive (and in any particular case we don't know whether an observation reflects a true effect or a false-positive effect), on average its magnitude will be greater in small subgroups than in large ones, reflecting the greater variability in small subgroups [37, pp. 912–913].

5.3.2 *Subgroup Analyses Determined a Posteriori*

Although subgroup analyses should normally be determined a priori, there is a limited role for subgroup analyses that have been determined a posteriori, provided certain conditions are satisfied. First, these should be clearly designated as hypothesis-generating rather than hypothesis-testing analyses. It follows from this that whilst

[7] On the assumption that the hypothesis tests are independent, this probability is derived in the following way: $1 - (1 - 0.05)^{20} = 0.64$, where 0.05 indicates the chosen significance level and 20 denotes the number of statistical comparisons. As noted earlier, testing interactions rather than subgroup effects involves fewer hypothesis tests and thus inflates the Type I error rate less markedly.

subgroup effects derived from such analyses should be presented with CIs, no statistical hypothesis tests should be performed (and the CIs should not be interpreted as such tests, but as expressions of the uncertainty associated with particular subgroup effects). Second, as with all other subgroup analyses, all subgroup analyses determined a posteriori that have been conducted should be reported. Unfortunately, the reporting of such analyses can be poor [38, 39] and the conclusions drawn are not always credible [40]. Table 5.2 provides some general guidelines on the conduct and interpretation of subgroup analyses.

Table 5.2 Points of guidance on subgroup analyses

Concept	Guidance
Pre-specification	Subgroup analyses should normally be determined a priori, before examining the data, and based on clinical or theoretical considerations, or on relevant findings from previous studies. Not only the subgrouping variable but also the way in which the subgroups will be defined (e.g., through categorisation of a continuous variable) should be prespecified. Ideally, the subgrouping variable should be used to stratify randomisation ([36]; see Table 5.3). It is advisable to restrict a subgroup analysis to the primary outcome variable [19, 29].
Avoidance of improper subgroups	As an improper subgroup is defined in terms of a variable measured after randomisation has taken place, it is potentially misleading, as the categorisation of participants within the subgrouping variable may be influenced by the treatment being tested [5].
Unstable subgroups	This refers to the possibility that if three or more subgroups are defined at baseline in terms of scores on a scale, the definition of some of these subgroups may change, through regression of extreme scores to the mean, by the time the subgroup analysis has been conducted on the observed treatment effect (see the main text for an example).
Statistical power	If statistical testing is intended, consideration should be given to the appropriate sample size, given that the testing of interactions requires many more participants than the testing of a main effect of the same magnitude [29]. Statistical testing of underpowered interactions—leading to false negatives—may cause readers to mistakenly assume that a treatment is of similar effectiveness within subgrouping variables [40]. However, whilst interactions should not be statistically *tested* unless the study has been appropriately powered, they may still be *estimated* and reported in combination with CIs, provided that they are regarded as purely exploratory and for the purpose of hypothesis generating rather than hypothesis testing.

(continued)

Table 5.2 (continued)

Concept	Guidance
Method of analysis	Ideally, for the purpose of statistical testing, an interaction term should be used, rather than separate hypothesis tests within specific subgroups (e.g., women and men) [18–22]. If separate tests are nonetheless conducted within subgroups, the resulting p-values should be interpreted cautiously. In this connection, Assmann et al. point to the possibility of within-subgroup effects arising by chance, simply through sampling variation, and argue: "Reliance on subgroup p values is misleading. If the overall result is significant, almost inevitably some subgroups will and some will not show significant differences depending on chance and the smallness of subgroups. Conversely, if an overall difference is not significant, some subgroups may have a bigger observed treatment difference by chance, which may even reach significance" [41, pp. 1067–1068].
Parsimony	Conducting too many subgroup analyses through statistical tests of interaction will increase the Type I error rate unacceptably, and even if no such testing is employed, interpretation of a large number of subgroup analyses will be difficult [23, 42]. If a number of subgroup analyses that have been determined a priori are nonetheless performed through statistical testing, an adjustment for multiple comparisons may be advisable [29]; alternatively, a statement can be given of the expected number of false positives if unadjusted p-values are presented [43]. By analogy with thinking on planned comparisons in analysis of variance [44, 45], if a strictly limited number of a priori interactions are to be tested, and the tests are applied to the interactions rather than to the subgroup effects within each interaction, it could be argued that a correction for multiple comparisons is not required, given that the interactions represent prior hypotheses. Moreover, using such adjustments will decrease the power of the interaction tests, which may already have low power [40]. Wittes [37] argues that the tendency of small subgroups to produce very variable effects, which may thereby be spurious, can be mitigated by restricting the statistical testing of subgroup effects to those of sufficient size to have a prior nominal statistical power of at least 40% or 50%.
Full reporting	Results of all subgroup analyses should be reported; selective reporting of just those that are statistically significant or just those considered to be clinically meaningful is misleading [41].
A posteriori subgroup analyses	If subgroup analyses that are determined a posteriori are performed, they should be treated with circumspection [20].

(continued)

Table 5.2 (continued)

Concept	Guidance
Plausibility	The plausibility of an interaction may be strengthened if, not only the interaction itself but also, its direction has been prespecified, and it is consistent in relation to other closely related outcomes in the trial [16]. For example, a trial of chemotherapy in advanced lung cancer defined a subgrouping variable in terms of smokers versus non-smokers [46]. Subgroup analyses using this categorisation were consistent for both survival and time to treatment failure: in each case there was a more favourable outcome among non-smokers. Identifying a biological or other mechanism underlying an observed interaction may also increase its credibility, though Howick [27] warns that we should not dismiss outright an observed interaction simply because one cannot conceive of a mechanism that might explain it.

Criticisms on statistical grounds of subgroup analyses—especially when conducted a posteriori—are in principle valid, but some of these criticisms seem to assume that the use of subgroup analyses is confined to assessment of statistical significance. Estimates of treatment effects from exploratory subgroup analyses usually provide the best available evidence from the study concerned of the relative effects of treatments within categories of patients, irrespective of their statistical significance.[8] Provided these estimates are interpreted circumspectly in terms of their provisional nature and with due regard to their likely inaccuracy, it seems unwise simply to forgo such information, especially in view of the costs involved in the study from which it is acquired; currently, definitive multi-centre clinical trials commonly cost well over £1m. The main return on this investment is an answer to the core research question as to the overall relative effectiveness of two of more interventions, but other information from the study, such as that from subgroup analyses, may be useful in its own right, or as the basis for a hypothesis to be tested formally in a subsequent study. As Friedman et al. note: "Clinical trials require considerable time and effort to conduct and the resulting data deserve maximum evaluation" [47, p. 431]. Furthermore, as Petticrew et al. [48] have argued, it could be morally remiss not to conduct subgroup analyses (provided that these are done robustly), as the information that they provide may allow interventions to be delivered with due regard to their relative benefit to different population groups, and thereby with greater equity.

[8] The extent to which subgroup effects are informative will depend in some measure on the size of the subgroup. Yusuf et al. advise that one should "rely more on the overall results to indicate the likely 'true' effect in a particular subgroup, rather than on the actual observations, unless the trial was specifically designed to have sufficient power within subgroups of interest" [5, p. 94]. We would suggest, however, that the adequacy of the sample size here should be determined in terms of the accuracy of the estimated subgroup effect rather than in terms of a hypothesis test.

Key Learning Point 5.2 Subgroup analyses should normally be few in number, prespecified, restricted to the primary outcome variable and performed by means of an interaction term in the statistical model. Any subgroup analyses conducted a posteriori should be treated as hypothesis-generating, and not subjected to statistical hypothesis testing.

5.3.3 How Much Do Subgroup Analyses Tell Us?

Subgroup analyses can produce findings that are at a greater level of specificity than the overall trial findings. However, this specificity remains somewhat limited. For example, it may be apparent from a subgroup analysis that the effect of an anti-inflammatory drug depends upon whether it is administered to men or women, and this information is certainly informative, but in daily practice the clinician will be dealing with a patient presenting with multiple intersecting characteristics that may be clinically relevant, in addition to the patient's sex. For example, if testing two drugs for rheumatoid arthritis, the important question is not just whether the treatment effect is greater in women than in men, for example, but whether it is greater in women who are seropositive rather than seronegative, have normal rather than impaired kidney function, and so forth. As Dahabreh et al. point out, "each patient has innumerable characteristics, and therefore can belong to an indefinite number of different subgroups. It follows too that there is an indefinite number of ways to disaggregate a trial population into subgroups" [49, p. 2186].

One way to address the above limitation is to simultaneously test more than one subgrouping variable in a multiway interaction. For example, in the assessment of treatment effectiveness a two-way interaction test involving gender and treatment allocation could be refined to include a three-way interaction test that differentiates between seropositive and seronegative patients, thus allowing for comparisons of subgroup effects across different combinations of gender and blood test result rather than purely according to gender. However, this is liable to create a model that is complex, both statistically and in terms of its interpretation. There are also implications for sample size; as noted earlier, testing a two-way interaction for significance can be expected to require four times as many participants as testing a main effect of the same magnitude, and the same factor is likely to apply if adding a three-way interaction of the same magnitude, resulting in sixteen times the sample size as for the main effect [33].[9] Additionally, an estimate of the effect of an intervention at a

[9] The variances associated with estimates of a main effect, a two-way interaction and a three-way interaction are $\sigma^2/2n$, $2\sigma^2/n$ and $8\sigma^2/n$, respectively [32]. Here, the variance for the individual patient responses, which is assumed to be the same in each treatment group, is given by σ^2 and the sample size is given by n.

particular level of a subgroup in such an analysis may be based on a small fraction of the trial sample and is therefore likely to be very inaccurate. For example, if one of the interventions is received by only a small number of patients who are female, seropositive and with normal kidney function, then this will adversely affect the accuracy, and hence the reliability, of the estimated interaction. Some of these difficulties are avoided if subgrouping variables are created based on risk, rather than a set of baseline variables taken one at a time. In particular, Dahabreh et al. [49] propose that risk scores for the trial outcome could be arrived at through a multivariable regression model and subgroups could be created accordingly.

A potentially fruitful alternative approach is to interpret a subgroup effect in the context of information derived from *outside* the trial. This might be the results of previous empirical studies, or the clinician's own judgment. Such an approach is in keeping with the principles of Evidence-Based Medicine, where it is argued that conclusions should be based on clinical experience as well as on research evidence:

> The practice of evidence based medicine means integrating individual clinical expertise with the best available external clinical evidence from systematic research [50, p. 71].

A different solution to the problem of interpreting aggregate treatment effects is to move away fully from the RCT and adopt a design that focuses directly on the individual patient, such as a single-subject ($n = 1$) study [51–53]. Here, one or more interventions are tested sequentially on a single participant, usually with alternating treatment and withdrawal phases.[10] In this way, the participant is used as his or her own control in the same way as in a crossover trial [54], but the unit of analysis is the individual patient, not the group. The findings are then potentially transferable to other similar patients, and Bayesian methods can assist in this process [53]. Thus, aggregating the findings from multiple $n = 1$ studies may be informative as to heterogeneous treatment effects [55]. However, this design is unsuitable for some diseases—e.g., those in which treatment may induce a permanent change, thereby rendering the withdrawal phases inert. Additionally, because they normally involve treatment withdrawal, single-subject studies give rise to specific ethical issues [56].

Two final points bear mentioning. First, a statistically nonsignificant result from a subgroup analysis (provided that the analysis has been adequately powered) is as informative as a statistically significant one—it is as useful to know that there is a lack of evidence for variable treatment effects across subgroups as it is to know that it is likely that the treatment's application must be tailored to specific such subgroups. Second, as Rothwell [29] has argued, in many cases where subgroup analyses have

---

[10] If more than one treatment is assessed in a single-subject study, such as in an ABACA design (where A represents a no-treatment phrase and B and C represent the phases in which two different treatments are administered) this allows us to determine the effect on a particular participant of each of these treatments; as noted earlier, this information cannot be extracted from a parallel-group RCT, as a participant only receives one of the treatments under study, and his or her response to the other treatment is a counterfactual and, as such, unobserved.

resulted in conclusions subsequently shown to be inappropriate, these analyses have not been conducted optimally, such as by failing to define the analysis before starting the trial or having an insufficient sample size to estimate an interaction with sufficient accuracy.

5.4 Transferability of Findings from the Study Context to a Different Context

5.4.1 External Validity

In addition to the difficulties outlined above in our discussion of subgroup analyses, there are difficulties in transferring findings from the study context to a different context: issues of external validity. It may be helpful if we return to the idea of treatment moderation here [15]. A treatment moderator is a variable that indicates *for whom* or *under what conditions* a choice between different treatments differentially affects outcome [57]. Accordingly, our earlier discussion of subgroup effects can be thought of in terms of moderators of treatment effect that may differ across categories within a subgrouping variable. Turning to external validity, this can also be understood in terms of treatment moderators. If the nature and/or the effect of such moderators differs between the study sample and the target population, this will restrict our ability to extend findings beyond the context of the study.[11]

Deaton and Cartwright note that external validity may indicate that a conclusion as to the *presence* of a treatment effect can be extended beyond the context of an RCT, or it may go further and indicate that the specific estimated *magnitude* of this effect can also be similarly extended [60]. They also distinguish between *extrapolation* (whether a result can be taken to be replicable in another particular situation or setting) and *generalisation* (whether the result can be taken to be replicable in all relevant situations or settings, or at least widely). An additional distinction is advanced by Justice et al. [61] between *reproducibility* (whether findings can be taken to be replicable in other patients from the same population but not included in

[11] It is worth distinguishing moderation from mediation. A treatment mediator has to do with the pathway through which a treatment effect occurs: it identifies *how* or *why* one treatment may be effective [57]. Dunn et al. [58] provide an example from the field of psychological medicine. In the COMMAND trial [59], cognitive therapy was examined as a means of reducing the effect of command hallucinations. The aim of the therapy was to reduce the perceived power of the hallucinatory voices, on the assumption that this would lower the risk of complying with these voices when they told the patient to harm themselves or others: the perceived power of the voices is a proposed mediator of the effect of cognitive therapy as it suggests the means through which this therapy reduces the risk of harmful effects.

the study) and *transportability* (whether findings can be taken to be replicable in a different but plausibly related population).[12]

As we shall see later, judgments as to external validity may be arrived at through formal statistical considerations—in relation to the principles of sampling theory—but they may additionally be guided by an understanding of the context to which findings might be applied, based on clinical experience and background theoretical knowledge.

5.4.2 Internal Versus External Validity

There is, quite correctly, a very strong concern for the internal validity of an RCT. As noted at the beginning of the chapter, internal validity has to do with the extent to which the study is free from confounding and other inferential biases and thereby allows differences in outcome to be attributed solely to the interventions being tested. However, the methodological features of an RCT that ensure internal validity tend to restrict external validity [10]. In a similar vein, Cartwright [63, 64] points out that in order to *clinch* rather than merely *vouch for* the causal inference that RCTs seek to make, their scope tends to be narrow.[13] This means that seeking greater certainty regarding the causal link in an RCT (between differences in the treatments tested in the trial and the associated treatment effect) requires a narrower focus, which may restrict the trial's external validity. Rothwell [65] argues that research into RCTs, and guidelines for assessing RCTs, have placed more emphasis on internal than on external validity, but, as Deaton and Cartwright argue, the latter type of validity is at least as important as the former: "it cannot be that knowing how to use results is less important than knowing how to demonstrate them" [60, p. 10].

Several studies have examined the external validity of the results of RCTs, across various clinical and geographical settings, by comparing trial participants with those who were not recruited or who declined consent, concluding that the study sample differed in important respects from the target population to which findings were intended to apply [66–73]. A review of RCTs of pharmaceutical therapy [74] found

---

[12] Compare Dekkers et al.'s [62] similar proposed distinction between *external validity* (study results are valid for patients other than those included in the sample, but who are in a treatment setting the same as that of the study) and *applicability* (study results are valid for patients to whom these results can be generalised, but who are in a different treatment setting from that of the study). Here, Dekkers et al. are using the term 'external validity' in a more specific sense than its common use as an umbrella term for inference beyond the context of the study to other settings and involving alternative patients. However, we will use the term in this latter, broader sense.

[13] Here, 'clinching' refers to a causal conclusion that can be guaranteed through deductive logic, but at the cost of certain restrictive assumptions. In contrast, 'vouching for' implies less certainty in a given causal conclusion, but with less restrictive assumptions underlying such claims and, as a result, broader scope.

that included participants had a lower risk profile (e.g., less likely to have severe clinical impairment or to have a poor prognosis) than the general clinical population, and that, in particular, older patients and those with comorbidities were commonly excluded. A similar finding regarding age emerged from reviews of trials in heart failure [67] and diabetes [75]. A particular concern is that upper age limits in RCTs may be arbitrary [76]. The population ageing evident in many countries makes this concern more acute, as increasing numbers of older people may thereby be excluded from medical research based on such age limits.[14]

In addition to the core element of randomisation, other specific design measures are normally adopted to secure internal validity in RCTs; some of these are given in Table 5.3. However, through these strategies, internal validity may be obtained at the cost of external validity. Each of them has the potential to create an environment within the trial that differs from that of the real-world clinical situation to which findings will be applied. A tightly defined intervention may not reflect the somewhat 'looser' way in which it will likely be employed in daily practice. Likewise, the most robust measurement strategies for RCT outcome measurement may not reflect everyday clinical strategies used to monitor patient outcomes. In some situations, clinicians and patients who are blinded to treatment may behave and interact in a different way from when treatment is known, reflecting the fact that blinding itself can be a source of performance bias (as defined in KLP 2.6, Chap. 2). For example, communication in an RCT between doctor and patient may have to take a less natural form when one or both parties have to be unaware of the treatment that the patient is receiving as part of the study. A further issue relates to the degree of expertise or specialisation of clinicians participating in a trial. Rothwell [77] discusses the ACAS trial of endarterectomy [78] and notes that only surgeons with a very high safety record were allowed to take part, such that the beneficial effect of surgery was attributable to a large degree to the low operative risk under the care of these practitioners. The findings of studies such as this that involve treatments that are practitioner- or technique-dependent may thereby not represent routine practice.

Furthermore, even if we are satisfied that all inclusion and exclusion criteria are satisfied, the type of patient who agrees to participate in an RCT may differ from the type of patient who declines to do so. For example, in a review of RCTs in heart failure, Lloyd-Williams et al. [66] note that those who consented to participate in such RCTs differed from non-consenters in terms of sex (males were more likely to consent), age (older people were less likely to consent) and concurrent medication (those taking ACE inhibitors were more likely to consent). The strength of patients' preferences for particular treatments is important in relation to willingness to participate. Olschewski et al. [90] report a trial of breast-conserving radiotherapy versus mastectomy in which, owing to strong treatment preferences, only 10% of women with breast cancer

[14] We can see here the counterpart to the fallacy of division referred to earlier. This is the fallacy of composition—assuming that what holds for some of the parts (here, the patients recruited to the trial) also holds for the whole (here, all patients of the sort eligible for recruitment and in turn, the target population) [7].

Table 5.3 Procedures for optimising internal validity in RCTs

Procedure	Description
Stratified randomisation	If potential confounding variables are incorporated in the randomisation process as stratification variables, this will remove or minimise their confounding influence (see Table 5.1). Stratified randomisation involves dividing the subgrouping variable into two or more strata and then randomising equal numbers within each stratum to the arms of the trial, thereby ensuring a more balanced distribution of the variable across the arms than if simple randomisation had been used. For example, stratifying by the presence or absence of a particular comorbidity will ensure that equal numbers of patients with and without the comorbidity are allocated to each arm of the trial, thus ensuring balance on this potential confounder.
Inclusion and exclusion criteria	Inclusion and exclusion criteria that are tightly defined can limit patient heterogeneity. This will make baseline imbalance in patient characteristics less likely and reduce the variance associated with a treatment effect, allowing it to be estimated with greater accuracy.
Operationalisation of interventions	If interventions are specified in detail in the trial protocol, this serves to keep the arms of the trial distinct in terms of what treatment participants receive and helps to prevent contamination (whereby participants in one arm of the trial may receive an intervention, or a component of an intervention, that is intended for participants in the other arm only).
Clinimetric properties of outcome variables	Outcome measures with good clinimetric properties, e.g., sound measurement validity (which, *ipso facto*, includes good reliability) and good sensitivity to change, serve to reduce both systematic and random measurement error in the outcome variables of the study.
Control of cointerventions	Restricting clinicians' use of other concomitant interventions for patients included in the trial helps in attributing any treatment effect specifically to the interventions being tested. If participants in one or both arms of the study have received a variety of other treatments, it will be hard to disentangle the effects of these from the effects of the treatments being tested.
Blinding (or masking)	Preventing participants and/or researchers from knowing which intervention has been allocated helps to prevent performance bias. This is the tendency for knowledge of treatment assignment to affect the way in which clinicians deliver, and/or participants respond to, the trial interventions [79] (see also KLP 2.6 of Chap. 2). Blinding also prevents information (or ascertainment) bias, whereby knowledge of treatment assignment may influence the measurement or recording of outcomes [80]. There are differing views on whether 'blinding' or 'masking' is the preferable term [27, 81].

(continued)

Table 5.3 (continued)

Procedure	Description
Statistical adjustment	Potential baseline differences in variables that are prognostically important can be dealt with by covariate adjustment, i.e., inclusion of these variables in the statistical model, so that their influence is adjusted for in the analysis. This will not only counteract their confounding influence but also increase the power of the statistical test on the outcome variable [82, 83]. It is generally agreed that variables used in covariate adjustment should be specified a priori, rather than being chosen on the basis of p-values arising from hypothesis tests on the baseline distribution of these variables across the study groups or, equivalently, the treatment arms [83–88]. Stratification variables should be included in the statistical model not so much with a view to adjustment (as they are, by definition, balanced across the treatment arms), but because any variable that is a constraint on randomisation should be represented in the analysis. A failure to include stratification variables will cause standard errors and Type I error rates to be incorrect, with the important effect of statistical power being attenuated [89]. Because a covariate or stratification factor included in the model will use up one or more degrees of freedom (which has an effect equivalent to reducing the sample size of the study by the number of 'lost' degrees of freedom), this will potentially reduce the power of the statistical test. However, except in very small studies, including a covariate or stratification factor will have the opposite effect of increasing the accuracy of the effect size estimate, and hence the power of the statistical test, so that, overall, the test will have greater power than if the variable were not included.

were willing to be randomised; the remaining women chose whichever of the two interventions they preferred and remained outside the trial.

It is important to recall that attempts to maximise the internal validity of an RCT have potential implications for its external validity, and some sort of balance should be struck between these competing demands. In recognition of such issues, pragmatic clinical trials [91] attempt to minimise the extent to which the context of the trial differs from that of clinical practice, by ensuring that inclusion criteria are not too narrow and exclusion criteria are not too restrictive, and by operationalising the interventions to be tested in a way that is comparable to their use in everyday practice. An example of reconciling the demands of internal and external validity within a pragmatic RCT is given by Godwin et al. [92]. In a study of home blood pressure monitoring, they allowed the monitoring device to be used flexibly by participants, provided that it was used at least once weekly. In this way, the monitor was used in the trial in a way likely to reflect its use in everyday clinical practice, but its use was sufficiently specified to allow its effectiveness to be tested in the trial.

Key Learning Point 5.3 Measures taken to maximise the internal validity of RCTs (such as use of strict inclusion and exclusion criteria or detailed operationalisation of interventions) may restrict their external validity. Nevertheless, steps can also be taken, especially in pragmatic trials, to achieve an appropriate balance between the demands of internal and external validity by seeking to minimise differences between the context of the trial and that of clinical practice.

Of course, findings that are not internally valid are not *worth* generalising,[15] and efforts to preserve internal validity should not be neglected. However, it does not follow from this that a high level of internal validity does anything to secure a similarly high level of external validity [93].[16]

5.4.3 Other Challenges in Achieving External Validity

We can identify other issues that may affect the external validity of RCT findings. For example, there are two ways in which the procedure used to recruit RCT participants affects how the study findings can be extrapolated or generalised. First, there are statistical issues. Randomisation is the cardinal feature of an RCT. However, whilst participants are randomly *allocated* (except in cluster-randomised trials), they are rarely randomly *selected*, especially if incident patients are recruited from a small number of centres, by advertisement at clinics, or through the media. This makes the empirical representativeness of participants in an RCT, and thereby, the generalisability of any conclusions, questionable.[17] The potential non-representativeness of the study sample also casts doubt upon the statistical conclusion of the trial. Hypothesis testing invokes a population in which the null hypothesis is initially assumed to be true. While, in practice, data are used from a single sample, assumed to be drawn from the above population, the testing of the null hypothesis assumes a theoretical distribution of estimates from repeated samples from this population. If the *study population* (the sampling frame for the study) does not represent the

[15] See, for example, the following comments: "...it is important to bear in mind that validity should not be compromised in an effort to achieve generalizability, because generalizability can be inferred only for a valid result" [93, p. 38]; "There is no point in worrying about whether a treatment works the same or differently in men or women until it has been shown to work in someone" [94, p. 795].

[16] This relates to the different logic underlying internal validity from that underlying external validity [95]. The former is a (hypothetico-)deductive logic, as articulated by the philosopher Karl Popper, whereas the latter is an inductive logic [96]. Whilst a conclusion to be generalised from an RCT must be internally (deductively) valid, this is a necessary but not a sufficient condition for it to be externally (inductively) valid.

[17] A corresponding argument applies to the selection of clusters in a cluster-randomised trial.

target population (the population to which the conclusions of the study are to be applied), the theoretical distribution of sample estimates may be incorrect; the trustworthiness of the statistical conclusion reached on the null hypothesis—whether it is rejected or not—will correspondingly be questionable. The latter problem will be compounded if, in addition, the null hypothesis on which a decision is made is tested on a sample that does not represent the study population. Owing to such lack of representativeness, the null hypothesis tested using the sample data will differ in its specification from the null hypothesis intended, which was framed in terms of the target *population*—so there is a sense in which the wrong hypothesis will have been tested.

In contrast to the design and statistical features considered above, other ways in which recruitment to an RCT, and participants' subsequent experience of trial processes, affect external validity can be more psychological in nature. Entering an RCT, as opposed to receiving routine clinical care, is characterised by various assumptions, expectations or misunderstandings, defined in Table 5.4. These phenomena serve to make the participant's experience within the trial somewhat different from that of normal clinical practice, and their treatment response may be influenced accordingly. For example, resentful demoralisation may cause participants receiving what they perceive as the disfavoured intervention to provide outcome data that are biased by this perception. Compensatory rivalry may also bias outcome data in participants receiving what they perceive to be a disfavoured intervention, but in a different way: as a means of redressing this disadvantage, they may report more favourable outcomes. In addition, compensatory rivalry may lead to contamination. If some participants seek to obtain some of the specific components of the intervention to which they have not been randomised, but regard as preferable, this will undermine the trial's ability to assess two distinct treatments. For example, patients in a group receiving only advice for their back pain may be aware that the other intervention being tested involves exercises in addition to advice. To redress this perceived disadvantage, they may work on their own fitness programme.

The factors discussed above suggest that the context in which an RCT occurs, and the experimental and psychological conditions under which it is conducted, may differ notably from those of clinical practice, making it hard, therefore, to transfer the treatment response noted in the trial to everyday practice. Of course, adopting a pragmatic approach to RCTs, whereby key elements in the trial are kept as close as possible to real-world practice, will help to minimise this tendency.

Issues of external validity may revolve around practitioners as well as patients. A multicentre trial of carotid endarterectomy [102] suffered poor accrual, recruiting just 893 patients in a two-year period, during which 140,000 patients across Canada and the United States underwent the intervention being evaluated. Such a low rate of recruitment suggests a mismatch between trial participants and the broader clinical population. Barnett et al. [103] suggest reasons for this shortfall in accrual relating to clinicians' attitudes and beliefs. Their comments warrant quoting at length:

> Some practitioners have told us frankly that their practices include such a major vested interest in the diagnostic and operative procedures associated with this treatment that

> they are not prepared to alter present practice procedures. The regulatory constraints of managed medical plans discourage referrals from busy practice centers to academic units. Many consulting physicians and surgeons… have reported to us that if they were to join a randomised trial, their referring practitioners would send their patients to another center where their expressed wishes for their own treatment of choice would be followed, rather than having them submitted to a randomization process that might not result in the treatment of their prejudice. In the opinion of a great many members of the profession, the procedure of carotid endarterectomy is a "standard treatment," and they are not prepared to reverse this thinking and cast it in the category of an experimental treatment modality for any degree of stenosis [103, pp. 1049–1050].

A final issue to do with the external validity of RCT findings relates to changes in practice. If, as part of the natural evolution of clinical practice, the way in which the intervention under investigation was applied in a trial and/or the comparator interventions against which it was tested have changed in the time since the results were

Table 5.4 Assumptions, expectations and misunderstandings associated with participating in an RCT

Term	Description
Resentful demoralisation	Participants who perceive themselves to have been allocated to the less desirable treatment in the trial may become disheartened and this may adversely affect their response to the intervention that they are receiving [97]. For example, participants who are receiving standard medication for their depression, and who know that other participants in the study are being given the latest drug therapy, may feel that they are not receiving state-of-the-art care and exhibit a poor response for this reason.
Compensatory rivalry	In contrast to resentful demoralisation, this term describes the phenomenon whereby participants who believe they are receiving the less desirable treatment respond better in an attempt to offset this perceived disadvantage [97]. They may do this by engaging more fully or conscientiously with the treatment to which they have been assigned, by seeking to access elements of the more favourable treatment to which they have not been assigned, or by unconsciously reporting more positive scores on a self-report measure.
Therapeutic misconception	Despite having received information clarifying the scientific purpose of the trial, participants could assume that the treatment that they receive reflects their individual clinical needs rather than the process of randomisation [98]. This allows participants to make sense of an unfamiliar research environment by reinterpreting it in terms of the more familiar clinical environment [99].
Therapeutic misestimation	This describes the way in which trial participants may overestimate the benefits and/or underestimate the harms associated with taking part in a trial [100].
Unrealistic optimism	This represents the way in which trial participants may overestimate the likelihood of therapeutic benefit for themselves relative to other similar participants in the trial [101].

published, their relevance to current practice may be doubtful. Some authors therefore suggest that an alternative to the RCT is required to assess interventions that are in a state of rapid evolution whereby RCT findings could rapidly become outdated as practice develops. In such a situation, Truog [104] suggests that a prospective observational study may be more informative than an RCT, as in an observational study choice of treatment may be left to the clinician and outcomes from patients treated with the interventions of interest may be assessed alongside changes in practice relating to their implementation. Bluhm has mounted a similar argument for the value of observational studies with specific respect to research in chronic diseases, pointing in particular to the difficulty of reconciling the conflicting demands of internal and external validity in a heterogeneous patient population:

> Because studies achieve external validity by maximizing variability and internal validity by minimizing variability, studies tend to be designed to serve one end or the other (most commonly internal validity). In the case of chronic disease research, the variability among the study population makes it difficult both to achieve homogeneity among the patients in the study and to assemble a representative sample [105, p. 115].

Patients with chronic diseases are heterogeneous in terms of differing clinical features, fluctuations in disease severity, comorbidities, changes that are indicated in ongoing clinical management, and the presence or absence of adjuvant therapy. Bluhm argues that a non-randomised study would better accommodate this heterogeneity than the more structured framework of an RCT. In this way, "the object of the study and the source of information about effectiveness would be actual clinical practice" [105, p. 117].[18] Trialists might counter that pragmatic trials, if appropriately designed, can address some of the difficulties that Bluhm points to. Nonetheless, his argument highlights importantly the need to consider alternative methods of determining treatment effectiveness.

Key Learning Point 5.4 The external validity of RCTs relates to how participants are sampled and then screened for inclusion, but it is also affected by practitioner willingness to recruit patients to a study, practitioner behaviour within the study, and by participants' psychological responses to being in a trial.

Thus far in this chapter, we have focused principally on the empirical extrapolation or generalisation of RCT findings. It has been suggested that if the conditions of the

[18] The natural experiment, in which an intervention to be tested occurs in the natural course of events and is uninfluenced by the investigator, except possibly in relation to its timing [106], sets the evidence from the study very firmly in the context to which it applies. However, this design is of greater relevance to public health interventions than to those of clinical practice at the level of the individual patient.

RCT—such as the type of patients recruited and the way in which the interventions in the trial are defined—can be made to closely approximate those of the sphere of practice for which findings are intended, such generalisation will be improved.[19] So, the solution to the problem of external validity appears to be one of achieving greater *statistical* generalisability of the quantitative estimates derived from the trial data. However, there is a more basic difficulty. The context of the trial can indeed be compared to that of clinical practice in terms of the characteristics of the participants, the expertise of the clinicians, the operationalisation of the interventions, and so forth. But there will be a multitude of other ways—many of them unknown and unspecifiable—in which these two contexts may differ.[20] By virtue of the fact that many of these differences are unknown or unspecifiable, they cannot be measured or controlled. Accordingly, whilst the RCT may produce valid and robust conclusions in the specific context in which it was conducted, the validity of these conclusions in other contexts will be uncertain.[21] The fundamental question here is posed by Cartwright: "What relevance has an RCT result in the target situation for predicting what will happen there when the cause is implemented, as it will be, without the experimental constraints and in an environment where many other causes may be at work as well?" [109, p. 60]. Of relevance here is a distinction made by Basu [110] between *circumstantial causality*, that is, a conclusion regarding cause and effect that holds for a specific situation at a specific time, and *universal causality*, which would claim that this conclusion applies in other situations beyond that of the study. In terms of this distinction, the RCT is strong in establishing circumstantial causality but is less able to establish universal causality: "RCT discoveries never graduate from something 'was a cause' of something else to something 'is a cause'" [110, p. 460].

Seeking to demonstrate external validity solely by empirical means is therefore inadequate. An alternative, but complementary, approach is based more on judgment than on statistics. Here, one uses existing clinical, or other, knowledge to determine whether findings can reasonably be applied to patients and contexts outside the trial. Such judgments are, of course, based on the data resulting from the trial, but they also draw on other sources of understanding derived from theoretical knowledge, reasoning and clinical experience. This is particularly relevant in relation to Justice et al.'s [61] notion of transportability—applying findings to clinical populations

---

[19] This will allow generalisability to be improved a priori, in terms of the design of the study and how elements within it are operationalised. In addition, if data are available at the individual patient level from completed trials, generalisability to a defined target population can be established a posteriori through forms of statistical weighting [107, 108]. Such weighting involves statistically adjusting the trial data in relation to known differences between the characteristics of the sample and those of the target population.

[20] Such differences might lie at almost any level: genetic, physiological, clinical, psychological, social, cultural, institutional or environmental.

[21] In the rare event that the participants in an RCT are a purely random sample of the study population, there would be a strong statistical case, based on established principles of sampling theory, for claiming representativeness in respect of the characteristics of these individuals; the strength of this claim would increase asymptotically as sample size increased. However, this would not account for potential contextual or temporal differences. Furthermore, it would not settle the question as to the relationship between the study population and the target population.

that differ in certain respects from that in which the trial was based but that are nonetheless broadly similar. Rothman et al. indicate that such a decision cannot rely solely on statistical considerations but must also focus on whether it is reasonable or plausible to apply the findings in this way in the light of existing theoretical or clinical knowledge and logical reasoning:

> If the generalization of study results is literally limited to the characteristics of those studied, then causal inferences cannot be generalized beyond those subjects who have been studied and the time period during which they have been studied… The process of generalization in science involves making assumptions about the domain in which the study results apply [111, p. 146].

They discuss the specific case of prior pathophysiological knowledge:

> … although most of the decisive data connecting smoking to lung cancer was derived from observations on men, no one doubted that the strong effects observed would carry over at least approximately to women, for the lungs of men and women appear to be similar if not identical in physiologic detail. On the other hand, given the huge sex differences in iron loss, it would seem unwise to generalize freely to men about the effects of iron supplementation observed in premenopausal women [111, p. 147].

Cartwright expresses the same caution regarding generalisation less formally but just as effectively:

> I have heard one famous advocate of RCTs insist that for the most part Frenchmen are like Englishmen – we do not need to duplicate RCTs in France once we have done them in England. He of course realizes that this may not be true across all treatments and all effects. And we know that superficial similarities can be extremely misleading. I carry my pound coins from Britain to the US and put them into vending machines that look identical to those in Britain, but they never produce a packet of crisps for me in the US [109, p. 66].

Irrespective of the extent to which findings are generalisable, statistically or otherwise, an additional problem lies at the level of reporting. If the description of study participants in the reports of RCTs is inadequate—as suggested by Brænd et al. [112], van Deudekom et al. [113] and Malmivaara [114]—it will be hard for clinicians to apply the findings of such studies to their patients. Equally, if what Fortin and Smith [115] refer to as "system factors" (e.g., the clinical setting in which the trial took place or the extent to which the treatment protocol in the trial differed from routine practice) are poorly described, clinicians will be inadequately informed on the relevance of findings to their own practice. Following CONSORT reporting guidelines for RCTs [87] is important here; see also https://www.equator-network.org/.

Key Learning Point 5.5 The external validity of an RCT relates to empirical generalisation in terms of the principles of sampling theory. However, it is also important to take into account theoretical knowledge, logical reasoning and clinical experience in order to make sense of the likely broader applicability of trial findings.

5.4.4 Assessing External Validity

External validity—whether viewed empirically or more contextually—needs to be evaluated if we are to make sense of, and implement, the results of RCTs. Dekkers et al. [62] propose a three-step approach to evaluating the external validity of an RCT. First, we can assess external validity in relation to eligibility criteria. The participants included in the trial may not accord with the stated eligibility criteria for membership of the target population to which findings are intended to apply. In terms of our earlier distinction between study and target populations, this mismatch may occur because the study sample does not represent the study population from which it was drawn and/or because this study population does not represent the target population to which the findings are to be applied. Second, we can consider temporal, ethnic, socio-economic and geographical factors. Such differences between the study sample and the target population may affect the external validity of the findings from an RCT, though the extent to which they do so is likely to be context-specific, as reflected in the comments of Rothman et al. quoted above [111]. For example, Dekkers et al. [62] argue that patients' responses to antihypertensive medication may be related to their ethnicity whereas responses to aspirin in patients with acute ischaemic stroke are less likely to demonstrate such a relationship. Independently of the issue of whether the study sample meets the intended eligibility criteria for the trial, Dekkers et al. [61] also refer to the issue of whether the findings of the study can be applied to target populations defined by eligibility criteria that are at least partly different from those of the study. For example, the mean age or the sex composition of the new target population may differ, or it may exhibit different comorbidities or dissimilar chronicity.

Box 5.1 provides some questions that may usefully be asked when assessing the external validity of RCT findings.

Box 5.1 Some questions to ask when assessing the external validity of findings from an RCT

- Are the demographic characteristics of participants (e.g., sex, age, sociodemographic profile, ethnicity) similar to those of the target clinical population?
- Are the clinical characteristics of participants (e.g., severity or chronicity of symptoms, functional limitation, presence of comorbidities) similar to those of the target clinical population?
- Do any mechanisms underlying a treatment support, or question, the external validity of trial findings? For example, if there is evidence that the pharmacological effect of a drug is invariant with age, this may make it easier to apply a trial's findings across patients of different ages. For other treatments, however, their mechanism of action may serve to preclude generalisation beyond the strict case definition of the study. Howick et al. [116] give the

examples of macrogols in childhood constipation and metformin in diabetes, respectively, for these two situations.

- Is there a high ratio of ineligible to eligible patients, such that many patients who had the target condition were nonetheless excluded from the study?
- Is there a high proportion of eligible patients who did not consent to participation in the trial?
- Is the case definition employed in the trial (i.e., the diagnostic criteria used to determine the eligibility of patients) similar to that likely to be used in routine practice?
- If a matched-subjects design has been used, many potential participants may have been excluded if there were difficulties in identifying enough matched subjects who were willing to participate [117]. If so, do such exclusions affect application of the findings from the trial?
- Are the clinical centres taking part in the study similar to those in which most patients are likely to be seen?
- Have the treatments tested in the study been operationalised in a way that reflects routine practice, or have they been 'over-optimised' in terms of their frequency, duration or intensity?
- If the study was carried out some time ago, has clinical practice changed in the interim in ways that would affect the application of the findings?
- If the study was carried out in another country, with a different health care system and/or different modes of clinical practice, might this affect application of its findings to a local context?
- Has the degree of control over the design of the study (e.g., with respect to blinding or confounding variables) made the context in which the interventions were delivered in the trial different from those of routine practice?
- If the tested intervention is technique- or expertise-dependent, are the participating clinicians representative of the broader clinical population likely to treat the patients in question?
- In an attempt to avoid confounding, have participating clinicians been restricted from using interventions other than those being tested in the trial (cointerventions) in a way that may not reflect routine clinical practice?
- Are the outcome variables used to measure treatment effectiveness ones that would reflect the way in which outcomes and treatment progress would normally be measured in routine practice?
- If the outcome variables do indeed differ from those used in clinical practice, could the nature of the corresponding measurements, or the mode or frequency of their administration, have influenced, and thus confounded, participants' responses to the study interventions?
- In the case of chronic conditions, does the period over which outcomes were monitored in the trial represent the duration with which patients and/or clinicians would be concerned for the condition in question?

Key Learning Point 5.6 Assessing the external validity of an RCT is crucial to determining its relevance to clinical practice. A range of potential influences on external validity—indicated in Box 5.1—should be considered.

5.5 Conclusion

In equipping medical and other health professional students for clinical research and practice, educators should make it clear that although RCTs may be regarded as the gold standard design when assessing treatment effectiveness, the application of RCT findings, both within and beyond the sample, is subject to important limitations. Whilst RCTs are designed to generate aggregate conclusions that can be applied broadly to patients who meet the case definition used for the study (the target population), we have seen that if the sample is unrepresentative of patients who are encountered in routine practice, this aim is imperfectly met. Moreover, even if the study sample corresponds appropriately to the target population, applying the aggregate findings from the trial across different categories of what may be a heterogeneous population is not straightforward, particularly if participants in the study exhibit a wide range of responses. Such heterogeneity in responses may be prominent in pragmatic trials with broad inclusion criteria [14].

Some attempts to offset these limitations bring their own problems. For example, a shift to more pragmatic trials, with broader entry criteria, may allow greater external validity of trial findings, but may make it even harder to apply these findings to specific patients or categories of patient. Conversely, trials with more restrictive entry criteria may provide targeted information on a relative homogeneous study population but leave the key clinical question unanswered in other groups of patients within the target population, particularly in conditions that exhibit high clinical heterogeneity [118].

Some of the difficulties of drawing inferences, either in terms of generalising beyond the study sample or in relation to subgroups within the study sample, can be addressed through refocusing certain aspects of RCT methodology, such as by ensuring that attempts to optimise internal validity do not unduly restrict the ability to apply the study's findings to clinical practice. Equally, efforts should be made to ensure that any analyses aimed at examining subgroup effects are performed correctly and interpreted judiciously. Shifting some of the emphasis from hypothesis testing, with its binary logic of either rejecting or not rejecting the null hypothesis, to processes of estimation using CIs would reduce the extent to which RCTs appear to 'ignore' certain information.

In the light of current advances in personalised medicine, a closer focus on differences in treatment effects across subcategories of patients would allow trial findings to be applied more accurately and appropriately, provided that such analyses are interpreted cautiously, for example in relation to Type 1 error inflation and whether they were specified a priori. Some of the pitfalls associated with subgroup analyses should not cause them to be neglected: "If reporting on subgroups is tempting but treacherous, failing to report on them seems unscientific and incurious" [37, p. 913].

Attempts to gauge the external validity of RCT findings should not be restricted to purely statistical considerations within sampling theory but should also draw upon other information and clinical judgment to determine whether findings can plausibly be applied to other contexts or clinical populations.

Finally, when RCT methodology is presented or discussed, it should be presented alongside alternative designs that could potentially be applied to the question being posed—for example, observational studies or single-subject studies—so that the different types of information that they provide can be compared and evaluated.

References

1. Murad MH, Asi N, Alsawas M, Alahdab F. New evidence pyramid. Evid Based Med. 2016;21:125–7.
2. Lancaster GA, Dodd SR, Williamson PR. Design and analysis of pilot studies: recommendations for good practice. J Eval Clin Pract. 2004;10:307–12.
3. Greenland S, Brumback B. An overview of relations among causal modelling methods. Int J Epidemiol. 2002;31:1030–7.
4. Höfler M. Causal inference based on counterfactuals. BMC Med Res Methodol. 2005;5:28.
5. Yusuf S, Wittes J, Probstfield J, Tyroler HA. Analysis and interpretation of treatment effects in subgroups of patients in randomized clinical trials. JAMA. 1991;266:93–8.
6. Longford NT. Selection bias and treatment heterogeneity in clinical trials. Stat Med. 1999;18:1467–74.
7. Walton DN. Informal logic: a handbook for critical argumentation. Cambridge: Cambridge University Press; 1989.
8. Dong J. Simpson's paradox. In: Gail MH, Nenichou J, editors. Encyclopedia of epidemiologic methods. Chichester: Wiley; 2000. p. 852–3.
9. Tu Y-K, Gunnell D, Gilthorpe MS. Simpson's paradox, Lord's paradox, and suppression effects are the same phenomenon—the reversal paradox. Emerg Themes Epidemiol. 2008;5:2.
10. Kravitz RL, Duan N, Braslow J. Evidence-based medicine, heterogeneity of treatment effects, and the trouble with averages. Milbank Q. 2004;82:661–87.
11. Subramanian SV. The "average" treatment effect: a construct ripe for retirement. A commentary on Deaton and Cartwright. Soc Sci Med. 2018;210:77–82.
12. Cook TD, DeMets DL. Introduction to statistical methods for clinical trials. Boca Raton FL: Chapman & Hall/CRC; 2008.
13. Montgomery AA, Peters TJ, Little P. Design, analysis and presentation of factorial randomized controlled trials. BMC Med Res Methodol. 2003;3:26.
14. Kraemer HC, Frank E, Kupfer DJ. Moderators of treatment outcomes: clinical, research and policy importance. JAMA. 2006;296:1286–9.
15. Cohen J, Cohen P, West SG, Aiken LS. Applied multiple regression/correlation analysis for the behavioral sciences. 3rd ed. Mahwah: Lawrence Erlbaum; 2003.
16. Sun X, Briel M, Walter SD, Guyatt GH. Is a subgroup effect believable? Updating criteria to evaluate the credibility of subgroup analyses. BMJ. 2010;340:c117.
17. Bakhai A, Fox Z, Wang D. Subgroup analysis. In: Wang D, Bakhai A, editors. Clinical trials. A practical guide to design, analysis and reporting. London: Remedica; 2006. p. 265–71.
18. Pocock SJ. Clinical trials: a practical approach. Chichester: Wiley; 1983.

19. Pocock SJ, Assmann SE, Enos LE, Kasten LE. Subgroup analysis, covariate adjustment and baseline comparisons in clinical trial reporting: current practice and problems. Stat Med. 2002;21:2917–30.
20. Schulz KF, Grimes DA. Multiplicity in randomised trials II: subgroup and interim analyses. Lancet. 2005;365:4657–61.
21. Fletcher J. Subgroup analyses: how to avoid being misled. BMJ. 2007;335:96–7.
22. Sun X, Briel M, Busse JW, You JJ, Akl EA, Mejza F, Bala MM, Bassler D, Mertz D, Diaz-Granados N, Vandvik PO, Malaga G, Srinathan SK, Dahm P, Johnston JC, Alonso-Coello P, Hassouneh B, Walter SD, Heels-Andsell D, Bhatnagar N, Altman DG, Guyatt G. Credibility of claims of subgroup effects in randomised controlled trials: systematic review. BMJ. 2012;344:e1553.
23. Brookes ST, Whitely E, Egger M, Smith GD, Mulheran PA, Peters TJ. Subgroup analyses in randomized trials: risks of subgroup-specific analyses; power and sample size for the interaction test. J Clin Epidemiol. 2004;57:229–36.
24. Feinstein AR. The problem of cogent subgroups: a clinicostatistical tragedy. J Clin Epidemiol. 1998;51:297–9.
25. Greenhalgh T. How to read a paper: the basics of evidence-based medicine and healthcare. 6th ed. Chichester: Wiley-Blackwell; 2019.
26. Greeenland S. Confounding. In: Gail MH, Nenichou J, editors. Encyclopedia of epidemiologic methods. Chichester: Wiley; 2000. p. 254–61.
27. Howick J. The philosophy of evidence-based medicine. Chichester: Wiley-Blackwell; 2011.
28 Wang D, Clayton T, Bakhai A. Confounding. In: Wang D, Bakhai A, editors. Clinical trials. A practical guide to design, analysis and reporting. London: Remedica; 2006. p. 295–304.
29. Rothwell PM. Subgroup analysis in randomised controlled trials: importance, indications, and interpretation. Lancet. 2005;365:176–86.
30. Barnett AG, van der Pols JC, Dobson AJ. Regression to the mean: what it is and how to deal with it. Int J Epidemiol. 2005;34:215–20.
31. Ogrodniczuk JS, Piper WE, Joyce AS, McCallum M. Effect of patient gender on outcome in two forms of short-term individual psychotherapy. Psychother Pract Res. 2001;10:69–78.
32. Piantadosi S. Clinical trials: a methodologic perspective. 2nd ed. Hoboken, NJ: Wiley; 2005.
33. Heo M, Leon AC. Sample sizes required to detect two-way and three-way interactions involving slope differences in mixed-effects linear models. J Biopharm Stat. 2010;20:787–802.
34. Kontopantelis E, Sperrin M, Mamas MA, Buchan IE. Investigating heterogeneity of effects and associations using interaction terms. J Clin Epidemiol. 2018;93:79–83.
35. Lee KL, McNeer F, Starmer CF, Harris PJ, Rosati RA. Clinical judgment and statistics: lessons from a simulated randomized trial in coronary artery disease. Circulation. 1980;61:508–15.
36. Cui L, Hung HMJ, Wang SJ, Tsong Y. Issues related to subgroup analysis in clinical trials. J Biopharm Stat. 2002;12:347–58.
37. Wittes J. On looking at subgroups. Circulation. 2009;119:912–5.
38. Hernández AV, Boersma E, Murray GD, Habbema JD, Steyerberg WW. Subgroup analyses in therapeutic cardiovascular trials: are most of them misleading? Am Heart J. 2006;151:257–64.
39. Fan J, Song F, Bachmann MO. Justification and reporting of subgroup analyses were lacking or inadequate in randomized controlled trials. J Clin Epidemiol. 2019;108:17–25.
40. Burke JF, Sussman JB, Kent DM, Hayward RA. Three simple rules to ensure reasonably credible subgroup analyses. BMJ. 2015;351: h5651.
41. Assmann SF, Pocock SJ, Enos LE, Kasten LE. Subgroup analysis and other (mis)uses of baseline data in clinical trials. Lancet. 2000;355:1064–9.
42. Cook DI, Gebski VJ, Keech AC. Subgroup analysis in clinical trials. Med J Aust. 2004;180:289–91.
43. Wang R, Lagakos SW, Ware JH, Hunter DJ, Drazen JM. Statistics in medicine—reporting of subgroup analyses in clinical trials. New Engl J Med. 2007;357:2189–94.
44. Rutherford A. ANOVA and ANCOVA: a GLM approach. 2nd ed. Hoboken NJ: Wiley; 2011.

45. Baguley T. Serious stats: a guide to advanced statistics for the behavioral sciences. Basingstoke: Palgrave Macmillan; 2012.
46. Thatcher N, Chang A, Parikh P, Pereira JR, Ciuleanu T, von Pawel J, Thongprasert S, Tan EH, Pemberton K, Archer V, Carroll K. Gefitinib plus best supportive care in previously treated patients with refractory advanced non-small-cell lung cancer: results from a randomised, placebo-controlled, multicentre study (Iressa Survival Evaluation in Lung Cancer). Lancet. 2005;366:1527–37.
47. Friedman LM, Furberg CD, DeMets DL, Reboussin DM, Granger CB. Fundamentals of clinical trials. 5th ed. Heidelberg: Springer; 2015.
48. Petticrew M, Tugwell P, Kristjansson E, Oliver S, Ueffing E, Welch V. Damned if you do, damned if you don't: subgroup analysis and equity. J Epidemiol Community Health. 2012;66:95–8.
49. Dahabreh IJ, Hayward R, Kent DM. Using group data to treat individuals: understanding heterogeneous treatment effects in the age of precision medicine and patient-centred evidence. Int J Epidemiol. 2016;45:2184–94.
50. Sackett DL, Rosenberg WMC, Gray JAM, Haynes RB, Richardson WS. Evidence based medicine: what it is and what it isn't. BMJ. 1996;312:71–2.
51. Richards SB, Taylor RL, Ramasamy R, Richards RY. Single subject research: applications in educational and clinical settings. San Diego, CA: Singular Publishing; 1999.
52. Lillie EO, Patay B, Diamant J, Issell B, Topol EJ, Schork NJ. The n-of-1 clinical trial: the ultimate strategy for individualizing medicine? Per Med. 2011;8:161–73.
53. Duan N, Kravitz RL, Schmid CH. Single-patient (n-of-1) trials: a pragmatic clinical decision methodology for patient-centered comparative effectiveness research. J Clin Epidemiol. 2013;66:S21–8.
54. Senn S. Cross-over trials in clinical research. 2nd ed. Chichester: Wiley; 2002.
55. Gabler NB, Duan N, Liao D, Elmore JG, Ganiats TG, Kravitz RL. Dealing with heterogeneity of treatment effects: is the literature up to the challenge? Trials. 2009;10:43.
56. Sim J. The ethics of single system (n = 1) research. Physiother Theory Pract. 2004;10:111–22.
57. Kraemer HC. Messages for clinicians: moderators and mediators of treatment outcome in randomized clinical trials. Am J Psychiatry. 2016;173:672–9.
58. Dunn G, Emsley R, Liu H, Landau S, Green J, White I, Pickles A. Evaluation and validation of social and psychological markers in randomised trials of complex interventions in mental health: a methodological research programme. Health Technol Assess. 2015;19(93):1–115.
59. Birchwood M, Peters E, Tarrier N, Dunn G, Lewis S, Wykes T, Davies L, Lester H, Michail M. A multi-centre, randomised controlled trial of cognitive therapy to prevent harmful compliance with command hallucinations. BMC Psychiatry. 2011;11:15554.
60. Deaton A, Cartwright N. Understanding and misunderstanding randomized controlled trials. Soc Sci Med. 2018;210:2–21.
61. Justice A, Covinsky KE, Berlin JA. Assessing the generalizability of prognostic information. Ann Intern Med. 1999;130:515–24.
62. Dekkers OM, von Elm E, Algra A, Romijn JA, Vandenbroucke JP. How to assess the external validity of therapeutic trials: a conceptual approach. Int J Epidemiol. 2010;39:89–94.
63. Cartwright N. Are RCTs the gold standard? BioSocieties. 2007;2:11–20.
64. Cartwright N. A philosopher's view of the long road from RCTs to effectiveness. Lancet. 2011;377:1400–1.
65. Rothwell PM. External validity of randomised controlled trials: 'to whom do the results of this trial apply?' Lancet. 2005;365:82–93.
66. Lloyd-Williams F, Mair F, Shiels C, Hanratty B, Goldstein P, Beaton S, Capewell S, Lye M, McDonald R, Roberts C, Connelly D. Why are patients in clinical trials of heart failure not like those we see in everyday practice? J Clin Epidemiol. 2003;56:1157–62.
67. Masoudi FA, Havranek EP, Wolfe P, Rathore SS, Steiner JF, Ordin DL, Krumholz HM. Most hospitalized older persons do not meet the enrollment criteria for clinical trials in heart failure. Am Heart J. 2003;146:250–7.

68. Yuasa H, Kurita K, Westesson PL. External validity of a randomised clinical trial of temporomandibular disorders: analysis of the patients who refused to participate in research. Br J Oral Maxillofac Surg. 2003;41:129–31.
69. Petersen MK, Andersen KV, Andersen AT, Søballe K. "To whom do the results of this trial apply"? External validity of a randomized controlled trial involving 130 patients scheduled for primary total hip replacement. Acta Orthop. 2007;78:12–8.
70. Millard JD, Muhangi L, Sewankambo M, Ndibazza J, Elliott AM, Webb EL. Assessing the external validity of a randomized controlled trial of anthelminthics in mothers and their children in Entebbe, Uganda. Trials. 2014;15:310.
71. Paul M, Bronstein E, Yahav D, Goldberg E, Bishara J, Leibovici L. External validity of a randomised controlled trial on the treatment of severe infections caused by MRSA. BMJ Open. 2015;5:e008838.
72. Canevelli M, Trebbastoni A, Quarata F, D'Antonio F, Cesari M, de Lena C, Bruno G. External validity of randomized controlled trials on Alzheimer's disease: the biases of frailty and biological aging. Front Neurol. 2017;8:628.
73. Jones GT, Jones EA, Beasley MJ, Macfarlane GJ, on behalf of the MUSICIAN study team. Investigating generalizability of results from a randomized controlled trial of the management of chronic widespread pain: the MUSICIAN study. Pain. 2017;158:96–102.
74. Kennedy-Martin T, Curtis S, Faries D, Robinson S, Johnston J. A literature review on the representativeness of randomized controlled trial samples and implications for the external validity of trial results. Trials. 2015;16:495.
75. Saunders C, Byrne CD, Guthrie B, Lindsay RS, McKnight JA, Philip S, Sattar N, Walker JJ, Wild SH; Scottish Diabetes Research Network Epidemiology Group. External validity of randomized controlled trials of glycaemic control and vascular disease: how representative are participants? Diabet Med. 2013;30:300–8.
76. Briggs R, Robinson S, O'Neill D. Ageism and clinical research. Ir Med J. 2012;105:311–2.
77. Rothwell PM. Factors that can affect the external validity of randomized clinical trials. PLoS Clin Trials. 2006;1:e9.
78. Asymptomatic Carotid Atherosclerosis Study Group. Carotid endarterectomy for patients with asymptomatic internal carotid artery stenosis. JAMA. 1995;273:1421–8.
79. Yakoub D, Panesar SP, Athanasiou T. Methodological framework for evaluation and prevention of publication bias in surgical studies. In: Anathasiou T, Debas H, Darzi A, editors. Key topics in surgical research and methodology. Heidelberg: Springer; 2010. p. 429–40.
80. Schulz KF, Altman DG, Moher D. Blinding is better than masking. BMJ. 2007;334:918.
81. Schulz KF, Grimes DA. Blinding in randomised trials: hiding who got what. Lancet. 2002;359:696–700.
82. Egbewale BE, Lewis M, Sim J. Bias, precision and statistical power of analysis of covariance in the analysis of randomized controlled trials with baseline imbalance: a simulation study. BMC Med Res Methodol. 2014;14:49.
83. Kahan BC, Jairath V, Doré CJ, Morris TP. The risks and rewards of covariate adjustment in randomized trials: an assessment of 12 outcomes from 8 studies. Trials. 2014;15:139.
84. Roberts C, Torgerson DJ. Baseline imbalance in randomised controlled trials. BMJ. 1999;319:185.
85. Raab GM, Day S, Sales J. How to select covariates to include in the analysis of a clinical trial. Control Clin Trials. 2000;21:330–42.
86. Senn SJ. Statistical issues in drug development. 3rd ed. Chichester: Wiley; 2021.
87. Moher D, Hopewell S, Schulz KF, Montori V, Gøtzsche PC, Devereaux PJ, Elbourne D, Egger M, Altman DG. CONSORT 2010 explanation and elaboration: updated guidelines for reporting parallel group randomised trials. BMJ. 2010;340:c869.
88. European Medicines Agency. Guideline on adjustment for baseline covariates in clinical trials. London: European Medicines Agency; 2015.
89. Kahan BC, Morris TP. Improper analysis of trials randomised using stratified blocks or minimisation. Stat Med. 2012;31:328–40.

90. Olschewski M, Schumacher M, Davis KB. Analysis of randomized and non-randomized patients in clinical trials using the comprehensive cohort follow-up study design. Control Clin Trials. 1992;13:226–39.
91. Schwartz D, Lellouch J. Explanatory and pragmatic attitudes in therapeutical trials. J Chronic Dis. 1967;20:637–48.
92. Godwin M, Ruhland L, Casson I, MacDonald S, Delva D, Birtwhistle R, Lam M, Seguin R. Pragmatic controlled trials in primary care: the struggle between external and internal validity. BMC Med Res Methodol. 2003;3:28.
93. Hennekens CH, Buring JE. Epidemiology in medicine. Philadelphia, PA: Lippincott Williams and Wilkins; 1987.
94. Meinert CL. The inclusion of women in clinical trials. Science. 1995;269:795–6.
95. Sleigh JW. Logical limits of randomized controlled trials. J Eval Clin Pract. 1997;3:145–8.
96. Newton-Smith WH. The rationality of science. London: Routledge and Kegan Paul; 1981.
97. Cook TD, Campbell DT. Quasi-experimentation: design and analysis issues for field settings. Boston, MA: Houghton Mifflin; 1979.
98. Appelbaum PS, Roth LH, Lidz CW, Benson P, Winslade W. False hopes and best data: consent to research and the therapeutic misconception. Hastings Cent Rep. 1987;17(2):20–4.
99. Gray BH. Human subjects in medical experimentation: a sociological study of the conduct and regulation of clinical research. New York NY: Wiley; 1975.
100. Horng S, Grady C. Misunderstanding in clinical research: distinguishing therapeutic misconception, therapeutic misestimation, and therapeutic optimism. IRB. 2003;25(1):11–6.
101. Jansen LA. The problem of optimism in clinical trials. IRB. 2006;28(4):13–9.
102. North American Symptomatic Carotid Endarterectomy Trial collaborators, Barnett HJM, Taylor DW, Haynes RB, Sackett DL, Peerless SJ, Ferguson GG, Fox AJ, Rankin RN, Hachinski VC, Wiebers DO, Eliasziw M. Beneficial effect of carotid endarterectomy in symptomatic patients with high-grade carotid stenosis. N Engl J Med. 1991;15:445–53.
103. Barnett HJM, Barnes RW, Clagett GP, Ferguson GG, Robertson JT, Walker PM. Symptomatic carotid artery stenosis: a solvable problem. North American Symptomatic Carotid Endarterectomy Trial. Stroke. 1992;23:1048–53.
104. Truog RD. Randomized controlled trials: lessons from ECMO. Clin Res. 1992;40:519–27.
105. Bluhm R. The epistemology and ethics of chronic disease research: further lessons from ECMO. Theor Med Bioeth. 2010;31:107–22.
106. Dawson A, Sim J. The nature and ethics of natural experiments. J Med Ethics. 2015;41:848–53.
107. Stuart EA, Ackerman B, Westreich D. Generalizability of randomized trial results to target populations: design and analysis possibilities. Res Soc Work Pract. 2018;28:532–7.
108. Ackerman B, Schmid I, Rudolph KE, Seamans MJ, Susukida R, Mojtabai R, Stuart EA. Implementing statistical methods for generalizing randomized trial findings to a target population. Addict Behav. 2019;94:124–32.
109. Cartwright N. What are randomized controlled trials good for? Phil Stud. 2010;147:59–70.
110. Basu K. Randomisation, causality and the role of reasoned intuition. Oxf Dev Stud. 2014;42:455–72.
111. Rothman KJ, Greenland S, Lash TL. Validity in epidemiologic studies. In: Rothman KJ, Greenland S, Lash TL, editors. Modern epidemiology. 3rd ed. Philadelphia, PA: Walters Kluwer/Lippincott Williams & Wilkins; 2008. p. 128–47.
112. Brænd AM, Straand J, Klovning A. Clinical drug trials in general practice: how well are external validity issues reported? BMC Fam Pract. 2017;18:113.
113. van Deudekom FJ, Postmus I, van der Ham DJ, Pothof AB, Broekhuizen K, Blauw GJ, Mooijaart SP. External validity of randomized controlled trials in older adults, a systematic review. PLoS ONE. 2017;12:e0174053.
114. Malmivaara A. Generalizability of findings from randomized controlled trials is limited in the leading general medical journals. J Clin Epidemiol. 2019;107:36–41.
115. Fortin M, Smith SM. Improving the external validity of clinical trials: the case of multiple chronic conditions. J Comorbid. 2013;3:30–5.

116. Howick J, Glasziou P, Aronson JK. Can understanding mechanisms solve the problem of extrapolating from study to target populations (the problem of 'external validity')? J Roy Soc Med. 2013;106:81–6.
117. Sim J, Wright C. Research in health care: concepts, designs and methods. Cheltenham: Stanley Thornes; 2000.
118. Travers J, Marsh S, Williams M, Weatherall M, Caldwell B, Shirtcliffe P, Aldington S, Beasley R. External validity of randomised controlled trials in asthma: to whom do the results of the trials apply? Thorax. 2007;62:219–33.

Julius Sim is Professor of Health Care Research in the School of Medicine and a member of the Biostatistics Group, at Keele University. He is also Director of the Keele hub of the NIHR Research Design Service, West Midlands. Julius has an academic background in sociology, philosophy and statistics and is a Chartered Statistician. He has extensive experience of teaching statistics to undergraduate medical students and statistics and research methods to postgraduate students in medicine, psychology, social science and health care. Julius' research interests lie principally in clinical trials methodology, qualitative methods and research ethics and he has published widely in these areas. He is a current and past member of several clinical trial oversight committees and of major grant awarding committees in England and Wales.

Gillian Lancaster is Professor of Medical Statistics in the School of Medicine and Director of the Clinical Trials Unit, at Keele University. The main focus of her research is developing patient-/caregiver-reported outcome measures and assessment tools primarily for use with children and young people in clinical trials or programmatic evaluations (e.g., Alder Hey Triage Pain score, MDAT, ADNAT and CLCF). She is part of a working group convened by the World Health Organization (WHO) to develop Indicators for Infant and Young Child Development (IYCD) and Global Scales for Early Development (GSED) for population monitoring. She is currently Editor in Chief of the BMC journal *Pilot and Feasibility Studies* and co-author of CONSORT extension guidelines for pilot and feasibility trials. Gillian is a strong advocate of statistics education and was Director of the Lancaster Postgraduate Statistics Centre for 9 years (a HEFCE-funded Centre of Excellence for Teaching and Learning).

Martyn Lewis is Reader in Medical Statistics in the School of Medicine and a member of the Biostatistics Group, at Keele University. His academic background is in mathematics, statistics and health economics. He has been a successful co-applicant on numerous national grants for musculoskeletal, and other, health research, totalling over £20 million, and has co-authored over 150 peer-reviewed research articles published in leading international medical and methodology journals. He is an experienced senior advisor and supervisor, including through membership of several clinical trial independent oversight committees and having a key role in operational decision making for Keele Clinical Trials Unit. Martyn also has an active teaching and research portfolio, where his interests are mainly focused on clinical trials methodology.

Chapter 6
What Can Comparisons of Randomised and Non-Randomised Studies Tell Us?

Daniel Steel and Andrew Jones

Abstract A strand of research in Evidence-Based Medicine (EBM) compares results of randomised and non-randomised studies to empirically assess the methodological superiority of randomised controlled trials (RCTs). This chapter describes how early comparisons of randomised and non-randomised studies found that non-randomised studies tend to overestimate positive effects, while more recent reviews suggest this is not the case. These comparisons have influenced debates about RCTs for decades, initially supporting RCTs as the preferred methodology in EBM and later bolstering critiques of RCTs' "gold standard" status. However, learning from these comparisons requires careful exploration of auxiliary hypotheses about how the results of RCTs and non-randomised studies would be expected to differ, on the assumption that RCTs are in fact superior. Two such auxiliary hypotheses, labelled *directional bias* and *random bias*, are discussed. While more recent reviews call *directional bias* into question, *random bias*, though testable, has yet to be thoroughly investigated.

6.1 Introduction and Overview

A central component of Evidence-Based Medicine (EBM) since its inception has been an emphasis on the methodological superiority of the randomised controlled trial (RCT) for assessing the efficacy of medical interventions [1, 2]. Another traditional theme of EBM is that empirical demonstrations of effects should be prized above theoretical explanations of how or why an intervention should work [2]. Put together, these two features of EBM suggest that the methodological superiority of RCTs should itself admit of some empirical demonstration. One way to attempt this

D. Steel (✉)
School of Population and Public Health, University of British Columbia, Vancouver, Canada
e-mail: daniel.steel@ubc.ca

A. Jones
Institute for the History and Philosophy of Science and Technology, University of Toronto, Toronto, Canada

M. MacDougall (ed.), *A Medical Educator's Guide to Thinking Critically about Randomised Controlled Trials: Deconstructing the "Gold Standard"*,
https://doi.org/10.1007/978-3-031-25859-6_6

is by means of systematic comparisons of RCTs and non-randomised studies to check for differences that might be indicative of the theoretically expected methodological advantages of randomisation. In fact, a literature of such comparisons dates back to at least 1977 [3]. As detailed below, the first generation of reviews in this genre suggested that non-randomised studies frequently overestimate treatment efficacy [4], but more recent reviews have failed to find consistent differences in the results of randomised[1] and non-randomised studies [5–7]. These more recent findings have been seen as a source of worry among some who affirm the methodological superiority of RCTs [8] and as supporting evidence for those who claim that RCTs do not merit their privileged status [9, 10]. This chapter aims to clarify this debate by carefully examining what the hypothesised methodological superiority of RCTs would predict regarding differences in results between randomised and non-randomised studies.

To begin, we can ask whether comparisons are at the level of individual studies, or in the aggregate. Some discussions of comparisons between randomised and non-randomised studies seem to have disagreements between individual studies in mind. For example, Howick writes, "Randomized trials sometimes contradict the results of observational studies. These empirical studies can be taken as evidence that the biases ruled out by randomization (self-selection bias and allocation bias) have important effects" [11, p. 54]. However, we should expect the results of individual randomised and non-randomised studies to sometimes differ even if RCTs were *not* superior. Differences in results of individual studies can arise from a variety of sources, such as differences in the samples of research subjects, differences in the treatment, differences in outcome variables, from the choice of comparison groups or controls, and so on [12]. Indeed, differing results can occur among RCTs for similar reasons [13]. That randomised and non-randomised studies sometimes produce divergent results, therefore, is uninformative about the relative methodological merits of the two study designs.

Consider comparisons of aggregate (i.e., pooled or averaged) results between randomised and non-randomised studies. Aggregate comparisons can occur at two levels of abstraction, either single-question comparisons, or multiple-question comparisons. For a single-question comparison, one searches literature for RCTs and non-randomised studies that address the same medical question, such as whether anticoagulant drugs reduce fatality from myocardial infarction [3]. Data from the RCTs are then pooled, as are data from the non-randomised studies. Effect sizes, which might be measured by risk ratios, odds ratios, or hazard ratios [7], are estimated from the pooled data for the two study designs and then compared. For example, one might ask whether the pooled effect size for non-randomised studies is within the aggregate 95% confidence interval for the RCTs. Sometimes an individual article reports more than one single-question comparison. Thus, Benson and Hartz [5] compare RCTs and observational studies on 19 medical questions and find that the pooled effect estimate from observational studies is outside the 95% confidence interval for the

[1] Throughout this chapter, in referring to randomised studies, we do so with a focus specifically on RCTs.

corresponding RCTs in only two cases. However, they do not pool results across medical questions, so their article presents 19 single-question comparisons.

In a multiple-question comparison, one performs a literature search for single-question comparisons on a variety of topics, and then pools their results to allow a general comparison of effect sizes for randomised and non-randomised studies. For example, Anglemeyer et al. [7] considered 14 reviews that compared RCTs and non-randomised studies on a total of 228 different medical conditions. They estimated a ratio of ratios (ROR; i.e., the RCT effect size divided by the non-RCT effect size) for each review, and then synthesised these ratios across reviews to arrive at a pooled ROR [7]. While single-question comparisons make apples-to-apples comparisons of randomised and non-randomised studies on specific medical questions, multiple-question comparisons synthesise data across conditions and thereby aim for more general conclusions. Thus, multiple-question comparisons attempt to answer such questions as whether non-randomised studies, in general, tend to overestimate treatment effects in comparison to RCTs, and if so, by how much. In what follows, claims about differences between randomised and non-randomised studies "in the aggregate" can pertain to both single and multiple-question comparisons.

Theoretical arguments for the methodological superiority of RCTs typically claim that they are less prone to bias than other study designs because they reduce allocation and selection bias and facilitate masking [11, 14, 15]. But it is unclear what such claims predict about aggregate differences in the results of randomised and non-randomised studies. Bias might result in overestimating or underestimating the treatment effect. So, greater bias in non-randomised studies might predict larger effect sizes or smaller effect sizes than those of randomised studies. Or it might predict no difference in aggregate effect size if the two directions of bias cancel one another out. These observations illustrate the necessity of a more careful exploration of hypotheses about how the results of randomised and non-randomised trials would be expected to differ, on the assumption that RCTs are in fact superior. Now let's consider two hypotheses on this topic, which we can call the directional and random bias hypotheses. The directional bias hypothesis (*directional bias*) asserts that non-randomised studies tend to be biased in a specific direction, either towards larger or smaller effect sizes. The random bias hypothesis (*random bias*) asserts that non-randomised studies are more biased but are just as likely to overestimate the effect size as underestimate it. By accepting *directional bias*, we would conclude that if RCTs provide accurate estimates of effect sizes, then effect sizes of non-randomised studies will be systematically greater, or less, depending on the direction of the bias, than those found by RCTs. However, by accepting *random bias,* we would expect that estimated effect sizes from non-randomised studies on a topic would be more heterogeneous (i.e., more likely to diverge from one another) than those of RCTs on the same topic. This is on account of the assumed greater bias in estimated effect sizes from non-randomised studies, that may vary from over- to under-estimating effect sizes from one study to the next.

Two main findings emerge from this chapter. The first is that reviews comparing results of randomised and non-randomised studies have been published for over 40 years, and the conclusions drawn from them have shifted during this time.

Initially, these reviews suggested that non-randomised studies tend to overestimate positive treatment effects in comparison to randomised studies, while more recent reviews often find little difference in the results of the two study types [7]. Reviews comparing randomised and non-randomised studies have influenced debates about RCTs, initially supporting the "gold standard" status of RCTs in EBM and then later calling that status into question.

The second main finding is that learning from comparisons of randomised and non-randomised studies requires making one's hypotheses more precise. The hypothesis that randomised and non-randomised studies tend to produce different results could be understood in more than one way, and thus fails to yield unambiguous predictions. Consequently, under the assumption of methodological superiority of randomisation it is important to specify just how results would be expected to differ. The distinction between *directional* and *random bias* explained above is one way to do this. The examination of literature to follow in this chapter suggests that *directional bias* has been disconfirmed, but *random bias* has yet to be thoroughly researched.

The remainder of this chapter is organised as follows. Section 6.2 provides a chronologically organised overview of literature comparing randomised to non-randomised studies. Section 6.3 introduces the concept of an *auxiliary hypothesis* and explains why auxiliary hypotheses are important for learning from comparisons of randomised and non-randomised studies. It then considers how two such auxiliary hypotheses, *directional bias* and *random bias*, as defined above, fare in light of existing reviews comparing randomised to non-randomised studies. The final section summarises the lessons that you can take away from this chapter.

6.2 Literature Comparing Randomised and Non-Randomised Studies

This section provides an overview of literature comparing randomised and non-randomised studies, organised chronologically into three periods. In the first, from 1977 to 1992, comparisons of randomised and non-randomised studies suggested that non-randomised studies produce larger effect sizes, a claim that was subsequently used to support the privileged role of RCTs in EBM [16]. In the second period, 1993–2004, reviews comparing randomised and non-randomised studies frequently yielded conflicting findings. By 2000, some reviews suggested that the two study designs typically produce similar results [5, 6]. Some critics of EBM, including the philosopher Worrall [10], interpreted these findings as challenging the "gold standard" status of RCTs. In our third period, 2005 to present, an influential Cochrane review concluded that, "on average, there is little difference between the results obtained from RCTs and observational studies" [7]. While this review has not ended controversy on the matter [17, 18], it has been cited in support of increased use of non-randomised "real-world evidence" [17].

Since systematic reviews comparing randomised and non-randomised studies have already been published [7, 19], the articles discussed below were located primarily by iteratively tracking down articles cited in previous reviews. Nearly all the studies discussed below can be found by searching the lists of reviews examined in Odgaard-Jensen et al. [19] and Anglemeyer et al. [7], searching the bibliographies of those reviews, and so on. More recent reviews were located by searching work that cited Anglemeyer et al. [7] on Google Scholar, with the rationale that, given its prominence, any subsequent review of consequence would cite it. The historical overview presented here, therefore, is not intended as a new systematic review in its own right. Instead, its aim is to present an historical narrative of how research on this topic has evolved over more than four decades, and how this evolution has influenced thinking about the status of RCTs.

6.2.1 Consolidating Consensus: 1977–1992

In 1977, Chalmers et al. [3] published an article that compared randomised and non-randomised controlled trials that tested whether anticoagulants reduce fatality rates for myocardial infarction. The authors found 32 relevant studies, 8 with random allocation and 26 with non-random allocation. The comparison revealed that fatality rates were lower in the treatment groups of the non-randomised studies than in the randomised studies, and that the non-randomised studies consequently found larger treatment effect sizes. Chalmers et al. attributed this difference to patient selection bias in the non-randomised studies [3, p. 1091].

Throughout the 1980s, subsequent reviews compared the results of randomised and non-randomised studies and similarly concluded that non-randomised studies overestimate treatment effects due to bias [4, 20–22]. For example, a study by Sacks et al. [22] compared the results of RCTs and trials using historical controls for six different therapies. In a historically controlled trial (HCT), a group of patients receiving a new treatment is compared with a control group of earlier patients who had received the standard treatment, the *historical controls*. Sacks et al.'s study revealed that 79% of HCTs, but only 20% of RCTs, found treatments to be effective. Since the HCTs and RCTs found similar results in the treatment groups and the differences were due to poorer outcomes among the historical controls, the authors suggested that bias in selecting the controls of the HCTs likely explained the differing effect sizes [22, p. 233]. For example, Sacks et al. suggested that while the historical controls often include all available records of patients with the condition, patients recruited for a trial are likely be those whom physicians think will do well with the new treatment [22, p. 237]. In a subsequent article, the same authors presented very similar results and used them as the basis for a statistical argument that HCTs are more prone than RCTs to inaccurately indicate positive treatment effects [4, p. 755]. Their approach was to estimate a range of sensitivities and specificities given the possible assumptions about the effectiveness of the six treatments. The sensitivity of a study is its probability of finding the treatment effective when it is in fact effective, while the trial's specificity is its probability of finding the treatment ineffective when it is in fact, ineffective. Hence, a study with low specificity is likely to mistakenly find an ineffective treatment to be effective. Sacks et al.'s analysis produced a non-overlapping range of estimates of *sensitivity* and *specificity* for HCTs and RCTs, such that RCTs were more specific but less sensitive while HCTs were more sensitive but less specific [4]. Consequently, Sacks et al.'s [4] results suggested that HCTs are more likely than RCTs to find a treatment effective when it was actually ineffective.

The consensus of reviews in the 1980s that non-randomised studies tend to overestimate treatment effects helped to reinforce the privileged status of RCTs within the emerging EBM movement [2]. For example, in clarifying the EBM position, Sackett et al. [16] stated that while non-randomised studies are relevant to certain research questions, they "routinely lead to false positive conclusions about efficacy," so that decisions about treatment efficacy should be based on randomised trials or meta-analyses of randomised trials.

6.2.2 Conflicting Reviews: 1993–2004

Reviews suggesting that non-randomised studies overestimate treatment effects continued to be published in the 1990s [23, 24]. Moreover, a secondary meta-analysis by Kunz and Oxman [25] reviewed 11 studies comparing randomised and non-randomised studies and found, "On average, failure to use randomisation or adequate concealment of allocation resulted in larger estimates of effect due to a poorer prognosis in nonrandomly selected control groups compared with randomly selected control groups" [25, p. 1185].

At the same time, however, some reviews began to appear that challenged the claim that non-randomised studies overestimate treatment effects in comparison to RCTs [26–30]. For example, Britton et al. [26] examined 18 reviews that compared the results of RCTs and non-randomised studies, and found that, despite the standard view that non-randomised studies overestimate effect sizes, differences between the results of randomised and non-randomised studies were not always statistically significant or in a consistent direction and that features of trial design besides methods of patient allocation can influence results.

In June of 2000, two highly influential reviews that explicitly challenged the "gold standard" view of RCTs were published side by side in the *New England Journal of Medicine* [5, 6]. As noted in the introduction, Benson and Hartz [5] compared RCTs and observational studies on 19 medical conditions, and found that the effect size estimate from the observational studies fell outside of the RCTs' 95% confidence interval in only two cases. They concluded that there is "little evidence that estimates of treatment effects in observational studies reported after 1984 are either consistently larger than or qualitatively different from those obtained in randomized, controlled trials" [5, p. 1878]. Concato et al. [6] compared meta-analyses of randomised trials with meta-analyses of non-randomised trials. In contrast to Sacks et al. [4], Concato et al. [6] compared RCTs to concurrent rather than historical case-controlled studies.

The authors found that the results of the non-randomised studies they considered were "remarkably similar" to the results of randomised trials on the same topics, and thus concluded that the results of high quality non-randomised studies "do not systematically overestimate the magnitude of the effects of treatment as compared with those in randomized, controlled trials" [6, p. 1887]. Concato et al. argued that this finding challenges "the current consensus about a hierarchy of study designs in clinical research" [6, p. 1890].

These reviews, however, were not without their critics. In an editorial, Pocock and Elbourne argued that the Benson and Hartz [5] and Concato et al. [6] reviews were based on a "highly selected sample" of studies [31, p. 1908]. Similarly, a 2001 editorial noted that Benson and Hartz [5] and Concato et al. [6] only considered "a very small portion" of randomised and non-randomised research [32, p. 880]. In addition, other comparisons of randomised and non-randomised trials continued to find that non-randomised studies overestimate treatment effects [33–35].

Despite this controversy, the reviews by Benson and Hartz [5] and Concato et al. [6] were used to support philosophical arguments that challenged the merits of randomisation and the view that RCTs are the gold standard of evidence. Worrall considered various arguments for randomisation, including the claim that non-randomised trials "are 'known' to exaggerate treatment effects" [10, p. 325]. According to Worrall, this claim presupposes that results of randomised trials are accurate, since studies showing that non-randomised studies yield larger effect size estimates than RCTs are "equally consistent with the [alternative] claim that RCTs consistently underestimate the true effect" [10, p. 326]. In addition, Worrall invoked the reviews by Kunz and Oxman [25], Benson and Hartz [5], and Concato et al. [6] as evidence that non-randomised studies do not in fact overestimate treatment effects in comparison to RCTs.

However, some of Worrall's [10] claims about literature comparing randomised and non-randomised studies are rather problematic. For example, in estimating the sensitivities and specificities of RCTs versus HCTs, Sacks et al. [4] were careful to avoid assumptions about which treatments are actually effective. Hence, their argument did not depend on the assumption that, when RCTs and non-randomised studies disagree, the RCTs must be right. On the other hand, while Kunz and Oxman's [25] review reported that non-randomised studies tend to produce larger estimated treatment effects than RCTs, their applications of the terms "overestimation" and "underestimation" when reporting on the direction of bias based on a comparison of the two types of study implicitly rested on the above assumption.

6.2.3 Elusive Differences: 2005–the Present

Several reviews in the period from 2005 to the present have continued to develop the claim that differing results between randomised and non-randomised studies are often explained by aspects of study design other than randomisation [36–39]. For example, a review carried out by Furlan et al. [38] found that when the compared randomised and non-randomised trials are more homogenous (e.g., "similar settings, population, interventions, and outcomes"), their results are more similar [38, p. 209]. In addition, several reviews compared randomised versus non-randomised studies on adverse effects of treatments [40–42]. Papanikolaou et al. [42] compared rates of major harms in large-scale ($n > 4{,}000$) RCTs and epidemiological studies for 15 medical interventions. They found a statistically significant ($p \leq 0.05$) difference in relative risks in only 2 out of the 13 cases for which the comparison was possible [42, p. 638]. They also compared absolute risks, and found a statistically significant difference in only one of the eight cases for which this comparison was possible [42, p. 638].[2] A review by Golder et al. [41] compared estimates of adverse treatment effects derived from meta-analyses of randomised trials with those derived from meta-analyses of observational studies. They pooled the odds ratios for both study types and concluded that "there is no difference on average in the risk estimate of adverse effects of an intervention derived from meta-analyses of RCTs and meta-analyses of observational studies" [41, p. 1].

But the most influential works in this period are arguably two Cochrane reviews [7, 19]. Odgaard-Jensen et al. [19] reviewed 18 systematic reviews, involving meta-analyses, that compared randomised and non-randomised studies as well as randomised studies with and without adequate allocation concealment. The authors concluded that findings of study types they examined sometimes differed; moreover, studies that were not randomised or that had inadequate allocation concealment sometimes yielded larger effect sizes than randomised studies with adequate allocation concealment [19, p. 2]. However, since Odgaard-Jensen et al. [19] did not pool effect estimates for the study types they examined, the import of their study for comparisons among them was unclear. Subsequently, Howick and Mebius [8] pooled the results of the studies included in the Odgaard-Jensen et al. review and found no statistically significant difference between study types.

The second Cochrane meta-review similarly concluded that, "on average, there is little evidence for significant effect estimate differences between observational studies and RCTs, regardless of specific observational study design, heterogeneity, or inclusion of studies of pharmacological interventions" [7, p. 2]. As noted in the introduction, Anglemeyer et al. [7] estimated a ratio of ratios (ROR; i.e., RCT to non-RCT effect sizes) to compare RCTs to observational studies across research questions. They found this ROR to be 1.08 (95% confidence interval 0.96–1.22), indicating that the aggregate effect sizes for randomised and non-randomised studies in the literature they reviewed are nearly identical [7, p. 2]. The authors proposed that other aspects of trial design besides randomisation should be considered when explaining diverging results [7, p. 2]. The Anglemeyer et al. [7] review has been cited in support of a call to give greater weight to non-randomised evidence [17, 43].

[2] The authors nevertheless claimed that, "usually (75% [6/8]) the randomized trials estimated larger absolute excess risks of harm than the nonrandomized studies did," on the grounds that estimates of absolute risks were larger in RCTs than epidemiological studies in six out of the eight cases for which the comparison could be made [42, p. 635]. However, since the difference in absolute risk was statistically significant in only one case, these data could also be interpreted as not providing evidence of a difference in absolute risk estimates between RCTs and epidemiological studies.

The Cochrane reviews did not set the matter to rest, however. A subsequent review by Hemkens et al. [18] compared mortality in observational studies relying on routinely collected health data to mortality in subsequently conducted RCTs. The authors concluded that, "Studies of routinely collected health data could give different answers from subsequent randomised controlled trials on the same clinical questions, and may substantially overestimate treatment effects" [18, p. 1]. However, Franklin et al. [44] argued that the statistical method of the Hemkens et al. review was biased and that better methods find no association between study design and outcomes when applied to the same data. In response, Hemken et al. published a correction to their original paper in which they reanalysed the data using an alternative statistical approach and arrived at similar results to their original paper [45]. The debate stimulated by the Hemkens et al. review illustrates how much has changed in literature on comparisons of randomised and non-randomised studies. Whereas asserting that non-randomised studies do not overestimate treatment effects was a cause of controversy in 2000, the opposite was true in 2016.

Key Learning Point 6.1

Reviews comparing results of RCTs and non-randomised studies date back to at least 1977. Initially, reviews were taken to provide evidence that non-randomised studies tend to overestimate treatment effects, which supported the RCT's position as the "gold standard." However, a number of more recent reviews have found that RCTs and non-randomised studies generally yield similar results, including a Cochrane review from 2014 [7]. These reviews have been cited by critics of the "gold standard" status of RCTs and by those who support a greater role for non-randomised studies in medicine.

6.3 More Precise Hypotheses Needed

Let's consider the question of what one can reasonably hope to learn from comparisons of the results of randomised and non-randomised studies. Some take a sceptical view of this matter, and argue that empirical comparisons are irrelevant to methodology. For example, Deaton and Cartwright write that "hopes for empirical tests about which methods are best" are "doomed" because the validity of methods can be deduced from underlying assumptions, "So which method is best in a given case depends on what we know or are prepared to assume in that case" [46, p. 89]. For example, the soundness of an observational study may depend on the assumption that the covariates have been adequately controlled for statistically, while an RCT may be unbiased if allocation is adequately concealed and treatment assignment masked [47]. In other words, Deaton and Cartwright claim that theoretical arguments based upon assumed premises suffice to determine the validity methods, thereby making empirical comparisons of methods irrelevant. However, Deaton and Cartwright's reasoning is problematic, because it is often difficult to know to what extent such justifying assumptions are satisfied in real-world cases, and our willingness to make assumptions should not be confused with the accuracy of the assumptions themselves. Consequently, theoretical arguments for a method—consisting of a deduction from assumed premises—may be insufficient to establish its superiority in practice, which suggests that empirical comparisons may be relevant to assessing the relative merits of methods [9]. Nevertheless, learning from these comparisons requires formulating more precise hypotheses about how randomised and non-randomised studies should differ if randomisation is the most effective means of reducing bias.

Considered in isolation, the claim that RCTs are less biased than non-randomised studies makes no predictions about how the results of the two study designs will differ, or even that systematic differences will be found. Greater bias might result in overestimating or underestimating an effect, and it is possible that these differences would cancel one another out in aggregate comparisons of randomised and non-randomised studies. This situation is unsurprising from the perspective of philosophy of science, wherein it has long been recognised that general theoretical claims typically do not make testable predictions in isolation, but only when conjoined with auxiliary hypotheses [48–51]. Auxiliary hypothesis are claims that, when conjoined with a hypothesis of primary interest, generate a testable prediction. For instance, consider how the claim that a person is infected with SARS-CoV-2 can be tested. One might proceed by observing symptoms, which would rely on background knowledge about the characteristic effects of COVID-19. Or one might use a polymerase chain reaction (PCR) test, which depends on knowledge of distinctive sequences of RNA in the viral genome. In either case, testing the target hypothesis that the person is infected with SARS-CoV-2 is possible only given other background knowledge. Philosophers of science use the term "auxiliary hypotheses" to refer to such background knowledge in order to highlight (a) its role in testing other claims that are more directly of interest ("auxiliary") and (b) that assumed background knowledge is sometimes mistaken ("hypotheses"). For present purposes, the first of these two points is the more important one. Just as the claim that a person is infected by SARS-CoV-2 can only be empirically tested with the aid of background knowledge, auxiliary hypotheses are needed to link the claimed methodological superiority of RCTs to predictions about differences in results of randomised and non-randomised studies.

We can appreciate the importance of auxiliary hypotheses in research that compares the results of medical studies by considering research on sponsorship bias. Sponsorship bias arises when research funded by an entity with a financial interest in the outcome tends to produce results that align with that interest. There are published systematic reviews, some of which include meta-analyses, comparing industry to non-industry sponsored research [52]. The relevant point here is that work in this genre is not merely based on the idea that, due to sponsorship bias, industry and non-industry studies on the same topic are expected to yield different results. Instead, such reviews are driven by a plausible auxiliary hypothesis about *how* they differ, namely, that industry funded studies are more likely to favour their sponsors' products. Moreover, several tactics for favouring a sponsor's product might be adopted, such as assigning too low a dose to a comparator drug in a head-to-head trial, not publishing negative findings, or selecting outcome variables after the data are in to make results appear more positive, each of which might yield more specific predictions about how industry and non-industry funded studies are likely to differ.

The example of sponsorship bias illustrates that auxiliary hypotheses can make a hypothesis more precise. In our case, the primary hypothesis of interest is that randomisation reduces bias. But testing this idea empirically by means of comparisons with non-randomised studies requires being more specific about the nature of the hypothesised bias. A good place to start this processs is with a null hypothesis, which we can understand here as the claim that non-randomised studies are, in general, no more biased in their estimates of treatment effects than RCTs. We can then consider two types of alternatives to this null hypothesis, *directional* and *random bias*. These are two possible ways of making the hypothesis that randomisation reduces bias more precise. Let's consider them in turn.

Directional bias asserts that greater bias in non-randomised studies leads to differences in the aggregate effects of non-randomised studies and RCTs. *Directional bias* itself can be broken down into two more specific hypotheses: either non-randomised studies generally tend to overestimate positive effects or they tend to underestimate positive effects in comparison to RCTs, where positive effects are understood as increased beneficial or reduced harmful effects of the intervention of interest. *Directional bias* would be vindicated, then, if one of these two sub-hypotheses were found to be true.

Let's examine the above two sub-hypotheses more closely. The hypothesis that non-randomised studies are biased towards positive effects was supported by the first generation of systematic reviews comparing randomised and non-randomised studies. The theoretical motivation for this hypothesis is perhaps related to the idea that non-randomised studies are less likely to adequately mask treatment assignment, suggesting larger effect sizes in non-randomised studies due to allocation bias or positive expectations on the part of patients or physicians. Another possible rationale for this hypothesis is that socio-economically advantaged individuals are more likely to be able to access novel treatments, and are more likely to have better health outcomes for several reasons that may be difficult to adequately control for statistically.

Although the possibility that non-randomised studies tend to underestimate positive treatment effects in comparison to RCTs has received less attention, there are also plausible scenarios in which it might arise. For example, patients who have already tried existing or less invasive therapies without success may be more likely to seek out and be prescribed experimental or more invasive interventions. Thus, without random treatment assignment, patients who are sicker or less likely to respond positively to treatments for other reasons may be overrepresented in the treatment group, creating a bias in the direction of underestimating positive treatment effects.

Directional bias predicts a systematic difference in effect sizes between randomised and non-randomised studies while the corresponding null hypothesis does not. As noted in the introduction, a prediction of the above sort can be made at more than one level of abstraction. The prediction might be that, for a specific medical question (e.g., the effectiveness of anticoagulants in preventing myocardial infarction), non-randomised studies overestimate beneficial treatment effects in comparison to RCTs. Or the prediction might be made in general, that is, across many medical questions. As we saw in Sect. 6.2.1, the more general interpretation of *directional bias* has been the primary focus of comparisons between randomised and non-randomised studies from early on. Sacks et al. [4] were concerned to compare the sensitivities and specificities of randomised and non-randomised studies across six different medical questions, while Sackett et al. [16] asserted that, in general, observational studies overestimate the effectiveness of treatments. Subsequent reviews similarly compared randomised and non-randomised studies across numerous medical questions [5–7, 19, 25]. The move towards a focus on this more general type of comparison is unsurprising, as it is particularly relevant to debates about whether RCTs should be viewed as the "gold standard". These debates concern the general status of RCTs in assessment of the effectiveness of treatments, not merely with respect to a particular treatment or condition. The Anglemeyer et al. [7] review was of interest, therefore, because it concluded so strongly that *directional bias*, interpreted in the more general sense, was disconfirmed by available evidence. Current defenders of *directional bias*, then, must hope that Anglemeyer et al. [7] was overturned by future reviews, or they must limit the scope of their claims to particular conditions or treatments.

Random bias can be considered with respect to a specific medical question or more generally, but if debates about the "gold standard" status of RCTs are at issue, then the latter, more general, interpretation is more relevant. However, the testable prediction made by *random bias* is much easier to grasp if one begins by considering comparisons of randomised and non-randomised studies on a single treatment. *Random bias* agrees with the null in not expecting differences in aggregate effect sizes across randomised and non-randomised studies on the same medical question. However, unlike the null, it predicts greater heterogeneity in results of non-randomised than randomised studies on the same medical question. It predicts, in other words, that divergent results should be more common among non-randomised studies than among RCTs. As Ioannidis et al. put it, "Popular wisdom has it that a 'gold standard' method should give more or less the same results when repeated several times, while a poor method would suffer from lots of variability" [31, p. 879]. Of course, there is no reason to expect that a method should generate similar estimates of effect size across different treatments or conditions. However, a statistical measure of heterogeneity among studies can be computed for studies on the same medical question, and then pooled across multiple medical questions. In this way, the heterogeneity of randomised and non-randomised studies can be compared more generally.

Only a few reviews consider the heterogeneity of randomised versus non-randomised studies, and usually only as an aside in the course of comparing effect size estimates between randomised and non-randomised studies [6, 30, 35, 42]. Moreover, among these reviews, some find RCT results to be more variable than those of observational studies [6, 30], while others report the opposite [35, 42]. In addition, a careful examination of *random bias* should control for confounders that might exaggerate the heterogeneity of RCTs in comparison to non-randomised studies in an analogous manner to that in which confounders can inflate treatment effects in individual RCTs (see, for example, Table 5.1 of Chap. 5). Relatedly, in comparison to observational studies, RCTs may be more likely to have smaller sample sizes or target niche patient groups. Consequently, the current literature is insufficient to render any firm judgment regarding *random bias*.

Key Learning Point 6.2

Learning from comparisons of RCTs and non-randomised studies requires making the hypothesis that randomisation reduces bias more precise. *Directional bias* and *random bias* represent two ways in which this can be done. *Directional bias* asserts that non-randomised studies tend to be biased in a specific direction, either towards overestimation or underestimation of treatment effects. *Random bias* asserts that non-randomised studies are more biased but are just as likely to overestimate treatment effects as to underestimate them. In accepting the latter hypothesis, we would predict that results from non-randomised studies are more likely to diverge from one another than results from randomised studies on the same topic. While findings reported in Key Learning Point 6.1 disconfirm *directional bias*, *random bias* awaits examination by rigorous meta-analyses.

6.4 Conclusions

Students of medicine and other health professions are generally familiar with theoretical arguments that RCTs are less biased than other methods for estimating treatment effects but are often unaware of the history of empirical comparisons of randomised and non-randomised studies. This situation is unfortunate because it can promote a blinkered and dogmatic understanding of the grounds for assessing methods. As the founders of EBM emphasised, theoretical rationale for a method—including the methods endorsed by EBM itself—is insufficient and should be considered in tandem with empirical demonstrations of effectiveness [2]. And as we have seen from the overview in this chapter, comparisons of randomised and non-randomised studies have played an important role in debates about the methodological merits of randomisation for more than four decades. Furthermore, the consensus of such studies has shifted from supporting towards questioning the claim that non-randomised studies overestimate positive treatment effects in comparison to RCTs. But the implications of this research remain unsettled, because *directional bias* is still debated while *random bias* awaits a thorough examination. Medical students, then, should know about the history of empirical comparisons between randomised and non-randomised studies. They should also be encouraged to see the relative merits of RCTs and other methods as the subject of an exciting, dynamic scientific inquiry in its own right, not something that can be definitively settled by theoretical reasoning about bias.

The narrative historical approach of the literature overview presented here has the advantage of being non-technical and accessible to readers without a background in statistics. But it is limited in that the presentation and interpretation of the sources is inherently somewhat subjective. Nevertheless, the central theme of the rise and fall of the consensus that non-randomised studies overestimate treatment efficacy in comparison to RCTs is readily apparent from a chronological arrangement of the sources.

Several further research projects are suggested by the findings described here. Rigorous meta-analyses on whether there is more heterogeneity among the results of non-randomised studies than RCTs, as predicted by *random bias*, would be an important contribution. Moreover, further research could ask why evidence shifted from initially supporting *directional bias* to later calling it into question. Perhaps this change is due to differences in methodology between older and more recent non-randomised studies. For instance, several earlier reviews [3, 4] compared RCTs to studies using historical controls, while later reviews were more likely to compare RCTs to studies with concurrent controls [6]. Another possibility is that, in earlier reviews, confirmation bias led to a selective focus on topics in which non-randomised studies produced larger positive effect sizes than RCTs.

So, what in the end can we learn by comparing the results of randomised and non-randomised studies? First, we can learn something about the current state of play in the literature that makes such comparisons. Current literature disconfirms the hypothesis that non-randomised studies are more likely than randomised studies to overestimate positive treatment effects, thereby undermining an argument made by early EBM proponents. Nevertheless, it would be premature to pronounce the demise of RCTs as the gold standard of medical evidence based on this current literature, as *random bias* remains a possibility. Second, and more philosophically, we can learn that theoretical arguments alone may be insufficient to establish the superiority of a method: careful empirical comparisons with alternative methods are also needed.

References

1. Sackett D, Brian Haynes RB, Guyatt GH, Tugwell P. Clinical epidemiology: a basic science for clinical medicine. Little Brown & Co.; 1985.
2. EBM Working Group. Evidence-based medicine: a new approach to teaching the practice of medicine. J Am Med Assoc. 1992;268(17):2420–5.
3. Chalmers TC, Matta RJ, Smith H, Kunzler AM. Evidence favouring the use of anticoagulants in the hospital phase of acute myocardial infarction. New Eng J Med. 1977;297:1091–6. https://doi.org/10.1056/NEJM197711172972004.

4. Sacks H, Chalmers TC, Smith H. Sensitivity and specificity of clinical trials: randomized v historical controls. Arch Intern Med. 1983;143:753–5.
5. Benson K, Hartz A. A comparison of observational studies and randomized, controlled trials. New Eng J Med. 2000;342:1878–86.
6. Concato J, Shah N, Horwitz RI. Randomized, controlled trials, observational studies, and the hierarchy of research designs. N Engl J Med. 2000;342:1887–92.
7. Anglemyer A, Horvath HT, Bero L. Healthcare outcomes assessed with observational study designs compared with those assessed in randomized trials. Cochrane Database Syst Rev. 2014. https://doi.org/10.1002/14651858.MR000034.pub2.
8. Howick J, Mebius A. In search of justification for the unpredictability paradox. Trials. 2014;15:480.
9. Solomon M. Just a paradigm: evidence-based medicine in epistemological context. Eur J Philos Sci. 2011;1:451–66.
10. Worrall J. What evidence in evidence-based medicine? Philos Sci. 2002;69:316–30.
11. Howick J. The philosophy of evidence-based medicine. Oxford: BMJ Books; 2011.
12. Lodi S, Phillips A, Lundgren J, et al. Effect estimates in randomized trials and observational studies: comparing apples with apples. Am J Epidemiol. 2019;188:1569–77. https://doi.org/10.1093/aje/kwz100.
13. Horwitz RI. Complexity and contradiction in clinical trial research. Am J Med. 1987;82:498–510. https://doi.org/10.1016/0002-9343(87)90450-5.
14. Armitage P. Fisher, Bradford Hill, and randomization. Int J Epidemiol. 2003;32:925–8. https://doi.org/10.1093/ije/dyg286.
15. Fisher RA. The design of experiments. Edinburgh Tweedale Court: Oliver & Boyd; 1935.
16. Sackett D, Rosenberg W, Gray J, Haynes B, Richardson S. Evidence based medicine: what it is and what it isn't. BMJ. 1996;312:71–2.
17. Franklin JM, Glynn RJ, Martin D, Schneeweiss S. Evaluating the use of nonrandomized real-world data analyses for regulatory decision making. Clin Pharmacol Ther. 2019;105:867–77.
18. Hemkens LG, Contopoulos-Ioannidis DG, Ioannidis JPA. Routinely collected data and comparative effectiveness evidence: promises and limitations. CMAJ. 2016;188:E158–64.
19. Odgaard-Jensen J, Vist GE, Timmer A, Kunz R, Akl EA, Schünemann H, Briel M, Nordmann AJ, Pregno S, Oxman AD. Randomization to protect against selection bias in healthcare trials. Cochrane Database Syst Rev. 2014. https://doi.org/10.1002/14651858.MR000012.pub3.
20. Chalmers TC, Celano P, Sacks HS, Smith H. Bias in treatment assignment in controlled clinical trials. New Eng J Med. 1983;309:1358–61. https://doi.org/10.1056/nejm198312013092204.
21. Colditz GA, Miller JN, Mosteller F. How study design affects outcomes in comparisons of therapy. I. Medical. Stat Med. 1989;8:441–54.
22. Sacks H, Chalmers TC, Smith H. Randomized versus historical controls for clinical trials. Am J Med. 1982;72:233–40. https://doi.org/10.1016/0002-9343(82)90815-4.
23. Bhansali MS, Vaidya JS, Bhatt RG, Patil PK, Badwe RA, Desai PB. Chemotherapy for carcinoma of the esophagus: a comparison of evidence from meta-analyses of randomized trials and of historical control studies. Ann Oncol. 1996;7:355–9.
24. Schulz KF, Chalmers I, Hayes RJ, Altman DG. Empirical evidence of bias: dimensions of methodological quality associated with estimates of treatment effects in controlled trials. JAMA. 1995;273:408–12.
25. Kunz R, Oxman AD. The unpredictability paradox: review of empirical comparisons of randomised and non-randomised clinical trials. BMJ. 1998;317:1185–90.
26. Britton A, McPherson K, McKee M, Sanderson C, Black N, Bain C. Choosing between randomised and non-randomised studies: a systematic review. Health Technol Assess. 1998;2:1–124.
27. Heinsman DT, Shadish WR. Assignment methods in experimentation: when do non-randomised experiments approximate answers from randomised experiments? Psychol Methods. 1996;1:154–69.
28. MacLehose RR, Reeves BC, Harvey IM, et al. A systematic review of comparisons of effect sizes derived from randomised and non-randomised studies. Health Technol Assess. 2000;4:1–154.

29. Reeves BC, MacLehose R, Harvey IM, Sheldon TA, Russell IT, Black AMS. Comparison of effect size estimates derived from randomised and non-randomised studies. In: Black N, Brazier J, Fitzpatrick R, Reeves B, editors. Health services research methods: a guide to best practice. London: BMJ Publishing Group; 1998. p. 73–85.
30. Shadish WR, Ragsdale K. Random versus non-random assignment in controlled experiments: do you get the same answer? J Consult Clin Psychol. 1996;64:1290–305.
31. Pocock SJ, Elbourne DR. Randomized trials or observational tribulations? N Engl J Med. 2000;342:1907–9.
32. Ioannidis JP, Haidich A, Lau J. Any casualties in the clash of randomised and observational evidence? BMJ. 2001;322:879–80.
33. Bhandari M, Tornetta P, Ellis T, Audige L, Sprague S, Kuo JC, et al. Hierarchy of evidence: differences in results between non-randomized studies and randomized trials in patients with femoral neck fractures. Arch Orthop Trauma Surg. 2004;124:10–6.
34. Guyatt GH, DiCenso A, Farewell V, Willan A, Griffith L. Randomized trials versus observational studies in adolescent pregnancy prevention. J Clin Epidemiol. 2000;53:167–74.
35. Ioannidis JP, Haidich AB, Pappa M, et al. Comparison of evidence of treatment effects in randomized and non-randomized studies. JAMA. 2001;286:821–30.
36. Ferriter M, Huband N. Does the non-randomized controlled study have a place in the systematic review? A pilot study. Crim Behav Ment Health. 2005;15:111–20.
37. Abraham NS, Byrne CJ, Young JM, Solomon MJ. Metaanalysis of well-designed non-randomized comparative studies of surgical procedures is as good as randomized controlled trials. J Clin Epidemiol. 2010;63:238–45.
38. Furlan AD, Tomlinson G, Jadadb AA, Bombardier C. Methodological quality and homogeneity influenced agreement between randomized trials and non-randomized studies of the same intervention for back pain. J Clin Epidemiol. 2008;61:209–31.
39. Melloni C, Washam J, Jones S, et al. Conflicting results between randomized trials and observational studies on the impact of proton pump inhibitors on cardiovascular events when coadministered with dual antiplatelet therapy: a systematic review. Circ Cardiovasc Qual Outcomes. 2015;8:47–55.
40. Edwards J, Kelly E, Lin Y, Lenders T, Ghali W, Graham A. Meta-analytic comparison of randomized and nonrandomized studies of breast cancer surgery. Can J Surg. 2012;55(3):155–62.
41. Golder S, Loke Y, Bland M. Meta-analyses of adverse effects data derived from randomised controlled trials as compared to observational studies: methodological overview. PLoS Med. 2011;8. https://doi.org/10.1371/journal.pmed.1001026.
42. Papanikolaou PN, Christidi GD, Ioannidis JPA. Comparison of evidence on harms of medical interventions in randomized and non-randomized studies. CMAJ. 2006;174:635–41.
43. Ross J. Randomized clinical studies and observational studies are more often alike than unlike. JAMA Intern Med. 2014;174:1557. https://doi.org/10.1001/jamainternmed.2014.3366.
44. Franklin JM, Dejene S, Huybrechts KF, Wang SV, Kulldorff M, Rothman KJ. A bias in the evaluation of bias comparing randomized trials with nonexperimental studies. Epidemiol Methods. 2017;6. https://doi.org/10.1515/em-2016-0018.
45. Sterne J. Commentary: does the selective inversion approach demonstrate bias in the results of studies using routinely collected data? BMJ. 362. https://doi.org/10.1136/bmj.k3259.
46. Deaton A, Cartwright N. Reflections on randomized control trials. Soc Sci Med. 2018;210:86–90. https://doi.org/10.1016/j.socscimed.2018.04.046.
47. Deaton A, Cartwright N. Understanding and misunderstanding randomized controlled trials. Soc Sci Med. 2018;210:2–21.
48. Duhem P. The aim and structure of physical theory. Wiener, P. translator. Princeton, NJ: Princeton University Press; 1954.
49. Kuhn T. The structure of scientific revolutions. 2nd ed. Chicago, IL: University of Chicago Press; 1970.
50. Lakatos I. Criticism and the growth of scientific knowledge. Cambridge: Cambridge University Press; 1970.

51. Popper K. The logic of scientific discovery. 1st English ed. London: Hutchenson & Co.; 1959.
52. Lundh A, Lexchin J, Mintzes B, Schroll JB, Bero L. Industry sponsorship and research outcome: systematic review with meta-analysis. Intensive Care Med. 2018;44:1603–12.

Daniel Steel is Associate Professor in the W. Maurice Young Centre for Applied Ethics in the School of Population and Public Health at the University of British Colombia. His research focuses on values and science in the context of environmental and public health issues. His current research includes a project funded by the Social Sciences and Humanities Research Council of Canada on the ethics of severe climate change. Daniel is also the author of *Philosophy and the Precautionary Principle: Science, Evidence and Environmental Policy* (2015 Cambridge University Press), and *Across the Boundaries: Extrapolation in Biology and Social Science* (2008 Oxford University Press).

Andrew Jones is a Ph.D. candidate at the University of Toronto's Institute for the History and Philosophy of Science and Technology. He received a BA and MA in Philosophy and Science and Technology Studies from the University of British Columbia. His research focuses on the philosophy of medicine and the history of psychiatry.

Chapter 7
Multi-Studies: A Novel Approach to Addressing Irreplicability in RCTs

Alexander Krauss

Abstract This chapter proposes a novel approach to addressing irreplicability encountered in randomised controlled trials (RCTs). Irreplicability affects, to different degrees, all fields of science. Some of the most common explanations for non-replicability are low sample size, low statistical power, p-hacking, publication bias and HARKing. Such issues face some studies but not others. Focusing on RCTs, this chapter highlights that, while it is well-recognised that many types of bias and other types of constraint generally face studies across the biomedical, behavioural and social sciences, they are not commonly discussed collectively. In response to this gap in the literature on RCTs, we shall consider a novel approach to addressing non-replicability. The chapter helps explain how a degree of inevitable variation between the outcomes of different studies is driven by a combined set of biases and constraints underlying the scientific *process* that relate to *methods, people, contexts, and time points* at which data are collected in studies. It outlines how causal evidence in such studies is inferred from estimated statistical results that are the outcome of multiple complex processes: sampling, randomising, blinding, controlling, delivering treatments, and so on. These processes involve many actors making many decisions at different steps when designing, implementing and analysing studies within a particular context and time period. Correspondingly, the chapter considers the influence of the combined set of constraints on the quality and replicability of RCTs, and how to better tackle this issue. This entails integrating comparability of groups, treatments, contexts, etc. within a single study—called here a *multi-study*. Implications are then drawn for practitioners.

A. Krauss (✉)
London School of Economics, Institute for Economic Analysis, Spanish National Research Council, Barcelona, Spain
e-mail: a.krauss@lse.ac.uk

M. MacDougall (ed.), *A Medical Educator's Guide to Thinking Critically about Randomised Controlled Trials: Deconstructing the "Gold Standard"*,
https://doi.org/10.1007/978-3-031-25859-6_7

7.1 Introduction

According to the American Statistical Association [1], a study is "reproducible if you can take the original data and the computer code used to analyse the data and reproduce all of the numerical findings from the study." Replicability refers to "repeating an entire study, independently of the original investigator without the use of original data (but generally using the same methods)" (ibid.; cf. Open Science Collaboration [2]). This chapter employs this distinction between reproducibility (using the original data) and replicability (collecting new data)—though there is a tendency in the literature to use these terms interchangeably, as is reflected in the *Nature* article we shall consider shortly. We shall focus on the constraints in replicating studies and how to improve replication.

Difficulties in replicating studies face, to different degrees, all fields of science [2–6]. In a 2016 survey in *Nature*, about 90% of the over 1500 scientists surveyed think that science is facing a "reproducibility" crisis and about 70% of these scientists report that they had attempted and failed to reproduce another scientist's findings [3]. Some of the most common explanations for non-replicability are low sample size, low statistical power, p-hacking, selective reporting, publication bias and pressure to publish, as well as HARKing (hypothesising after results are known) [3, 6, 7]. While such biases and constraints face some studies and not others, a broader combined set of complex biases and constraints generally face, to some extent, all studies across the biomedical, behavioural and social sciences. But they are not commonly discussed together in the literature.

In providing examples from randomised controlled trials (RCTS), this chapter explains how a degree of inevitable variation between the outcomes of different studies is driven by a combined set of biases and constraints underlying the scientific process, the methods used and the different conditions between people, contexts and time periods in which studies are conducted. It outlines how causal evidence in such studies is inferred from estimated statistical results that are the outcome of multiple processes. Such processes include selecting a sample, randomising, blinding and controlling the sample of participants, implementing treatments and monitoring the level of adherence among participants.

Research studies involve many actors making multiple decisions at the different stages of design, implementation and analysis, and within a particular context and time period. They are conducted by collecting data within a particular context (clinic, lab, field, etc.) with particular subjects and conditions, and the phenomena being studied often change across contexts and can evolve over time. Thus, some difficulties in replicating scientific research are, to some degree, unavoidable. Differences in how scientific processes and methodological techniques are carried out in different contexts with different subjects help explain why replication studies are generally not able to obtain the same results (with the same outcomes, effect size and degree of statistical significance) as in the original studies. The issue of non-replicability is therefore viewed here, not as a *crisis* that takes place for a given period of time but, in part, as a reflection of the complexities in conducting research and continually improving the quality of research.

This chapter provides a new perspective on irreplicability by identifying some key constraints and limitations in the conduct of RCTs and using these to propose a rigorous approach to research design. The new approach involves conducting a *single* study—called here a *multi-study*—among different groups of people (subjects), with different background traits and characteristics, at multiple time points, using different treatment groups and control groups, and within different contexts. By way of background, we shall consider the ten most cited RCT studies worldwide.[1]

A background on RCTs and causal bias is first provided (Sect. 7.2). We then consider the limitations, assumptions and biases facing RCTs (Sects. 7.3–7.4). It is only when the range of complex issues and problems facing trials are considered collectively that we are we able to draw inferences about the overall constraints on replicability of RCTs (Sect. 7.5). Most importantly, we can then consider how to reduce such complexity and improve replicability by improving the research design of studies (Sect. 7.6).

7.2 The RCT Method: Measuring Causation, and Reducing Human Biases

Experimental methods are the main means we use in the sciences to identify causal relationships in the world. In the biomedical, behavioural, and social, sciences, the RCT method is commonly viewed as the best method to estimate causal effect sizes and provide replicable results.

In developing RCTs, the methodological technique of randomisation was created to reduce the problem of confounding, discussed in Chaps. 2 and 5, by randomising the experimental groups in the sample to help balance independent variables. Blinding was then created to reduce biases such as observer effects and confirmation effects by withholding information that can, often unconsciously, change the behaviour of participants, experimenters and others involved in the experiment. Placebos were eventually developed to reduce the problem of participants' expectations about a treatment influencing the experiment's outcomes by giving a share of participants a placebo (Chap. 2). The RCT method is thus the result of a continual reworking or refinement of previously used methods or methodological techniques

[1] The examples provided throughout the chapter are all based on the top ten cited RCT studies worldwide in any scientific journal, with each cited by at least 6,500 or more articles as of 2016 based on searching Scopus (the largest database of peer-reviewed journals). They include world-leading trials on the topics of breast cancer and chemotherapy [17], colorectal cancer [18], stroke [25], postmenopause and estrogen [20], insulin therapy for critically ill patients [19], two separate trials on cholesterol and heart disease [23, 24] and three separate trials on diabetes [21, 22, 26]. While these trials relate to the fields of general medicine, biology and neurology, the insights outlined here are as useful for researchers and practitioners across many other fields, including psychology, neuroscience, economics and agriculture.

over time that aim to reduce human influences in the scientific process and allow for greater replicability.

Methods like RCTs are generally passed along to the next generation that has not been exposed to the debates that had to take place before those methods could be accepted. Debates on the best way to administer blinding, placebos, multiple treatments, etc. are, as outlined here, ongoing. By continually reworking and integrating such methodological features into the RCT design, the method has emerged by an iterative process over time. And we shall continue developing techniques to reduce the constraints and biases facing the method. While science may not always seem self-correcting when looking at scientific outputs—due to contradictory studies and hypotheses—science can be generally viewed as self-correcting when looking at the methods we develop and improve in a cumulative way over time.

For the two-armed, double-blinded, parallel group, individually randomised controlled trial design discussed in previous chapters, the intervention is the *only* causally relevant difference between the groups, while other conditions are similar. The effectiveness of the intervention is estimated statistically by comparing the average outcome in the intervention group with the average outcome in the control group at the sample level. As noted in Chap. 1, the resultant estimate is the effect size estimate for the intervention.

In Chaps. 2 and 5, the authors consider the concept of counterfactual. The unique power of RCTs to create good counterfactuals and produce reliable causal results is often thought to be found in the process of randomisation [8]. Use of this process is viewed by many as the defining trait for scientific rigour and increased replicability, and a number of the methodological issues that we shall consider later in this chapter are related to randomisation. First, however, we need to be clear about what is meant by *causal bias* in trials, as this type of bias influences the replicability of studies.

To identify the different aspects of a trial's design, implementation and analysis that can contribute to bias and reduced replicability, we need to understand the nature of biased causal inference and prediction. The leading account of biased causal inference in the RCT literature is grounded in the probabilistic theory of causality [9]. In that account, causal inference is not biased if "all factors that can produce or eliminate a probabilistic dependence between C [the cause] and E [the effect] are the same in both wings [arms of the trial] except for C" [8].

However, other sources of bias that are not directly captured in that common account influence our estimated causal outcomes and their predictive capacity and in turn, study replicability. We need a broader account that incorporates, not only causal bias in terms of imbalances *between* trial groups [8, 10] but also, imbalances *within* one of the trial groups or *across* the entire sample [11].

Key Learning Point 7.1

If it were practically possible, bias in causal inference from an RCT would be eliminated if all confounders were equally distributed *between* the treatment and control groups, if all subjects within the intervention and control groups received an identical intervention and control respectively, and if all study design, implementation and analysis features *across* the entire sample were properly selected and used to produce an undistorted estimated average effect of the treatment.

In what follows, this broader account of causal bias aims to improve RCTs by directing our attention to the range of biases at play that can each hinder replicability.

7.3 The Replicability of RCTs: The Influence of Sampling Bias, Compliance Bias and Lack of Blinding

The common view of RCTs is that they ensure an "unbiased estimate of the average effect" [12] of a treatment (as stated in highly cited textbooks across disciplines) [13–15]. Some limitations, assumptions and biases underlying RCTs cannot however generally be fully avoided—statistical and non-statistical techniques cannot entirely eliminate them. Three examples are sampling bias, compliance bias and bias due to lack of blinding, all of which affect the estimated causal results of RCTs and their replicability [7, 16].

Arguably, in any RCT, there is potential for some degree of unblinding of the various people involved in the trial that can lead to performance bias (Chap. 2). Among the ten most cited RCTs, to varying degrees, some did not involve full implementation of double-blinding [17–19] whereas others did but participants were later unblinded to support patient care, such as the management of adverse treatment effects [20–22]. In some cases, when participants require continual monitoring, it is not possible to conduct a blinded trial. For example, participants' blood glucose levels in trials may need to be regularly tested to adjust insulin doses [19]. In other cases, there is at least the potential for trial participants to check their vital status—for example, their cholesterol levels—outside the trial and then to discontinue the placebo to begin taking an active open-label treatment [23], which involves disclosing the name of the treatment to both the patient and health professionals involved in their care. Such issues of blinding and unblinding influence the reported causal outcomes and thus, replicability. Pre-registering studies and their research design (Chap. 1) can help to some extent in controlling mismanagement of blinding.

In any trial, there is potential for non-adherence to a prescribed treatment regimen among sufficiently many participants to influence the estimated effectiveness of an intervention. This influence—compliance bias—can arise, for example, from patients failing to fully comply with the regimen or deciding to stop taking a treatment altogether and there being a difference between patients who do or do not comply, leading to a confounding of the estimated treatment effect.

Other statistical biases that influence causal outcomes can arise during the trial, due to participants deciding to drop out during the trial or switch between treatment and control groups, or due to missing data for participants. These issues affect the ten most cited RCTs and their replicability. The more frequently such issues arise in each trial, the more problematic the causal claims become. Some of these issues are difficult to avoid and to fully detect while others become part of the statistical deviations. An example of such issues from the above RCTs is found in the estrogen trial [20], for which 42% of treated participants later discontinued the study drug, the vital status of 4% of participants was unknown (that is, data were missing) and 3% of participants passed away. Another difficulty in interpreting this trial's estimated causal outcomes and replicating the study is that 11% of participants in the placebo arm switched to the active treatment arm. The decision by participants to switch between groups, once they become familiar with a trial, is of importance for their current health status and quality of life and should not just be viewed in statistical terms.

In any trial, some share of recruited people refuse to participate, leading to sampling bias through an overrepresentation of certain study participants and in turn affecting the estimated causal results. Seven of the ten most cited trials do not report the share of people recruited who refuse to take part in the trial whereas, for the remaining three trials, the reported shares are high. In one of the cholesterol trials [24], 51% of those recruited for the trial appeared at the initial screening, after which only 4% of the recruited sample were randomised into the study. In the estrogen trial [20], 5% of all women initially screened consented to participate in the trial (and reported no hysterectomy), after which 88% of those who consented were then randomised into the study. In general, those people who choose to participate are more likely to have time, expect to benefit, consider the risk from participating to be low and possibly have greater demand for the intervention. As the intended sample gets smaller due to people refusing to participate, it becomes more likely that there are cases where the discrepancy between characteristics of individuals refusing to participate and those of the actual sample is substantial. Intention-to-treat analysis (Chap. 2) cannot in and of itself address the impact that both this issue and that of missing data for actual participants can have on biasing a trial's average estimated effects and hindering replicability.

In general, the causal claims of RCTs are commonly affected by such general issues as sampling bias, compliance bias and bias due to lack of blinding. Difficulties associated with replicating trials are thereby also in part the result of some degree of bias generally arising in conducting trials. In replicating studies, we cannot therefore attain the identical results but only aim for similar results, assessed by measures including effects in the same direction as in the original studies, and comparable effect sizes. Causal claims can however be influenced by a range of other methodological issues. Many of these are well recognised and discussed in isolation but need to be discussed collectively to assess the overall constraints on replicability of RCTs.

7.4 The Replicability of RCTs: The Influence of Other Issues

An important constraint influencing the replicability of RCTs is that they, even the top ten most cited trials [17, 25], at times consist of a few hundred individuals that are too restrictive to produce reliable results and can lead to sampling bias. An example is that the stroke trial [25], with 624 participants in total, reports that at three months after the stroke, 54 treated patients died compared to 64 placebo patients—with the main outcome thus being just a difference of 10 deaths. Such small trials, together with, at times, strict inclusion and exclusion criteria for participants and poor randomisation, often bring about important imbalances in covariates which bias the results.

Overall, larger sample sizes allow for greater statistical power and thus facilitate replication. But replicability, and the usefulness of results outside the trial context, are also constrained in small trials conducted in a particular context (a clinic, lab, geographic area etc.), as it is difficult to attain similar initial conditions for a new study. To increase reliability in estimated results, researchers ideally need larger trials that estimate larger effects across different contexts and studies (see Sect. 7.6).

An assumption in trials is that the important factors and conditions needed for the treatment to have an effect are all met. An intervention—whether a medication, an education programme or the like—may, for example, only work if recipients are nourished and healthy enough for the intervention to be effective, recipients take a sufficient amount of the prescribed intervention, practitioners are trained and experienced in delivering it effectively and the capacity of public institutions is adequate for its overall implementation, among a range of other influencing factors and conditions. No estimated causal results are thus affected solely by the target intervention, but rather, additionally, by many other background attributes, conditions and events that can take place at an individual, household, community and national level. These complex relationships, of course, influence the *internal* and *external validity* of RCT outcomes (see Chaps. 2 and 5 for details of these types of validity).

When talking about RCTs, there is a tendency to assume that proper randomisation has taken place in treatment allocation across groups, but this ideal is not invariably achieved. When trials are poorly randomised, covariates are poorly balanced between groups [8], which can in turn constrain the replicability of their estimated causal outcomes. This is the case for a number of the top ten cited RCTs. An example is the stroke trial [25]. This trial suggests that the outcome is explained by the treatment (reflected by a 4% lower mortality rate for treated patients (17%) after three months of a stroke compared to placebo patients (21%)). However, baseline data indicate that other factors that strongly affect the outcomes of stroke and mortality were not equally allocated: treated patients relative to placebo patients were 3% less likely to have had congestive heart failure, 8% less likely to have been smoking before the stroke, 14% more likely to have taken aspirin therapy, and 3% more likely to have had and survived a previous stroke. Viewing these differences in relation to the outcome of just a 4-percentage point difference in mortality, they point to alternative factors that might also explain the trial's main outcomes—in part or entirely. We cannot exclude the possibility that the effect of the treatment may even have been negative and one or more of these alternative causes may be driving the outcomes.

Another example of poor randomisation is the breast cancer trial [17] in which those in the treatment group (who received chemotherapy plus the study treatment) were 10 percentage points more likely to have already had chemotherapy prior to entering the trial relative to those in the control group (who only received chemotherapy). Because response to chemotherapy differs for those already exposed to it relative to those receiving it for the first time, it is difficult to claim that the study treatment was solely driving the results. Indeed, even when adhered to, simple randomisation does not guarantee covariate balance, and study design features, such as sample size and stratification may need to be carefully considered (Chaps. 1 and 2).

As noted in Chap. 2, another limitation that can arise in trials is failure to collect baseline and subsequent data for all relevant covariates that can influence outcomes. This limitation can stand in the way of appreciating, and correcting for, the influence of key characteristics, such as being younger, being more physically fit, following a healthier diet, having a more qualified health practitioner, having lower levels of need for the treatment, being at an earlier stage of the illness or condition to be treated, receiving better health care, attending a better health facility and so on. It can arise, in some cases, through difficulty in obtaining the required covariate data and in others, through a lack of awareness of their potential influence or importance. In such cases, the claim can be made that the treatment is driving the trial's estimated causal outcomes, without due regard for the need to collect and assess relevant covariates. For example, potential explanatory variables, including participant levels of fitness, physical activity and stress, were ignored in some of the ten most cited trials [18, 21, 23, 24]. This can contribute to a further degree of uncertainty in the internal validity of a study's causal conclusions and the replicability of the outcomes.

A further matter of concern is that a causal claim is for many trials a function of when a researcher chooses to collect baseline and endline data. Such time points can vary from one trial to another or indeed across participants in a single trial. Importantly, however, the estimated average causal outcome is a function of when the researcher assesses the treatment, whether every week, month, quarter or year, and so on, and failure to take this into consideration in the trial design can undermine both the replicability of the outcomes and their internal and external validity. Within half of the ten most cited RCTs, the length of follow-up was for some participants two or three times longer than for others; yet only the average effect size was estimated, despite the different amounts of treatment received by different participants [20–23, 26].

Also, one of the heart disease trials [23] estimated that the effect of the cholesterol-lowering drug seemed to begin, on average, after about a year and was later reduced. Interventions generally have different levels of effect over time. For example, they may work well in the shorter term but not once our body becomes accustomed to a certain medication. This brings an additional level of complexity to modelling, with factors influencing the effectiveness of a treatment varying over time.

Variation in estimated results due to when we decide to evaluate a treatment is to some degree inevitable for most trials. Thus, ideally, evaluation at multiple time points to improve our understanding of the evaluation trajectory, including time lags, ought to be factored into the design of RCTs.

An associated constraint in interpreting a trial's treatment effects is related to a placebo-only or conventional-treatment-only limitation. Of the ten most cited trials, five evaluated the tested treatment only against the current treatment, four evaluated only against a placebo, and one evaluated against both to be able to compare relative causal outcomes using the two comparators (current treatment as well as placebo). Variation in such design features generally leads to different causal estimates and conclusions. For RCTs in which the conventional treatment is the control, in the absence of a placebo, the benefit of the intervention can at times be attributed to a poor outcome brought about by the conventional treatment.

Another assumption made in evaluating RCTs is that *covariate values at baseline* remain constant throughout the trial. However, as has already been highlighted in Chap. 5, in some cases, these values change during the trial and therefore, we need to assess them, not only at baseline but also, at endline; otherwise, they can bias results and constrain replicability.

The longer the trial is, the more influential changes in covariates often become, due to the potential for these covariates then serving as confounders. For longer trials, confounding can arise, for example, due to changes in practitioners or clinic management or implementation of a national health policy, further to commencement of the trial. But changes in covariate values can also take place in shorter trials: If those in the control group are given the common treatment or nothing at all and for example 5% of those in the treatment group decide to combine the tested drug treatment with other forms of treatment such as additional exercise or better nutrition to improve their condition more rapidly but we only collect data on levels of exercise and nutrition at baseline, then we do not know if the tested drug treatment alone is driving the outcomes. Unless we can ensure that participants at the endline have the identical traits and conditions that they had at the baseline, we cannot claim that the outcome is just because of the treatment. Most RCTs face this concern, with none of the ten most cited RCTs including such endline data. This is because potential confounding arising from such changes after randomisation (and during trial implementation) is not being controlled for and trials are not always estimating *pure* treatment effects. Overall, we need to abandon the view that random treatment allocation at the outset can ensure what happens post-randomisation and can, by itself, lead to valid and replicable study results.

A related constraint for the validity of RCTs is that they are not commonly designed to evaluate the dynamic character of phenomena and the distribution of effects but rather only the average effect. The average effect can be positive even when some, even the majority of, subjects are not influenced or may even be negatively influenced by the treatment but a minority still experience large effects. In an ideal RCT, all participants would experience the same effect but, in practice, we face heterogeneity and outliers that can affect the estimated average causal results [27]. Outliers and heterogeneity of treatment effects exist in clinical trials because, for instance, people are of different age, gender and physical health, experience different conditions, treatment needs and responses, develop different levels of resistance to the treatment, and so on. They are, as such, not a resolvable statistical problem. They are a consequence of studying dynamic biological, behavioural and social phenomena that involve complex processes.

Such dynamic phenomena are continually changing—their scope, their intensity, their duration, etc.—and are better understood as *evolving causes* rather than precisely measurable, static causes, amenable to statistical analysis. Researchers, however, generally evade issues relating to complexity, heterogeneity and evolving causes, as trials are designed to just estimate the average treatment effect. Only knowing the average treatment effect constrains our understanding about the effects on different people and groups among the population and presents constraints on the external validity of the results (Chaps. 2 and 5); yet, if someone is deciding, for example, to undergo extensive surgery or take a particular lifesaving, but also life-threatening, treatment, such as chemotherapy, they will want to know if it will work for them, not just on average within a trial sample under study.

Another constraint is the extrapolation limitation that refers to trial conditions not being the same as those in the general population [8, 28]. This limitation, which refers to the external validity of a trial, is discussed in Chaps. 2 and 5. As noted in these earlier chapters, constraints on internal validity that also limit replicability, can, for example, arise due to demographic and clinical differences between trial participants and the broader population. These include those relating to restrictiveness of eligibility criteria, participant consent rates, and quality of trial facilities or practitioners, among other factors [8]. Internal validity can also be limited by the extent to which real-life institutional, regulatory, economic, educational and behavioural conditions are matched by trial conditions.

The internal and external validity of trial results are not always reported in detail. For some of the top ten RCTs, external validity is, however, partially discussed. For example, for the insulin therapy trial [19], participants were selected from one surgical intensive care unit, which suggests that the trial findings should not be applied to paients in medical intensive care units, or with, for example, demographic characteristics or illnesses, not represented in the study sample. Among the top ten trials, one of the diabetes trials [8] provides most detail on the study's applicability. The authors of the study concede that "The validity of generalizing the results of previous prevention studies is uncertain. Interventions that work in some societies may not work in others, because social, economic, and cultural forces influence [for example] diet and exercise." The authors of this trial state that the results could apply to about 3% of the US population. In general, when researchers do not explicitly discuss the potential scope of their results outside the trial context, practitioners are faced with the challenge of determining to whom they may apply.

Precise exporting of results from a trial population to a different population is not possible. When a positive causal outcome is estimated for a treatment in one context (in a given clinic, lab, field, etc.), adopting the treatment in another context rests on the assumption that the traits of participants and the other broader factors and conditions that helped bring about the estimated causal outcome in the trial context would be similar in the other context [8]. Yet, the distributions of the causes needed to enable an intervention to work (such as metabolic enzymes or existing economic policies) are not always very similar in the trial and the target context. What these causes amount to is often variable across particular domains and contexts. It is unlikely that replicating the same trial (or scaling it up) in another context or in the same context at a different point in time would produce the same average effect size estimate and that therefore for the stroke trial [25], we would arrive again at a 17% mortality rate at three months after the onset of stroke for those treated and 21% for those without treatment. Indeed, strictly speaking, since a trial's estimated causal results are relative to particular factors under particular conditions within a given sample at a given point in time and since causes can therefore evolve over time, we can generally only speak of a singular causal result based on a one-time causal relationship. This places constraints on the replicability of RCT results.

Overall, a range of limitations, assumptions and biases influence the estimated causal results of RCTs and the replicability of these results, with those discussed here not being exhaustive [16].

7.5 Overall Constraints on the Validity and Replicability of RCT Outcomes

The degree of validity and replicability of the estimated treatment effect from an RCT depends on the degree to which we are able to control the constraints, including the assumptions and the wide range of biases we face in designing, implementing and analysing trials.

Statistical techniques, such as intention-to-treat analysis or stratified randomisation, can reduce the effects of a number of individual biases (Chap. 2) but they cannot eliminate all biases in a given trial. Some of the studies discussed above also face several biases, including sampling bias, compliance bias and bias due to lack of blinding, which, when combined, present a considerable challenge for the replicability of RCT outcomes.

For some of the ten most cited trials, we are generally dealing with inaccurate effect size estimates and thus, less replicability. Many studies do, however, have solid research designs and robust effect size estimates. Yet, causal conclusions in studies should not be viewed separately from these implicit limitations, assumptions and biases. And the number of them underlying a trial's causal results and inferences can increase at each step: from how we choose our research questions and objectives, create our variables, select our sample, randomise, blind and control, to how we carry out treatments and monitor participants, collect our data and conduct our data analysis, deal with unknown factors, interpret our results, and so on.

Establishing which assumptions and biases are most or least important in affecting the estimated causal results is not feasible in general terms. We can only assess levels of importance within a specific study context and these depend on the extent to which the assumptions are met and the biases are managed. RCT studies generally result in some degree of bias (as do studies involving other designs). Relatedly, an important challenge for any given RCT is that of meaningfully estimating to what extent those sources of bias collectively explain the estimated effect size.

The complexity of the above challenge is influenced by the many decisions which researchers require to make as part of the study design. These include:

- which particular factors are the most important to stratify for during randomisation;
- how and in which location to select participants;
- which inclusion and exclusion criteria and outcome variables are appropriate;
- what length of time between collected baseline and endline data points is appropriate, etc.

Each decision influences the particular causal results estimated in a study. Important though it is, specification of randomisation as a study design feature cannot in and of itself guarantee causation, replicability and scientific objectivity of the outcome(s). It is, however, one step within the larger and complex process of designing, implementing and evaluating the same study.

Key Learning Point 7.2

This chapter assumes that irreplicability in RCTs can largely be explained in terms of the scientific process being a complex human process. This process involves many actors, including study designers, participants, data collectors, implementing practitioners and statisticians, among others, making many decisions at multiple steps when designing, implementing and analysing studies, with some degree of bias generally arising during this complex process. In general, effect size estimates from studies become less accurate, and results less replicable, with decreasing sample size, increasing covariate imbalance across trial groups (see earlier chapters) and increasing levels of bias from other sources, such as from lack of adequate blinding.

That trials face some degree of bias that affects their causal results is simply inevitable for studies to be conducted in the real world. And addressing one bias can at times mean introducing another bias. For example, making a sample more heterogeneous can help improve how useful results are after the trial but can also reduce precision in the trial's estimated results and hence it's internal validity (Chap. 5). It is generally unavoidable that a number of things do not go according to plan in conducting studies. By the time a study is carried out and completed, flaws and biases become increasingly apparent over time. We then have to make a judgement: are outcomes with some degree of bias sufficient to inform policy decisions and practitioners? Generally, they are and they are usually the best we have, but the judgement also depends on how useful the results are in practice and their level of replicability and validity compared with other studies using the same method or, at times, other methods. Yet, no single study should be the sole and authoritative source used to inform policy and our decisions. Different studies assessing the same treatment can help reduce the level of uncertainty in the estimated results of any single study. Meta-analyses of studies play an essential role in synthesising existing evidence across individual studies and often provide the most robust evidence we have. However, they also come with some limitations, such as the pooling together of biases from individual studies and the potential of overestimating positive treatment effects and underestimating negative treatment effects, since studies with negative results are less likely to get published.

7.6 Addressing Replicability Issues by Building Replicability into the Research Design of Studies: The Need for *Multi-studies*

Assessing the combined set of biases and constraints that can collectively affect the causal results of RCTs and seeking to address them collectively is an approach proposed here to tackling sources of irreplicability in RCT findings. Double-blinding, placebos and the like are not inevitable, nor are they the only methodological solutions to the problems they aim to address.

In a 2016 survey in *Nature*, about 90% of the over 1500 scientists surveyed reported that "more robust experimental designs" would improve reproducibility in science [5]. In light of the analysis in the current chapter, irreplicability can in part be explained by the results of conventional replication studies often not being meaningfully comparable to the results of the original studies, since study features generally change across time, location and study groups. The central implication and policy recommendation in this chapter is that to improve the replicability of studies we need to build replicability into the initial research design rather rely on conventional replication studies. This would help tackle many of the constraints on replicability collectively and before they arise. The recommendation here is that RCTs be designed so that individually they adopt as many design principles as possible aimed at reducing bias. This advice needs, of course, to be carefully balanced with the varying limitations on available funding, time and resources and the existence of rare disorders which limit recruitment of some patients to trials in real-world settings. With this caveat in view, the following list of procedures is recommended as a checklist for future trial design:

- *Compare various population groups through a large sample size*—to increase reliability of evidence, researchers ideally need large samples, if possible, of thousands of people, across a range of separate groups with different background traits and characteristics relating to age, gender, socio-economic status, ethnicity, level of need for the treatment, etc.
- *Compare conditions in multiple contexts*—to reduce context-specific conditions shaping evidence, we need to implement a study in multiple geographic locations, clinics, etc.

- *Compare new treatments against both current treatments and a placebo*—to better interpret and understand a study's estimated effect sizes we need to evaluate the tested treatments against both conventional treatments, where they exist, and a placebo, to be able to compare relative effects.
- *Compare different interventions*—we need to test different interventions against each other within a single study and, where applicable, multiple combined treatments, in unison—for example, comparing (i) increased exercise, (ii) improved nutrition, (iii) no smoking and (iv) a given medication in the form (i), (i + ii), (i + ii + iii), (i + ii + iii + iv), etc., to assess relative benefits.
- *Compare treatment effects at multiple time points*—we need to evaluate study results at multiple time points to better understand the outcome trajectory across baseline, multiple midline, and endline, data points.
- *Where baseline covariate values are subject to change over time within the trial period,* we need to also record the endline values—this is particularly important for longer studies.
- *Optimise blinding and wherever possible, use full blinding*—to minimise bias due to lack of blinding we need to get beyond common RCTs using double blinding and, when applicable, move to full blinding which implies that *nobody*—not just participants and practitioners, but also experiment designers, data collectors, data evaluators, etc.—would know before, during or after the study the group assignments for individual participants.

We gain more validity in our evidence, and generally learn most, by comparing results against each other. This involves conducting a single study among different groups of people with different background traits and characteristics, and at multiple time points, using different treatment groups and control groups, and within different contexts. This research design is called here a *multi-study.*

Key Learning Point 7.3

The results of replication studies have highlighted the existence of a range of factors influencing reported treatment effects. These include, for example, the study population, geographical location and the experimenters involved in the research. A *multi-study* would, however, involve design features which allow such factors to be represented across different groups, representative of different sample populations, different locations, different experimenters, and so on, while incorporating covariate adjustment within groups to correct for other factors which cannot be fixed across patients. Multi-studies have two key advantages: first, they control for the very features that are later commonly controlled for in conventional replication studies and, secondly, they also control for many of the biases and other constraints that can arise within any given study and that typically hinder replicability.

As presented in Key Learning Point 7.3, multi-studies have the potential to build replicability more effectively into a study design. By taking into consideration the complexity of human beings, including evolving human and contextual causes, such studies should improve our understanding of a given phenomenon through provision of multiple effect size estimates that can be directly compared against each other. They also represent an approach for enhancing the validity, and hence the quality and replicability, of RCTs in clinical research.

So, what is the advantage of reforming existing RCT design principles in favour of those of *multi-studies* rather than continuing to conduct conventional replication studies? Once we have integrated replicability into a study, we can reduce the need to replicate original studies in different settings to explore generalisability of findings. Instead, we can build the multiple study features into a single study that should produce more comparable results.

7.7 Conclusion and Implications

The ability to verify research results is a central feature of science and scientific methodology. This chapter has outlined how a combined set of important methodological limitations, assumptions and biases involved in estimating causal effects can constrain the replicability of studies. Differences in how the complex scientific process and methodological techniques are carried out in a different context with different human subjects help explain why replication studies are generally not able to obtain the same results as in the original studies. Evolving causes that underlie complex human processes can undermine the degree of coherence in results from clinical research, although it is important to bear in mind that a perfect match in effect size estimates and significance levels across studies is not something that our scientific methods and studies in different contexts can reasonably be expected to guarantee (Chap. 4). Much attention has been placed in recent years on a phenomenon referred to as the "replication crisis" as defined in Chap. 4. While much of the reasoning in support of its existence is open to debate, some of the relevant work has been helpful in exposing the weaknesses and complexities of science. This includes the need for better blinding, reviewing baseline data at endline, where this is meaningful, and creating multiple control and treatment groups, among other research design strategies. This is in addition to improving reporting guidelines for journal articles and pre-registration of studies [6].

While the need for such reform does not undermine the status of the RCT as the gold standard for primary studies in clinical research, it is helpful to be aware that individual RCTs are in some senses not as *random* as might be assumed in conventional Evidence-Based Medicine teaching. For example, they are not always able to ensure an even distribution of all measurable, non-measurable and unknown influencers, and as noted in Chap. 2, the selection process for recruitment to trials is typically not administered at random. RCTs are also not always as *controlled* as thought. For example, they are not designed well to control for participants refusing to participate, missing data, switching between groups or in the case of longer studies, for changes in covariates during the study.

Ultimately, we create the research designs we use—and they are not always limitation-free, assumption-free and bias-free. Yet we do not conduct rigorous studies because they guard against every bias and defect; we conduct them because they reduce levels of bias and address other experimental limitations. There is, however, continual room for improving the methods we use and the replicability of studies. It is because RCTs play a vital role in informing patient care and policy decisions that it is so important for researchers to take concrete steps to continually improve their design and identify ways to reduce the overall degree of bias. Researchers can thereby manage the impact of studies on patients' lives when they go through limitation by limitation, bias by bias and assumption by assumption and aim to address them as far as possible when designing, implementing and analysing studies.

Key Learning Points

7.4 A combined set of biases and other constraints, underlying the scientific *process* and *methods* used, and the different *time points* and conditions involving *people and study contexts*, together drive, to some degree, inevitable variation between study outcomes. This can in turn lead to difficulties in replicating studies.

7.5 A seemingly obvious approach in principle, therefore, in combatting irreplicability is to tackle many of the constraints of replicability collectively in the course of designing any single RCT, thus removing the need for assessment of replicability through later studies.

7.6 This involves conducting a *multi-study*—that is, a single study among different groups of people, with different background traits and characteristics, carried out at multiple time points using different experimental intervention groups and control groups, and within different contexts.

If clinical researchers and policymakers must insist on having a 'gold standard' in research it should not necessarily be tied to one particular research design. RCTs need to be combined with other research designs and methods, each with its own limitations, in order to gain a more holistic understanding of a given intervention. Different methods can provide further insight into different aspects of a treatment or phenomenon that the different methods are not designed to do alone. Using multiple research designs and methods (triangulation) to address the same question helps reduce biases and flaws that can underlie a given method, just as conducting multiple studies also reduces study-specific biases and flaws. Acquiring evidence that is supported across different methodologies increases the validity of our evidence. When we take the total evidence into account, we can thus make the most robust inferences.

To better understand the foundations and recognise the limitations of our leading scientific methods, we need to assess and continually reassess them. This critical lens enables us to continually identify better ways to improve how RCTs are conducted and thus, ultimately, optimise the impact they can have on patient care and treatment. The full potential of the RCT method has not been reached and we shall continue developing its methodological features over time to better address problems we have not yet solved and biases we have not yet eliminated. The rigorous approach to building replicability into the experimental design in *multi-studies* takes the focus away from traditional replication studies. This in turn can improve the validity of RCT outcomes and our understanding of the effectiveness of treatments, thus potentially improving their transferability to real-life settings.

Acknowledgements I am thankful for comments from Nancy Cartwright, John Worrall, Federica Russo, Carl Hoefer, Stephan Guettinger, Corinna Peters, Camille Lassale, Federica Malfatti and Saana Jukola. I received funding from the Marie Curie programme of the European Commission and the Ministry of Science and Innovation of the Government of Spain (grant RYC2020-029424-I).

References

1. American Statistical Association. Recommendations to funding agencies for supporting reproducible research. 2017. https://www.amstat.org/asa/files/pdfs/pol-reproducibleresearchrecommendations.pdf. Accessed 2 December 2023.
2. Open Science Collaboration. Estimating the reproducibility of psychological science. Science. 2015;349(6251):943–51.
3. Baker M. 1,500 scientists lift the lid on reproducibility. Nature. 2016;533.
4. Goldacre B. Make journals report clinical trials properly. Nature. 2016;530:7.
5. Allison D, et al. Reproducibility: a tragedy of errors. Nature. 2016;530:27–9.
6. Munafò MR, Nosek BA, Bishop DVM, Button KS, Chambers CD, Percie Du Sert N, et al. A manifesto for reproducible science. Nat Hum Behav. 2017;1(1):0021.
7. Bishop D. Rein in the four horsemen of irreproducibility. Nature. 2019;568:435.
8. Cartwright N. What are randomised controlled trials good for? Philos Stud. 2010;147:59–70.
9. Suppes P. A probabilistic theory of causality. Amsterdam: North-Holland; 1970.
10. Cartwright N. Are RCTs the gold standard? BioSocieties. 2007;2(1):11–20.
11. Krauss A. Assessing the overall validity of randomised controlled trials, International studies in the philosophy of science. 2021;34(3):159–182.
12. Kelly A, Lesh R, Baek J. Handbook of design research methods in education: Innovations in science, technology, engineering, and mathematics learning and teaching. New York: Routledge; 2014. p. 2.
13. Kane R. Understanding health care outcomes research. Jones & Bartlett; 2006. p. 118.
14. Gorard S, Taylor C. Combining methods in educational and social research. Open University Press, McGraw-Hill; 2004. p. 94.
15. Waters E, Swinburn B, Seidell J, Uauy R. Preventing childhood obesity: evidence policy and practice. Wiley-Blackwell: BMJ Books; 2010.
16. Krauss A. Why all randomised controlled trials produce biased results. Ann Med. 2018;50(4):312–322.
17. Slamon D, et al. Use of chemotherapy plus a monoclonal antibody against HER2 for metastatic breast cancer that overexpresses HER2. N Engl J Med. 2001;344:783–92.

18. Hurwitz H, et al. Bevacizumab plus irinotecan, fluorouracil, and leucovorin for metastatic colorectal cancer. N Engl J Med. 2004;350:2235–342.
19. Van Den Berghe G, et al. Intensive insulin therapy in critically ill patients. N Engl J Med. 2001;345:1359–67.
20. Rossouw J, et al. Risks and benefits of estrogen plus progestin in healthy postmenopausal women: principal results from the women's health initiative randomized controlled trial. JAMA. 2002;288(3):321–33.
21. Diabetes Control and Complications Trial Research Group (DCC). The effect of intensive treatment of diabetes on the development and progression of long-term complications in insulin-dependent diabetes mellitus. N Engl J Med. 1993;329(14):977–86.
22. Knowler W, et al. Reduction in the incidence of type 2 diabetes with lifestyle intervention or metformin. N Engl J Med. 2002;346(6):393–403.
23. Scandinavian Simvastatin Survival Study Group (SSSSG). Randomised trial of cholesterol lowering in 4444 patients with coronary heart disease: the Scandinavian Simvastatin Survival Study. Lancet. 1994;344(8934):1383–9.
24. Shepherd J, et al. Prevention of coronary heart disease with pravastatin in men with hypercholesterolemia. N Engl J Med. 1995;333(20):1301–7.
25. Marler J. Tissue plasminogen activator for acute ischemic stroke. N Engl J Med. 1995;333:1581–8.
26. Turner R. Intensive blood-glucose control with sulphonylureas or insulin compared with conventional treatment and risk of complications in patients with type 2 diabetes. The Lancet. 1998;352(9131):837–53.
27. Deaton A. Instruments of development: randomization in the tropics, and the search for the elusive keys to economic development. NBER; 2009.
28. Clarke B, Gillies D, Illari P, Russo F, Williamson J. Mechanisms and the evidence hierarchy. Topoi. 2014;33:339–60.

Alexander Krauss is an Assistant Research Professor with the Spanish National Research Council and a Research Associate at the London School of Economics. Prior to this, he was a Postdoctoral Fellow at University College London, after working for five years at the World Bank. Trained as an economist, his research is in economics, science of science, and philosophy of science. He is currently completing a book on what drives science and discoveries.

The manufacturer's authorised representative in the EU is Springer Nature Customer Service Centre GmbH, Europaplatz 3, 69115 Heidelberg, Germany. If you have any concerns regarding our products, please contact ProductSafety@springernature.com

Printed and bound by CPI Group (UK) Ltd, Croydon, CR0 4YY
01/10/2026
02225865-0001